CONFLICT

CONFLICT

The Evolution of Warfare from 1945
to Ukraine

GENERAL DAVID PETRAEUS
ANDREW ROBERTS

HARPER

An Imprint of HarperCollins*Publishers*

Published in 2023 in Great Britain by William Collins,
an imprint of HarperCollins Publishers.

The quotation on page 235 is reproduced with permission of Curtis Brown, London on
behalf of The Estate of Winston S. Churchill © The Estate of Winston S. Churchill.

Acknowledgment is made to the following for permission to reprint photographs:

SECTION 1: Sovfoto/Getty Images; Universal History Archive/Getty Images;
Bettmann/Getty Images; Keystone-France/Getty Images; Pictures From History/
Getty Images; Charles Hewitt/Getty; Stan Meagher/Stringer/Getty; Popperfoto/
Getty Images; Sovfoto/Getty Images; Underwood Archives/Getty Images; Central Press/
Getty Images; Ilan Ron/GPO/Getty Images; Adrian Brown/Alamy Stock Photo;
Bettmann/Getty Images; Associated Press/Alamy Stock Photo.

SECTION 2: Scott Wallace/Getty Images; Bettmann/Getty Images; Pictures
From History/Universal Images Group via Getty Images; KARIM SAHIB/AFP via
Getty Images; Trinity Mirror/Mirrorpix/Alamy Stock Photo; PA Images/
Alamy Stock Photo; Kevin King/Getty Images; JIM WATSON/Getty Images;
XM Collection/Alamy Stock Photo; Joe Raedle/Getty Images; Associated Press/
Alamy Stock Photo; MediaPunch Inc./Alamy Stock Photo; SOPA Images Limited/
Alamy Stock Photo; Global Images Ukraine/Getty Images; Zuma Press/
Alamy Stock Photo; American Photo Archive/Alamy Stock Photo.

FIRST EDITION

Library of Congress Cataloging-in-Publication Data has been applied for.

ISBN 978-0-06-329313-7

23 24 25 26 27 LBC 6 5 4 3 2

David Petraeus: "To the men and women in US, coalition and host-nation uniforms, as well as the diplomats, spies and development experts, with whom I was privileged to serve during nearly four decades in uniform and government service."

Andrew Roberts: "To Simon Sebag Montefiore and our forty-year friendship"

CONTENTS

INTRODUCTION

There are not many positive services which the historian can
do ... higher than that of tracing the causes of wars, describing
the means by which they are fought, ascertaining the reasons
that led to victory on one side or the other, describing the
effects, and estimating the conditions likely to produce
future wars in which they would be fought.

Professor Cyril Falls, Inaugural Address as
Chichele Professor of War at Oxford, 1946[1]

In the early hours of Thursday, 24 February 2022, President Vladimir Putin of Russia unleashed what he hoped would be swift and devastating attacks on Kyiv and other points in Ukraine, intended to topple its government by a *coup de main*. Although, as this book will show, military conflict has evolved significantly since 1945, Putin deliberately chose to wage Second World War-style hostilities, with devastating consequences for both Ukraine and Russia.

One intention of this book is to put Putin's actions in Ukraine and his methods of engagement into their proper historical context, but it has greater ambitions than just that. It will also attempt to show how militaries around the world have learned – or failed to learn – from each previous war when trying to fashion the means to fight the next, and it will investigate the personal qualities needed for successful strategic leadership.

It is important to establish what this book is *not*. It is not intended as a comprehensive history of all conflict since 1945, which would be nearly impossible in a single volume. Instead, it concentrates solely on conflicts that have contributed to the evolution of warfare. Nor is it a book about politics and why wars break out; rather, it is about what happens on battlefields once they have. If a conflict saw warfare evolve in some manner – such as in tactical concepts, or a crucial new weapon, or when defense became superior to offense (or vice versa) – then we have included it. The effect of increased accuracy in anti-tank weaponry in the Yom Kippur War, say, or the tactics of massed tank attack in the Gulf War, will find a place in this book, whereas we do not examine a series of lesser, often guerrilla, conflicts that were essentially fought according to much the same precepts as each other.

The Prussian military philosopher and theorist Carl von Clausewitz described warfare as politics by other means, and just as politics did not end in 1945, neither has warfare. Indeed, conflict has continued at least somewhere on the globe every year since the end of the Second World War. The twentieth century was the most brutal of all the many centuries of human existence; it is estimated that more people had already perished violently in the first half of the twentieth century than in all the previous centuries put together.

As for the twenty-first, within a month of the invasion of Ukraine, Russia had lost twice as many military personnel killed as the United States had in twenty years in Iraq, and by March 2023 the Russian death toll is thought to have reached five times the 13,300 soldiers that the USSR lost after a decade in Afghanistan in the 1980s. War is thus still very much worth studying.

The first eight of the following chapters form a chronological account of how warfare has evolved since the end of the Second World War. The ninth analyzes the Russo-Ukrainian War, highlighting those areas where it might provide clues to what war will look like in the future. Chapter

10 then draws lessons from both Ukraine and the earlier chapters about what we might expect in the wars of the twenty-first century. We concentrate upon the importance of being at the cutting edge of the latest military technology as well as on the critical roles played by leadership, training, morale, coalition-building, doctrine, the significance of highly professional non-commissioned officers and the importance of contesting the information sphere.

Strategic concepts have evolved faster since the Second World War than at any comparable period in history. A commander in that war was essentially using the same structures – corps, divisions, regiments, battalions – that Napoleon had employed in the early nineteenth century (albeit with the addition of much greater mobility, artillery and other indirect fire and airpower). Since then, however, warfare has evolved at a dizzying pace, particularly over the past two decades, and this book delineates how and why crucial changes have taken place – as well as the dramatic developments we can expect in the coming decades. Unlike any previous works on the subject, it has been written through the collaboration of a renowned battlefield commander and an established military historian, who have each brought to bear their different expertise and viewpoints.

In the aftermath of the Russian invasion of Ukraine, we recognized that there was very little literature that places that struggle in its *military history* context, despite the avalanche of often excellent works on its political, economic and geostrategic contexts. In a world that has expensively and painstakingly developed precision weaponry and smart bombs, the Russians deliberately chose to revert to a brutal, Second World War – and, in parts of the Donbas, First World War – style of fighting. With military strategy and tactics evolving hugely since 1945, where each conflict teaches lessons for the next in multifarious ways that we show in the following chapters, what can explain Russia's decision to fight a throwback war reminiscent of the Great Patriotic War – only this time with Russia as the aggressor rather than the victim?

Russia won untarnishable glory in 1945 for having provided the oceans of blood necessary to rid the world of the evil of Nazism. For every five soldiers killed fighting Nazi Germany on the battlefields of the

Second World War, four died on the Eastern Front. Yet ever since 1945, Russia has been drawing down on its credit for that great service to humanity, and never more so than in President Putin's unprovoked, reckless and unbelievably vicious invasion of Ukraine. By showing how warfare has evolved in different climes and with different weaponry and political situations over the decades since the death of Adolf Hitler, we hope to highlight just how strangely regressive is the present Russo-Ukrainian conflict. Warfare evolves; it does not ossify. Yet it is clearly also capable of being suddenly and shockingly thrown into reverse.

In each chapter we also provide examples of how, when done well, successful strategic leadership can transform even the most seriously disadvantageous situations for the better. Yet, when it fails, it can turn likely victory into certain defeat. Leaders, some of whom are politicians in this book while others are soldiers, must be able to master four major tasks.[2] Firstly, they need comprehensively to grasp the overall strategic situation in a conflict and craft the appropriate strategic approach – in essence, to get the big ideas right. Secondly, they must communicate those big ideas, the strategy, effectively throughout the breadth and depth of their organization and to all other stakeholders. Thirdly, they need to oversee the implementation of the big ideas, driving the execution of the campaign plan relentlessly and determinedly. Lastly, they have to determine how the big ideas need to be refined, adapted and augmented, so that they can perform the first three tasks again and again and again. The statesmen and soldiers who perform these four tasks properly are the exemplars who stand out from these pages. The witness of history demonstrates that exceptional strategic leadership is the one absolute prerequisite for success, but also that it is as rare as the black swan.

One

The Death of the Dream of Peace

1945–1953

A third world war may well prove beyond the limits of what civilized society can endure, perhaps even beyond the limits of our continued existence as a human world.

Jan Christian Smuts' speech to the United Nations,
1 May 1945

In the immediate aftermath of the Second World War, so terrible had been the suffering, with up to 60 million killed, that a dream arose that there might be peace on earth. This hope was perpetuated by the seeming end of ultra-nationalism and the birth of the United Nations, whose founding Charter stated that it hoped to "save succeeding generations from the scourge of war" by ensuring that nations would "practice tolerance and live together in peace with one another as good neighbors."[1] Almost a million people signed a register of "world citizens" and pledged to abolish war by abolishing nations and creating a United States of the World. In a similar bout of idealism, within a month of Japan's surrender, President Truman abolished the Office of Strategic Services, the United States overseas intelligence operation (and precursor to the CIA).

After the Nuremberg Trials, whose indictments had focused more on senior Nazis for waging aggressive war than on the Holocaust, the hope of mankind was that invasions and warfare might be abandoned as a way of solving international disputes. The universal cry of "Never again" applied as much to the practice of invading countries as to the monstrous crimes of the Nazi era. It was as noble as it was naive.

The immediate post-war period started off well with regard to actual cross-border warfare, although the two British partitions of the period – of the Indian subcontinent in 1947 and of Mandatory Palestine in 1948 – embedded severe and traumatic conflicts that endure to this day. Similarly, the Chinese Civil War led to an independent Formosa (modern-day Taiwan) which is still a potential global flashpoint. The late 1940s might not have seen actual cross-border invasions, therefore, but they sowed the seeds for festering resentment and tensions that even today might ignite into open conflict. As the nineteenth-century British jurist Sir Henry Maine put it, "War appears to be as old as Mankind, but peace is a modern invention."[2] Even two centuries after his observation, peace is an invention that still has teething troubles.

Far more powerful than the dreams of peace in preventing cross-border invasions in the late 1940s was the threat of nuclear annihilation, at least once the Soviet Union had tested "RDS-1," its first plutonium device, at the Semipalatinsk site in Kazakhstan on 29 August 1949. By the end of the first Cold War, the United States had conducted 1,032 nuclear tests and the Soviet Union a further 715. More than half of the Soviet tests were conducted at Semipalatinsk, a sequence that had severe consequences for the local population in the generations to come, including high cancer rates, genetic defects and birth deformities.

After 29 August 1949, the world faced for the first time the real possibility of Mutually Assured Destruction, usually referred to by its insightfully neat acronym. MAD changed the Cold War, because, as one military historian succinctly put it, "A total war between states possessing [atomic and nuclear weapons] carried the risk of total suicide, which imposed a new upper limit on the rationality of force."[3] Dr. Henry Kissinger, an early theorist of nuclear strategy, put it in equally bleak terms when he wrote that "Not long after Hiroshima and Nagasaki, the risks of fielding nuclear weapons became incalculable, the stakes disconnected from the consequences."[4] In light of this, the overriding US strategy became one of deterrence – what Kissinger would describe as "a psychological strategy of negative objectives."[5]

Although President Dwight Eisenhower had privately wondered with his advisers why tactical nuclear weapons couldn't be used "just exactly as you would a bullet or anything else" in the Taiwan Straits, he was far more circumspect in public.[6] In Geneva in July 1955, at the first Anglo-American–Russian summit since the end of the Second World War, he emphasized just how mutually assured the destruction of a nuclear war would be, through the sowing of radioactive isotopes by a carrying wind, devastating to life in all its forms across the northern hemisphere.[7]

This became even truer once the megatonnage possessed by the nuclear powers grew exponentially. Hiroshima had been destroyed by a 14-kiloton bomb and Nagasaki by one of 20, or the equivalent of 14,000 and 20,000 tons of TNT. These were minuscule compared to today's megaton warheads, which have reached the power of 1 million tons of TNT.[8] If exploded at the optimum height, a 1-megaton bomb would destroy every brick building within a 3½-mile radius and cause significant damage for up to 13 miles. Dry leaves would ignite as far as 11 miles away, with people suffering second-degree burns inside their houses. Downwind, radiation would swiftly kill victims or slowly cause cancer over a period of months – their skin burning, their bodies bleeding internally. As Lawrence Freedman and Jeffrey Michaels, historians of nuclear strategy, have pointed out, nuclear weaponry in the Cold War "was a palpably blunt instrument and would be used in a blunt manner."[9]

After the First World War, many assumed that a build-up of armaments was fundamentally destabilizing, but the Cold War served to disprove this notion: leaders with vast nuclear arsenals found that a nuclear threat concentrated the minds of world leaders and successfully disincentivized superpower-on-superpower conflict, albeit in favor of limited conflicts and proxy warfare. In 1952, the US tested its first hydrogen bomb (cheerfully nicknamed "Ivy Mike"), and the following

year found the Soviets successfully testing their counterpart, "Joe 4."
Not to be outdone on the world stage, Winston Churchill announced
that Britain had produced its own atomic bomb in February 1952 (with
a hydrogen bomb following in 1957), after which China acquired
nuclear capacity in 1967, Israel (probably) by 1967, France in 1968,
India in 1974, Pakistan in 1998 and North Korea in 2006.

It was not until 1972 that President Richard Nixon of the United
States and Leonid Brezhnev of the USSR both recognized that the introduction of multiple independently targeted re-entry vehicles (MIRVs)
meant that the number of warheads could be increased so dramatically
that it would overwhelm any antiballistic missile (ABM) defenses. The
result was the Strategic Arms Limitation Treaty that limited ABMs and
MIRVs to minimal numbers. What prevented the offensive from
completely overwhelming the defensive, and thus increasing the attrac-
tion of a surprise attack ("first strike"), was the invulnerability of
nuclear-armed submarines.

The deterrence effect of MAD depended upon the assurance that
both adversaries would retain enough nuclear weapons even after a
first strike to inflict unacceptable damage on the other.[10] In the 1980s,
rapid technological improvements in MIRV technology produced
increased accuracy and flexible targeting, meaning that targets could
be moved from massive urban areas to specific military structures. This
in turn led to the concept of a "controlled" or "limited" or "tactical"
nuclear war, of the sort that has been described as "a more subtle,
benign, and surgically accurate nuclear war; a possible extension of
conventional war" more reminiscent of the kind Eisenhower had envisioned in the Taiwan Straits, and Vladimir Putin likes occasionally to
threaten in Ukraine.[11]

In March 1983, President Ronald Reagan launched his Strategic
Defense Initiative project to explore a space-based system for antiballistic missile defense, but it was in its infancy in September when a
malfunctioning Soviet satellite early-warning system falsely reported
an American missile attack, which was in fact the result of sunlight
reflecting off the top of clouds.[12] Fortunately, common sense prevailed
at the Soviet local command level and no doomsday scenario took place.

In September 1987, a joint US–Soviet summit statement by Reagan and Soviet General Secretary Mikhail Gorbachev declared their "solemn conviction that a nuclear war cannot be won and must never be fought."[13]

Yet, if such a war *were* fought, much would depend on the personality of individual world leaders – and their reaction to news of an attack. As Freedman and Michaels put it, such a response might encompass "a reckless fury, lethargic submission, craven cowardice, or a firm and resolute action," the first and last of which could end in the annihilation of the human race.[14] Nor do those amount to a complete checklist. Some critics of nuclear power argue that life on earth will be extinguished by the greatest nuclear catastrophe, but this is not true: merely the lives of humans and other larger animals. Even before the last human in the last cave succumbed to nuclear-induced cancer, cockroaches, rats and other such species will have inherited the earth, and the evolutionary cycle will have started again among the cities that were turned into seas of glass. The blue planet would continue to orbit the sun, simply without us on it.

The upshot of this all too easily envisaged, tangible scenario has been the refusal to employ nuclear weapons since August 1945, which have therefore not killed anyone since Hiroshima and Nagasaki were bombed over three-quarters of a century ago (although the Cuban missile crisis of 1962 came close). In one sense, nuclear weapons have kept the peace by limiting wars: they have helped limit post-Second World War warfare to smaller, frequent but lesser armed conflicts, of which there have been, depending on how they are counted, between 150 and 300 since 1945.[15] "The fifty million who have died in war since that date," wrote the military historian John Keegan two decades ago, "have, for the most part, been killed by cheap, mass-produced weapons and small-calibre ammunition, costing little more than the transistor radios and dry-cell batteries that have flooded the world since that period."[16]

Despite the fact that the dangers of a nuclear exchange have largely produced caution among superpowers in international politics over the past seven decades, praise for MAD has been scarce. No fewer than eight Nobel Prizes have been awarded to individuals and organizations for their advocacy of nuclear disarmament, and only one proponent of nuclear deterrence has been honored – Thomas Schelling in 2005 – and that was in the field of economics.[17]

Historians long debated who was responsible for the Cold War that broke out only a matter of months after Japan's surrender and was well under way before it almost got hot at the time of the Berlin Airlift of June 1948 to May 1949. Since the opening of the Soviet archives in the early 1990s, however, it has become clear that, from the very moment the Second World War ended, Joseph Stalin was intent on extending Marxism-Leninism wherever he found a lack of Western resolve. Lenin had, after all, stated that a clash between communism and capitalism was inevitable and unavoidable, and the timing of a Cold War when Germany was weak and Europe impoverished seemed propitious to his successor, as did the relative disposition of military forces. "The power to create an unappeasably hostile world power and a complete confrontation was in the hands of one man," concluded Robert Conquest, the greatest of the Kremlinologists. The dream of peace did not die of its own accord; it was murdered by Joseph Stalin.

The diplomat George Kennan, writing from Moscow, recognized Stalin's intentions in the famous "Long Telegram" he sent in February 1946 which was subsequently published in the influential magazine *Foreign Affairs* under the pseudonym "X." That March, Winston Churchill also warned of Stalin's intractability in his Iron Curtain Speech at Westminster College in Fulton, Missouri. The world was made aware that a purveyor and instigator of terror was in control of a vast, monolithic army – and quickly growing an empire of his own.

Everything that followed, as pluralist politics were crushed in Eastern Europe, stemmed directly from Stalin's intention to degrade relations and fight an ideological war with the West: mass intimidation during Poland's elections in January 1947, King Michael of Romania's forced abdication in December 1947, the communist coup in Czechoslovakia in February 1948 and the Berlin Blockade four months later.

It was therefore a necessary act of reactive statesmanship for President Truman to establish the Central Intelligence Agency in

September 1947, and for his National Security Council to authorize it to perform covert actions two months later. Technically the attempted blockade of Berlin by Russia the following June was a *casus belli*, but Truman and the British premier Clement Attlee sensibly chose to interpret it more as one of intense provocation.[18] They responded by airlifting supplies into the city at huge expense in a remarkably efficient operation that lasted eleven months. The result of Russia's provocation was the precise opposite of what Moscow wished, a phenomenon that will recur regularly in this book. In April 1949, ten months into the Airlift, NATO was founded by twelve states united in their desire for collective security against Russian aggression. Seventy-three years later, as a result of Putin's invasion of Ukraine, both Sweden and Finland applied to join the by then thirty-country alliance.

The Chinese Civil War: the Generalissimo's failure of centralized command

Because of their doomsday qualities, nuclear weapons have provided an umbrella under which conventional warfare has often thrived rather than withered away. But their existence did not affect the Chinese Civil War, which is still easily the largest military engagement since the Second World War in terms of the number of people involved, a war that determined Chinese history for succeeding generations, and thus the modern experience of the whole world to this day.

The Second World War had seen China suffer cruelly during the military expansion of Imperial Japan into its territory in the preceding years, and when it ended, 80,000 Chinese soldiers of Chiang Kai-shek's Sixth Army were flown into Nanjing to reclaim their capital. Meanwhile, in Shanghai, poor peasant soldiers of the Ninety-Fourth Army were greeted as heroes by the wealthy Chinese citizens of the city: the liberated finely dressed in silk gowns, the liberators lucky to be wearing straw sandals.[19] Yet the Chinese Nationalists knew that their fight was not over, for an earlier foe was re-emerging from the ruins of the Japanese defeat. The Soviet Union had acquired control over all strategic points in Manchuria, a crucial area in the north, and it was not

Chiang Kai-shek (also known as Jiang Jieshi) whom they wanted to inherit it.

Mao Zedong had been fighting the Chinese Nationalists for nearly two decades. By the end of the Second World War he controlled hundreds of thousands of guerrilla fighters, mostly dispersed in the remote rural areas of northern China. The Japanese invasion had benefited Mao's guerrillas in their fight against Chiang Kai-shek's Guomindang Army, which had suffered significant losses at the hands of the Japanese. By 1945, Mao's National Revolutionary Army numbered well over a million fighters, and 100 million Chinese lived in areas controlled by the communists.

To China, today one of the world's two superpowers, the circumstances of its modern founding as a communist regime are important. A myth has developed, sedulously propagated by the Chinese Communist Party, that it won because the Chinese people were yearning for Marxism-Leninism. This is nonsense. In fact, the key to understanding the outcome of the war is to appreciate how, sometimes through no fault of his own, Chiang failed to perform properly the four tasks of a strategic leader set out in our Introduction, whereas Mao Zedong eventually came to master them all successfully.

In October 1945, soon after the Japanese surrender, Mao and Chiang Kai-shek met for the first time in twenty years at Chongqing, the latter's wartime capital; Mao was accompanied by the American Ambassador to China. Any hope that both parties might be able to cooperate following Mao's declaration that "all parties must unite under the leadership of Chairman Chiang to build a modern China" evaporated soon afterwards. Upon returning to his supporters, Mao informed them that he considered the declaration "a mere scrap of paper."[20] Fighting broke out in the late spring of 1946. Although they won some victories against the communists' northern bases, capturing 150 towns from January to March 1947, the Guomindang afterwards failed to win a single major victory for the whole of the rest of the war.

From 1937 to 1945 there had been two wars going on in China – the national war of resistance against the Japanese and the continuing Guomindang–communist civil war – in the former of which the

Guomindang had taken a far greater part, with concomitantly far greater losses.[21] During the war against Japan, in which between 14 and 20 million Chinese had died, China suffered appalling rural poverty and there had been a good deal of collaboration between the Chinese coastal urban elites and the Japanese, which weakened the state and its leaders, not least because corruption and inflation were also rife.[22]

Corruption was (and is) dangerous, not only because it diverts resources from the war effort but also because of the way it utterly demoralizes the troops and citizens who are its victims. "As he tasted the moment of victory," the historian Rana Mitter has pointed out, "Chiang Kai-shek looked out over ruin both foreign and domestic ... American disillusionment with the Chongqing government was fuelled by the wreck of the regime that ruled China. The nation had grand visions, but the reality was mass hunger, official corruption, and a brutal security state."[23]

This was largely because the Guomindang rather than the communists had to a considerable extent borne the brunt of fighting against the Japanese, and were almost exhausted by the time of Japan's sudden surrender on 1 September 1945.[24] Yet they were still the government of China, with all the powers of taxation, conscription and political patronage that went with that.[25] And Chiang, who had attended the Allies' Cairo conference in 1943 and met Churchill and Roosevelt, was still a recognized figure on the global stage. Thus the story of the Chinese Civil War is essentially one of Chiang and his senior commanders throwing away every advantage they had, while Mao survived until such time as he was able to launch devastating counter-offensives.[26]

Indeed, looking at the forces on paper, it is surprising that a civil war even broke out in 1945, let alone that it was lost by Chiang. The Guomindang had 2.6 million men under arms, whereas the communists had fewer than half a million, only half of whom had rifles.[27] The Guomindang controlled most of the key cities of China, whereas the communists were initially confined to their rural bases in the north. Yet the unruly nature of the Guomindang, which was riddled with competing egos and factions, kept the focus internal rather than on the common enemy, little guessing what horrors the communists would eventually wreak on their country.

In the light of Mao's impressive fighting force and territorial gains, the weakening of the Guomindang government and military and Stalin's crucial footholds in Manchuria, a clear opportunity arose for Mao to unite with the Russians and create a stretch of communist-controlled territory that swept from Mongolia across Manchuria and up to the Soviet Union. Hundreds of thousands of Chinese communist troops marched northwards to meet the Red Army, and in April 1946 the Soviets finally left Manchuria to the communist Chinese.

The Chinese Civil War is usually divided by historians into three great campaigns: the Liaoshen campaign for the control of Manchuria between May and November 1948, the Pingjin campaign for Beijing and Tianjin from November 1948 to January 1949, and the contemporaneous Huaihai campaign in north China. It was in the Huaihai campaign that the communists won the civil war, which until then could probably have gone either way. During those three campaigns alone, the Guomindang suffered over 1.5 million casualties to the communists' quarter of a million.[28]

If the first task of a strategic leader is to get the big ideas right, Chiang failed dramatically. He ostracized the civic leaders who had collaborated with the Japanese in the coastal cities at just the time when he desperately needed their support. He over-centralized power in himself and thus alienated important Guomindang leaders in Manchuria, Xinjiang and south-west China. He attacked the communists too soon after meeting Mao, and thus lost the support of key figures in the Truman administration. Even though he took territory in late 1946 and early 1947, he failed to hold it. His policy towards the powerful local warlords* who controlled much of China was to fight rather than conciliate them, and he made several important tactical errors in late 1948.[29]

---

* In the early part of the civil war, some villages and localities armed themselves to oppose both sides – something also seen in other civil wars from England to Russia.

Chiang's strategy of concentrating forces in cities and scattering them along railway lines made them vulnerable to highly mobile communist guerrillas. In a country as vast as China, supply lines were always in danger, and those of the thinly spread Guomindang forces were cut regularly. Chiang's decision to extend his lines of communication to try to recapture the north-east where the communists were receiving support from the USSR was badly misjudged.[30] This litany was all the more surprising because he had been an impressive commander when fighting Japan. Perhaps because of this, he was venerated by his staff who, an historian records, "tended to vie with each other for his favor. This limited their independence and speed of action, since they waited on his word rather than using their initiative."[31] This is not a phenomenon confined to the Chinese Civil War, of course. However much it might be comforting to hear one's opinions parroted back, a general's staff requires some professional naysayers if it is to be effective.

By contrast, Mao understood both the strategic and tactical situations he faced and he got the big ideas right. He grasped the importance of remaining agile, rather than, in his own words, "make holding or seizing a place our main objective."[32] He had studied *The Art of War* by Sun Tzu, the Chinese theorist of war of c.550–500 BC, and learned the advantages of avoiding direct confrontations if a strategy of indirect maneuvering was preferable. Despite the initial weakness of his positions, his reading of Sun Tzu had also taught him how to use strength and evaluate risk, practice deception and whenever possible do the opposite of what the enemy is expecting.[33]

It helped that Mao had ultimate control over policy and strategy, with dictatorial powers of which Chiang, who was nonetheless accused of being a dictator, could only dream. Moreover, Marxist ideology was interpreted very flexibly at the local level, because unlike the Guomindang the communists were willing to do deals with virtually anybody under any circumstances; they even employed the 200,000 Manchuguo troops who had fought for Japan. Any actual opposition to communist rule was dealt with summarily; no fewer than 150,000 soldiers were executed during the war, often after torture.[34]

The communists implemented a ruthless policy of rural land redistribution, enthusiastically beating and executing 1 million local landlords and landowners after their "People's Courts" delivered guilty verdicts. With some 90 percent of the Chinese population made up of rural peasantry, coercive recruitment to the communist Army was made sweeter by promises of land and liberation. For a Chinese landlord, ownership of as little as two-thirds of an acre constituted a death sentence. This was not the spontaneous acclaim of the Chinese peasantry for the precepts of the Marxist-Leninist dialectic so much as a land grab, and often the settling of ancient local grievances.[35] That said, as one historian has pointed out, "Both the Nationalists and the Communists resorted to horrendous strategies, including scorched earth policies, flooding vast tracts of land and engaging in urban terror campaigns, murderous purges and the use of starvation as a military tactic."[36]

In this case, as we shall see often over the coming years, the United States hesitantly and inconsistently supported one side (the Guomindang), while the Russians supported the other (the communists). Proxy warfare of this nature would soon became so ubiquitous in the post-Second World War period that direct superpower intervention was to become the exception rather than the rule.[37] This use of proxies to fight the superpowers' wars opened them both to the charge of being hypocritically willing to fight to the last Chinese, or Angolan, or Nicaraguan – or latterly Ukrainian.

The Americans withdrew from China in January 1947, although they continued to provide the Guomindang with advisers and equipment until mid-1948. By then, Chiang had lost over a million men, and was clearly losing the war. Historians will continue to debate whether President Truman "lost China," but the accusation would follow him for the rest of his career – adding to the triumvirate of Truman's huge, haunting decisions in East Asia, which already encompassed the nuclear annihilation of Hiroshima and Nagasaki, and would soon include American involvement in the Korean War. Historically, however, foreign intervention in civil wars rarely affects their outcome, as the Spanish and Russian Civil Wars had already shown.

Over the two years from August 1945, the communist People's Liberation Army (PLA) quadrupled in size by forcibly conscripting the peasantry in captured areas and accepting Guomindang deserters into its ranks. The Guomindang were initially armed with 3 million rifles and 200,000 machine guns captured from the Japanese, but rather than abandon positions that were impossible to defend they tended to send in further elite forces which were later surrounded and annihilated, leading to further demoralization.[38] "Make wiping out the enemy's effective strength our main objective," ordered Mao. "In every battle, strive to wipe them out thoroughly and do not let any escape from the net."[39] On those occasions when they were offered a choice between annihilation and changing sides, many Nationalist soldiers chose the latter.

Desertion from the Guomindang Army was perfectly understandable considering how brutally its soldiers were treated. Some 40 percent of conscripts deserted during basic training, while another 20 percent died of starvation.[40] Once in the field, a desertion rate of 10 percent per unit per month was not uncommon. If the man next to you deserted, you were beaten and denied rations. Sometimes troops had to be roped or even chained together to prevent desertions during marches, and they could be tethered together at night too. "Troops were allowed to relieve themselves only at prescribed times and it had to be done collectively."[41] It was almost a laboratory experiment for demoralization. The brutalization was infectious: many Guomindang troops had formerly belonged to warlords' bandit gangs for whom, as an historian puts it, "Looting, rape, torture, execution and the burning of villages were the usual – not the exceptional – behavior."[42]

Hyperinflation also helped to destroy Nationalist morale, as military pay became next to worthless. Officers began to steal the wages of their subordinates, sell military supplies and equipment and shake down civilians.[43] They also used bribery to get desk jobs away from the front line, where they could make money by selling troops' rice rations to local dealers. "Too often," notes an historian, "soldiers existed on handfuls of rice which they carried compressed in their pockets."[44] Officers' private enrichment at their soldiers' expense is a recurrent phenomenon in badly run armies.

Capitalizing on the inflation-induced poverty ravaging China, the communists laid siege to city after city, starving the populations into capitulation, the most horrific example of which was the five-month siege of Changchun, the largest city in Manchuria. Lin Biao, commander of the PLA's Manchuria Field Army, ordered it to be turned into "a city of death." The civilians of the city – cut off from all other Nationalist-held areas nearby – resorted to eating grass, bark and the human flesh that was eventually bought and sold on a black market. Some 160,000 died of starvation, while heavy artillery hammered the city at all hours. "Changchun was like Hiroshima," wrote one PLA lieutenant colonel in a memoir that was later banned. "The casualties were about the same. Hiroshima took nine seconds; Changchun took five months."[45] By the time the siege ended, every leaf and blade of grass had been consumed.[46] Daughters had been sold for minute amounts of food, the stuffing of pillows had been ingested by starving families and entire households were found frozen as if in sleep, having starved to death.

Later in the war, city after city capitulated to the communists without resistance: none wished to become another Changchun. The small guerrilla skirmishes of the early part of the war had transformed into vast, merciless arenas of attrition; Changchun represented medieval warfare with anti-aircraft guns. "In Manchuria alone," writes the historian Frank Dikötter, "the communists recruited or conscripted approximately one million men. In battle after battle, Chiang's best government troops were destroyed."[47]

An example of Chiang's inability to impose his will on his own army was seen during the battle of Jinzhou in October 1948. A month before, Mao had ordered Lin Biao to capture that key city, which would seal off Manchuria from any Guomindang relief forces for Changchun and Shenyang – "To close the door and beat the dog," as he put it.[48] From high ground close to Jinzhou, PLA heavy artillery made the city's sole airfield inoperable before Guomindang forces could be sent from Shenyang to reinforce it. Despite that, Chiang still hoped to use the assault on Jinzhou to pin Lin Biao down while reinforcements arrived from Shenyang and northern China. "The plans made and the orders given were sound," General David Barr, Chiang's American adviser, later averred, "and, had they been obeyed, the results would probably have been favorable."[49] Lin Biao himself agreed, later suggesting, "We prepared a feast for one table, but now we have two tables of guests – what are we to do?" Mao ordered him to attack the city before any more guests could arrive.

But it was not enough to get the strategic ideas right. The second task of a strategic leader highlights the importance of effectively communicating the big ideas to subordinates, and the third is to oversee their effective implementation. Chiang was able to communicate his counter-attack proposals to his powerful subordinates such as General Wei Lihuang, the commander in Shenyang, and General Fu Zuoyi, who commanded in northern China, but neither was willing to dilute his own forces to help save Jinzhou, so the implementation of the strategy was not driven through in time. Chiang had to spend no fewer than nine days flying back and forth between Shenyang and Beijing trying to persuade his overmighty subjects to come to the city's aid, losing valuable time in negotiation.[50]

By contrast, Lin Biao's army of a quarter of a million men used the time profitably, digging miles of trenches surrounding Jinzhou, shutting off the water supply and all communications into the city, and subjecting it to a 900-gun bombardment. On 14 October, his forces breached the city wall and they took Jinzhou by nightfall, capturing 400 trucks and enough ammunition for 60,000 troops.

The loss of Jinzhou proved disastrous for Chiang Kai-shek: six days later Changchun surrendered too, and on 2 November the communists took Shenyang itself. Ten days later the Liaoshen campaign ended in Guomindang defeat. In January 1949, Xuzhou fell to the communists and the Huaihai campaign ended in a defeat too. Beijing surrendered to the communists without resistance on 22 January, and Mao proclaimed the People's Republic of China on 1 October. Chiang fled to Taiwan on 10 December 1949, after which no formal peace treaty, or even armistice, was ever signed.

In January 1950 China and the USSR signed a security agreement which laid the seeds for the Korean War less than six months later. The Chinese Civil War had cost a staggering 6 million lives, and left profound consequences that last to this day.[51] Most importantly, Taiwan has evolved into a thriving democratic outpost of anti-CCP belief and activity whose sovereignty is threatened by an increasingly assertive China. This is arguably the most sensitive and dangerous situation in the world today. The United States has stated that, while it does not support independence for Taiwan, it would come to its aid if it were invaded, making the island a possible cause of direct conflict between the world's two superpowers.

The Chinese Civil War was important for the evolution of warfare for a number of reasons, but most significantly because it demonstrated to the world that guerrilla warfare undertaken according to Maoist military principles by smaller forces could ultimately be successful against a Western-backed government. The infiltration of communist guerrillas into French-held Vietnam predated Mao's victory, but the Chinese Civil War provided the Vietnamese communists with a template for what to do next, first against the French and then against the Americans.

The Korean War: getting the big strategic idea right

On Sunday, 25 June 1950, North Korean forces attacked south of the 38th Parallel border* into South Korea, with 135,000 troops, T-34 tanks and Yak warplanes. Kim Il-sung, the North Korean dictator, had been given the green light by Stalin the previous January, and with a ten-to-one superiority in a narrow sector, fighting with the advantage of surprise, his Army attacked in Blitzkrieg style redolent of the Second World War.[52] Just as with Blitzkrieg, the Army was followed closely by the country's secret police, which murdered POWs and suspected political opponents.[53] The Army of South Korea, whose President Syngman Rhee was an authoritarian American puppet, lacked heavy weaponry and has been described as "little more than a lightly armed gendarmerie."[54] It retreated quickly before the onslaught.

---

* The border across the Korean peninsula was the first in its history, and had been drawn arbitrarily at the 38th Parallel line of latitude north of the Equator to serve the interests not of the Koreans but of the US and USSR.

The United States had only 500 troops in Korea at the time, and as late as a week before the invasion CIA reports had indicated that any attack that might have been planned had been called off. Its concluding judgment – that North Korea would simply rely on propaganda and subversion to undermine the South rather than on force – was woefully misguided.[55] The invasion was the first overt military assault across an internationally recognized border since 1945, and a powerful and self-confident United States quickly sought to counter it. By 30 June, after the Truman administration had decided that the attempted *coup de main* must not be allowed to stand, the United States had deployed thousands of troops to South Korea.

The operations covered in this book rarely equate in size with those of the Second World War. Even the US-led attack in the Gulf in 1991 and Putin's invasion of Ukraine in 2022 were carried out by roughly the same number of troops that landed in Normandy on D-Day, but the latter operation was backed up by a million troops within a month of the landings. Yet they were all small in comparison with the Red Army's contemporaneous Operation Bagration, which numbered in the millions and which killed, wounded or captured 510,000 Germans. The only post-war conflict to approximate these vast figures were the Chinese Civil War and Korea.

Kim Il-sung had one advantage that all totalitarian leaders tend to share: they can undertake a truly surprise attack, with no preparation needed to win over the opinion of the general public or dissenting politicians. As the military historian Sterling Pavelec notes, surprise attacks open wars so often because they can leverage immediate – if usually short-term – advantage over the enemy.[56] They were used by Hitler against the USSR in Operation Barbarossa in June 1941, by Japan against the United States that December and, as we shall see, by the Israelis in the Six-Day War in 1967, by the Arabs in the Yom Kippur War in 1973, by the Argentines against the Falkland Islands in 1982, by the Iraqis against Kuwait in 1990 and by al-Qaeda on 9/11, among many other examples. As the senior Pentagon official Paul Wolfowitz once observed, "Surprise attacks happen so often that the surprising thing is that we are still surprised by them."

Yet surprise attacks have pitfalls, as Kim Il-sung and several other dictators have found. In particular, they tend to shock the enemy into a more active response than would a slow build-up. They also need adequate reinforcement if they enjoy early success, and that support might not always be developed because of the need to maintain the element of surprise. A final disadvantage is that they leave no one in any moral doubt as to who was the aggressor. Kim Il-sung's surprise attack launching the Korean War left the United States and the United Nations convinced that North Korea needed to be stopped from swallowing the South – and if possible punished for the attempt.

President Truman, still wounded by the accusation that he had "lost" China to the communists in 1949, felt he could not allow the same damaging results to manifest themselves in Korea – especially as Kim Il-sung was supported militarily, financially and politically by both China and Russia. Two days after the North Korean invasion, therefore, he stated unequivocally that "The attack on Korea makes it plain beyond all doubt that Communism has passed beyond the use of subversion to conquer independent nations and will now use armed invasion and war."[57] Dwight D. Eisenhower, then the President of Columbia University and soon to assume command of NATO forces, agreed, saying, "We'll have a dozen Koreas if we don't take a firm stand."[58] Korea was thus a new evolution of conflict in the immediate post-Second World War period: it was a war of warning that stemmed directly from what statesmen had learned from the Munich Agreement of 1938 about appeasing dictatorships.

In stark contrast to later episodes in the twentieth century, the response of the United Nations was powerful. The timing of North Korea's invasion had been unfortunate for Kim Il-sung because the Soviet Union was boycotting the United Nations for allowing the Guomindang to occupy China's seat on the Security Council, rather than giving it to the Mao regime. The boycott allowed the Western powers to pass Resolutions 82 and 84, authorizing a United Nations force to go to South Korea's aid if the North did not withdraw to the Parallel. With these two resolutions, the Security Council turned the Korean War into a United Nations military operation.

The United States would become by far the largest contributor of troops from outside the peninsula, providing 88 percent of the 341,000 UN soldiers and bearing the greatest financial cost, in return for controlling the operation.[59] Alongside South Korea and the United States, troops from Australia, Belgium, Canada, Colombia, Ethiopia, France, Greece, Luxembourg, the Netherlands, New Zealand, the Philippines, Thailand, Turkey and the United Kingdom saw ground action. Eight of those countries sent warships, and five sent warplanes. Transport ships and planes were provided by eight countries, including Norway, and medical units by seven countries, including Denmark, India, Italy and Sweden.[60] The offer of three infantry divisions from the Guomindang was sensibly turned down, as it would have been likely to inflame China.

On 30 June, Truman authorized the naval blockade of North Korea and the dispatch of large numbers of ground troops. "The Korean War occupies a unique place in history," the military historian Max Hastings wrote, "as the first superpower essay of the nuclear age in the employment of limited force to achieve limited objectives."[61] Having been stopped from expanding any further in Europe and denied any role in the pacification of Japan, Stalin turned to Asia to advance the proletarian revolution. He believed after the communist victory in China that he could prod and harry the capitalist West without provoking too heavy a backlash.[62] In May 1950, he had confidently informed Chairman Mao that the Americans were "not ready at present for a big war ... together we will be stronger than the USA and England, while the other European capitalist states ... do not present serious military forces."[63] In this he had miscalculated badly, but 5 million people were to die before that became clear.

Such underestimation of the West's willingness to engage, believing it to be too decadent, has been all too common in post-war history. "Terrorist clerics, godless Marxists and other enemies of the West share few beliefs," writes the commentator Janan Ganesh. "One is that free societies have an innate flakiness: a sort of will to impotence."[64] Yet this theory, which was also shared by the Axis powers between 1936 and 1941, has constantly been proved wrong, most recently in the West's willingness to ship enormous quantities of lethal weapons to Ukraine and to impose significant sanctions and export controls on Russia.

Each of the three American commanders in Korea – Douglas MacArthur, Matthew Ridgway and Mark Clark – were all veritable stars of the Second World War.[65] The personality of generals has always been extremely important in determining the conduct and success of warfare, but few modern commanders demonstrate this truth to the extent of General Douglas MacArthur, whose reputation has undergone ups and downs reminiscent of a Coney Island rollercoaster. For all his undoubted intelligence and great personal bravery – he was one of the most highly decorated American officers of the First World War – and his successful island-hopping strategy against Japan in the Second, the credit for which he must share with Admiral Chester Nimitz, there were many drawbacks to his character that would probably preclude him from holding general rank today.[66]

Despite his stellar, if highly self-curated, reputation, the then seventy-year-old MacArthur – the "American Caesar" as one of his many biographies was entitled – suffered from a series of character flaws that had already emerged before and during the Second World War and were to re-emerge even more powerfully in Korea. Principal among these were hubris and vanity. Even by the standards of the mid-twentieth century, MacArthur was a monster of egotism. "MacArthur is the type of man who thinks that when he gets to heaven God will step down from the great white throne and bow him into his vacated seat," wrote Harold Ickes, the Interior Secretary of the Roosevelt administration. Marshal Foch wrote in his book *Precepts and Judgments* in 1919 that "The power to command has never meant the power to remain mysterious," but MacArthur took the opposite phenomenon to its limits, putting out scores of press releases covering his doings in minute detail. Of the 142 communiqués he sent from Corregidor during the Second World War, 109 spotlighted himself.

Yet mere egotism alone would not have prevented MacArthur from reaching the upper ranks of today's armies.[67] His relationship with the sixteen-year-old half-Scottish, half-Filipina vaudeville star Isabel "Dimples" Rosario, who called him "Daddy," would have given modern colleagues pause for thought, although it was his secret acceptance of $500,000 in "recompense and reward" from the Filipino President Manuel Quezon on 3 January 1942 – worth around $18 million in today's money – that would have sunk his career if it had come out any earlier than it did, in 1979.

MacArthur's major problem in Korea was that he ultimately got the big strategic idea for the conflict wrong, thus failing in the performance of the first and most important task of a strategic leader. His big idea was that it would be relatively easy to destroy the North Korean Army with superior American firepower, and that it did not matter if China sent an army across the Yalu River to North Korea's aid, as he could always destroy that too. Nor did he have senior lieutenants who questioned these two fundamentally mistaken beliefs. It did not help that he conducted the war from Tokyo, over 600 miles away, and hardly ever visited Korea.

Rarely has the lesson of war been better proven that although a commander does not need to be on the front line, a commanding general does at least need to be in the same country as the theatre of operations. The study of large-scale maps, Lord Salisbury once said, drives men mad, and MacArthur was at such a distance from the conflict that, in historian Michael Neiberg's words, he presented "an extreme case of the inability to make the adjustment from total war to limited war."[68] Unfortunately, such an assessment would only become obvious as the war progressed.

Seoul, the capital of South Korea, fell to Kim's forces on 29 June, only four days into the conflict. The US–British–South Korean coalition armies were forced down the peninsula to the Pusan Perimeter, while the US Navy blockaded the North. Coalition reinforcements started to arrive on 1 July under Major General William F. Dean, but when they first came into contact with the North Korean Army north of Osan on 4 July, they were forced to retreat. MacArthur was appointed supreme commander of UN forces in Korea three days later.

The Korean War was fought by Second World War generals, and so it tended to conform to that previous war's strategies and tactics, but there was new equipment. Jet planes had entered service at the very end of the Second World War, and in Korea they predominated. Another evolution saw the use of helicopters, which were only in their very infancy before 1945 but by Korea were used to evacuate the wounded. Thereafter, helicopters were to become a crucial factor in warfare, both for moving large numbers of troops swiftly and (in general) safely and after Korea also as offensive platforms in the form of attack helicopters.

By 1 September 1950, the North Koreans had attacked across the Naktong River, and there was a real fear that the Pusan Perimeter, forming the south-eastern 10 percent of the Korean peninsula on the Straits of Tsushima, might be breached, whereupon the war would have been as good as lost for the coalition. Two weeks later, however, MacArthur pulled off one of the great feats of modern warfare, Operation Chromite, a large and strategically brilliant amphibious attack on Inchon, a key strategic port 100 miles behind enemy lines. Some 83,000 troops of the US X Corps and South Koreans landed in a textbook replay of the kind of operation that had been so successful against the Japanese armies in the Central Pacific.[69]

It is undeniable that brilliance and folly sometimes go hand in hand, as they occasionally did with Douglas MacArthur. It also sometimes takes charismatic generals built on the giant scale – as MacArthur undoubtedly was, for all his faults – to conceive and pull off such an ambitious operation as Chromite, at which lesser figures might have balked. In the Second World War, General George Patton occasionally set himself almost impossible tasks and deadlines, and accomplished them, highly conscious that history was watching him.

Yet there are always downsides when a general's ego is epicentral to an operation. The success of Operation Chromite only boosted MacArthur's self-regard, not least because General Omar Bradley, the Chairman of the Joint Chiefs of Staff and another giant of the Second World War, had initially opposed it, stating that "large-scale combined amphibious operations ... will never occur again."[70] The naval staff were equally opposed, pointing out the 32-feet high tide, one of the largest in the world, and the two easily mined narrow approaches. MacArthur responded by pointing out that it was Inchon's very defensibility that would ensure the element of surprise. "By seizing Seoul I would completely paralyze the enemy's supply system coming and going," he argued in a bravura performance to the Joint Chiefs.[71]

Although military history is full of amphibious assaults from the Siege of Troy onwards, they have always been a highly risky maneuver, because of the danger of troops being flung back into the sea while at their most vulnerable: the moment of transferring from sea to land. Operation Chromite was conducted without any practice that might have alerted the enemy, and it was a triumph. It was to be followed over the next three-quarters of a century by a large number of successful amphibious assaults, including Operation Musketeer during the Suez Crisis in November 1956, Operation Starlite in Vietnam in August 1965, Operation Motorman in Ulster in July 1972, the Turkish invasion of Cyprus in July 1974 and on three occasions in the Falklands War in April, May and June 1982. The Al-Faw Peninsula in Iraq has witnessed two amphibious assaults, from Iran in 1986 and from Britain's 3 Commando Brigade in 2003. For all the technical difficulties in moving troops from sea to land and supporting them, amphibious assaults have been seen in struggles as far removed geographically and operationally as the Sri Lankan Civil War, the Gulf War, the Comoros Islands, Kenya and Yemen. The fact that any future Chinese invasion of Taiwan would be difficult and risky does not mean that it would be historically rare.

The Inchon beachhead was consolidated after only two days, and all of South Korea was liberated in fifteen, with Seoul recaptured on 26 September. Many North Koreans escaped MacArthur's trap, however, retreating along the peninsula's eastern coastal road. Here came the second great act of hubris of the war, after Kim's initial decision to invade, but at the time its underlying assumption was shared by the administration. On 30 September 1950, George C. Marshall, the newly appointed US Secretary of State, told MacArthur that "We want you to feel unhampered tactically and strategically to proceed north of the 38th Parallel," but without explaining what he was supposed to do there.

The hubris of invading North Korea was therefore that of the Truman administration, but it was more than reflected in MacArthur, who believed that he could reunite the peninsula under Syngman Rhee in just three months. On 1 October, the day after being given the go-ahead by Marshall, MacArthur sent South Korean troops streaming across the 38th Parallel, ignoring clear warnings from China that it would intervene if North Korea were invaded. MacArthur's subsequent demand that he be allowed to bomb Chinese targets in North Korea was refused. His belief that nuclear weapons were "merely quantitative increments to the arsenal of war," and his request that he should be given permission to employ them if necessary, were not picked up early enough by the Truman administration as signs of his growing instability.[72] His stated intention was "to drive to the Chinese border, dismantle North Korea, and unify the peninsula under Syngman Rhee," and he gave no guarantees that he would even stop at the Chinese border.[73]

MacArthur made a series of other errors in underestimating both the likelihood and the consequences of Chinese intervention. He pushed his forces northwards on one of the widest fronts in modern warfare, made an amphibious landing at Wonsan that did not succeed in trapping significant numbers of the enemy, and divided and overextended his forces despite the oncoming severe winter and the fact that central Korea's lateral valleys meant that coalition units were unable to support each other.

On 12 October, Mao ordered his "volunteers" – actually the Chinese XIII Army Group – to start slipping across the Yalu River from Manchuria. This they did for thirteen nights in numbers that initially reached 130,000, and eventually over a quarter of a million men.[74] This extraordinary undertaking was managed without MacArthur's intelligence officers noticing anything at all. The Chinese used camouflage, eschewed wireless communication and carried everything by hand – they were, in the words of the American brigadier-general Samuel Marshall, "a phantom that cast no shadow" across 300 miles.

So confident were the American generals in their own success that, in the words of Max Hastings, "their senses were deadened to a fresh consideration."[75] This failure to spot an entire Chinese army crossing the Yalu River has been described as "probably the greatest intelligence lapse in postwar American military history."

The Chinese crossing of the Yalu had been under way for three nights when MacArthur met President Truman on Wake Island on the early morning of Sunday, 15 October 1950. MacArthur's choice of meeting place forced his Commander-in-Chief to travel a total of 14,400 miles across seven time zones.[76] "Have to talk to God's right-hand man tomorrow," Truman noted the day before he met MacArthur for the first time.[77] "Two men can sometimes learn more of each other's minds in two hours, face to face, than in years of correct correspondence," was the justification that Admiral Arthur Radford, the Commander of the Pacific Fleet, made for the meeting, at which he was present.[78]

There was to be a good deal of mythologizing on both sides about the meeting afterwards, and it was later alleged that Truman minded not being saluted and was put out by MacArthur's open-necked khaki shirt, but at the time the encounter was friendly.[79] "I've been waiting a long time meeting you, General," said Truman. "I hope it won't be so long next time, Mr. President," replied MacArthur.[80] It was only later, once the relationship had comprehensively soured, that accounts became embellished or even invented wholesale. Nor was it helped by the often unreliable later reminiscences of some of the seventeen aides and advisers present. In a ninety-six-minute meeting, no fewer than five people took notes and there was a shorthand stenographer, so we have an accurate report of what was said.

There was no agenda; Truman asked questions and MacArthur answered them at length, telling the President that the enemy capital Pyongyang would fall within a week (which it did), after which North Korea's "formal resistance" would end by Thanksgiving, and the US Eighth Army would be back in Tokyo by Christmas. When Truman asked what chance there was of either Russia or China directly assisting North Korea on the battlefield, MacArthur replied, "Very little ... We are no longer fearful of their intervention."[81] He thought that China could not put more than 60,000 men across the Yalu River, and they would be unsupported by any air force.[82] MacArthur had characteristically overpromised and underdelivered in giving specific dates for victory to Truman, in contravention of a much more sensible approach for generals when making predictions to presidents (which is to underpromise and overdeliver).

A great deal of ground was covered in these good-natured discussions, although when the meeting ended at 9:05 a.m. MacArthur turned down Truman's offer of lunch later that day, which Bradley thought insulting to the President, whether intentionally or not. At the time, however, the *New York Times* described Truman as "like an insurance salesman who had at last signed up an important prospect."[83] The official communiqué commented on "the very complete unanimity of view" of the meeting – which was no more than accurate – and described MacArthur as "one of America's great soldier-statesmen."[84] Truman awarded MacArthur a Distinguished Service Medal on the runway before returning home.

In its immediate aftermath, the meeting was deemed a success, but only because the war looked like it was about to be won. In his crucial assessment of the likelihood of Chinese and Russian intervention, however, MacArthur had completely miscalculated. Personal relations between the President and his theatre Commander, strong at Wake Island, collapsed not because of any perceived rudeness on MacArthur's part, but because he had got a vital aspect of the war entirely wrong. Personal relations between the military and politicians can be good, bad or indifferent, but nothing makes up for the difference between victory and, as in this case, near defeat. Of course, MacArthur was not alone in his assessment of China, which the CIA also believed would not intervene. "On this one MacArthur and the rest of us were all wrong," Assistant Secretary of State Dean Rusk later admitted.[85]

On 25 October, the Chinese launched their counter-attack in the battle of Onjong 40 miles south of the Yalu River. By falsely designating his troops as volunteers, Mao hoped to avoid a wider, direct war with the United States. Similarly, when in early November Stalin also provided Soviet MiGs, their Russian pilots wore North Korean uniforms and their planes had the markings of the North Korean People's Army Air Force. Soviet military leaders, fearful of their pilots being exposed as Russian, ordered them not to enter South Korean airspace lest they were downed and captured.

By 27 October, the US Eighth Army had been halted by the enemy advance, and three days later South Korea's 6th Division was routed by the Chinese at Yongdu. In this phase of the war, the Chinese "countered the tactics of George S. Patton with those of Stonewall Jackson," with light infantry moving through the mountainous terrain in winter and outmaneuvering the better-equipped, road-bound UN forces.[86] An analogy can be made with the Japanese attack in Malaya in early 1942, when the invading troops avoided roads and instead carried their equipment through the jungle.

By early November 1950, the coalition was in retreat, and MacArthur found himself fighting a very different war from the limited one that he, Truman and the Joint Chiefs had envisaged. The concept of limited, proxy wars might seem unheroic to some, but it has saved the United States from entering into a direct conflict with a superpower for over seventy years. Yet MacArthur wanted to extend the Korean War to include a naval blockade of China, the bombing of bases in Manchuria, the use of Guomindang battalions from Taiwan and later the option to use tactical nuclear weapons. "The new intellectual ground of limited war proved difficult to navigate," records Michael Neiberg of MacArthur. "The total war mindset of 1941–1945 proved difficult to set aside."[87]

The Joint Chiefs of Staff understandably feared that such a full-scale confrontation with a Soviet ally might lead to a clash with Russia in Europe. On 8 November, the first jet-aircraft dogfights in history were seen in an air battle above Sinuiju. As the US F-80s and especially the F-86s were superior to Russian-flown MiG-15s, the coalition retained air superiority, which under normal circumstances should have been a decisive factor, but was not in Korea because superiority never turned into dominance.

The Chinese First Wave Offensive of 25 October to 24 November resulted in 27,827 American casualties, over one-third of whom were killed or missing in action.[88] On 24 November, the day after the Thanksgiving that MacArthur had promised the President would herald the North Korean military's *Götterdämmerung*, MacArthur launched a counter-offensive, telling his staff within earshot of reporters that "If this attack is successful, I hope we can get the boys home for Christmas."[89] It was a vainglorious remark of the kind every senior commander should avoid at all costs. The offensive having failed after the battle of the Chosin Reservoir, in which UN troops (nicknamed "the Chosin Few" and "the frozen Chosin") were encircled by over 100,000 Chinese troops, coalition forces withdrew further south, with Siberian winds whipping down the peninsula.

At 3 p.m. on 28 November, Truman and Bradley categorically refused MacArthur permission to extend the war into China by means of B-29 bombing and a naval blockade, which David McCullough considers "stands among the triumphs of the Truman administration."[90] Nonetheless, a week later the Chinese and North Koreans retook Pyongyang, and by mid-December both armies were back at the 38th Parallel. All of MacArthur's recent efforts had been for naught. Furthermore, the North Korean and Chinese forces weren't sated.

Lieutenant General Matthew Ridgway, the Deputy Chief of Staff of the US Army, had commanded the 82nd Airborne Division in Sicily, Italy and Normandy in the Second World War, and afterwards the XVIII Airborne Corps. On 22 December, he received a call while enjoying a cocktail at a friend's house, informing him that he had been appointed as the new ground commander of the US Eighth Army, his predecessor, Walton "Bulldog" Walker, having been killed in a jeep accident. With enemy forces, which now numbered almost half a million, crossing the 38th Parallel on 28 December as the coalition army retreated southwards through the snow and ice, the situation he inherited could hardly have been worse – not least because his commander Douglas MacArthur had by then become too loose a cannon to be entrusted with the higher direction of the war. (Remarkably, no member of the US Joint Chiefs had visited Korea between July 1946 and July 1950.)

American losses had increased alarmingly; 14,650 Americans had already died in the war, most in November and December 1950 – some 799 on 30 November alone.[91] In opinion polls in November and the following January, half of Americans said it had been wrong to enter the war, and more than half wanted nuclear bombs used on enemy military targets to bring it to an end. With his Commander-in-Chief denying him any such course of action, MacArthur turned defeatist. "The command," he stated, "should be withdrawn from Korea just as rapidly as it was tactically possible to do so."[92] Plans were made to evacuate the UN forces to Okinawa, mainland Japan and the Philippines, as MacArthur worked on a narrative to blame the defeat on Truman's refusal to allow him to widen the war.

In Ridgway, by contrast, the United States had a commander who was soon on the ground and who had no political ambitions. An extraordinarily capable man with a strategic mind, he brilliantly executed the four tasks of a strategic leader during his time in Korea. Most importantly, he quickly concluded that the main problem was leadership, rather than anything to do with the enemy. Walker had retreated too precipitously, and an evacuation of the US Marines from eastern Korea had been badly handled, but the hastiness of the Chinese thrust would give Ridgway opportunities for counter-attack as the communists extended their lines of communication going south in much the same way as MacArthur had going north.

Dismissing the common trope of the superiority of the communist fighting man, Ridgway put his trust in the sheer lethality of advanced American weaponry.[93] He laid down four strong defensive lines all the way back to Pusan, and, in contrast to MacArthur, remained resolutely optimistic about final victory. It is hard to avoid the conclusion that, with a very few specific exceptions such as Inchon, Douglas MacArthur's performance in Korea provides an excellent example of how not to command, whereas Ridgway's more competent, hands-on and determined approach proved better suited to the tough, attritional nature of the war.

On 4 January 1951, Chinese and North Korean forces retook an evacuated Seoul. Later that month the coalition returned to the offensive, retaking Inchon on 10 February. A huge communist offensive in central Korea using human-wave frontal assault tactics was then repulsed with massive loss of (enemy) life. In April and May alone, the communists suffered 160,000 casualties.[94] These attacks were reminiscent of the kind that had been undertaken by the Russians at Stalingrad or the Japanese in the Pacific during the Second World War – utterly disregarding the human cost in the hope of overwhelming the enemy. It is a tactic that can be more easily undertaken by totalitarian powers.

A French general of the First World War once said that "Three men and a machine gun can stop a battalion of heroes," and however bravely the Chinese human waves charged onwards, Western firepower wreaked devastation among them.[95] At Kap Yong in April 1951, for example, an attack on a well-entrenched Canadian battalion supported by artillery killed 800 Chinese in twenty minutes, for the loss of twenty Canadians.[96] Meanwhile, B-29s pummelled relentlessly the elongated communist supply lines.

General MacArthur's increasingly public disloyalty and defeatism resulted in President Truman relieving him of his command on 11 April 1951, after a letter from MacArthur critical of the President's concept of limited war became public. When the news of MacArthur's sacking came through over the radio at one divisional command post, there was first a terrific wind that levelled tents, followed by a hailstorm that lashed the countryside, followed again by a driving snowstorm. "Gee," one soldier exclaimed, "do you suppose he really is God, after all?!"[97] MacArthur's ill-disguised involvement in American domestic politics was a cardinal mistake. Generals often have to undertake political roles as part of their jobs in theatre, but in democratic countries they should never interfere directly in their own country's politics. Although Clausewitz was correct to say that war is politics by other means, generals ought to confine themselves to the politics of the place where they are fighting rather than back home.

Despite his remarkable success with Operation Chromite, MacArthur had consistently misunderstood the strategic theatre of the war and failed to develop the appropriate approach. Instead it was Ridgway who would turn around the coalition's fortunes.

On 22 April, the Eighth Army reached the 38th Parallel. "The enemy forces are so closely knit together," reported the communist commander Peng Dehuai, "that there exists no gap and tactically they advance and retreat steadily and entrench themselves at every step."[98] Once he had reached the 38th Parallel, Ridgway was ordered to hold it and essentially to fight for a stalemate. The antithesis of MacArthur, Ridgway grasped the situation and undertook his task by maintaining a long continuous defensive line across the peninsula which was supported by massive firepower, and which the Chinese could not break through. Just as in the First World War, when the line of trenches stretching from Switzerland to the English Channel made large outflanking maneuvering impossible, the Korean War settled down to an attritional slogging match which lasted for two more years, during which the Chinese and North Koreans continued to suffer huge losses by attacking in close formation.[99]

Generals often need to be diplomats too, and on 25 October 1951 Ridgway took part in armistice discussions at Panmunjom. There is, of course, a serious danger in undertaking constant public peace negotiations during a war, since it necessarily undermines morale among those fighting. No one wants to be the last person to die in a war that is about to end. This problem extends disproportionately powerfully to democracies, which cannot hide and cover up peace negotiations in a way that totalitarian powers can and do. Regardless, nothing came of these negotiations, and in March 1952 the communists falsely accused American forces in Korea of using germ warfare, a classic example of the "Big Lie" so beloved of Joseph Goebbels. In China today, the Korean conflict is referred to as "the War to Resist America and Aid Korea," and is presented entirely as a defensive response to a threatened American invasion.

A Korean armistice – crucially, not a peace – was finally signed at Panmunjom on 27 July 1953. MacArthur might have told Congress in April 1951 that there was no substitute for victory, but in this case an uncomfortable but workable compromise did emerge. The Korean War showed that one of the most important ways that warfare had evolved since 1945 was that one is rarely rewarded with the satisfaction of the suicide of the enemy dictator in his bunker in Berlin, or a formal surrender signature ceremony aboard the USS *Missouri* in Tokyo Harbor. Wars end more messily in the modern world, and there are several acceptable substitutes for victory if it turns out – usually due to domestic political considerations rather than anything that has taken place on the battlefield – that outright victory is beyond reach.

Over 400,000 North Korean and Chinese soldiers died during the Korean War, alongside around 1.5 million civilians – although communist estimates of losses are notoriously unreliable and it might have been many more. The United States lost over 36,000 killed – 33,471 battlefield deaths and 2,827 non-battle – while South Korea suffered around 140,000 military deaths and the loss of around 1 million civilians.[100] Kim Il-sung survived to found a Stalinist dynasty which his grandson rules to this day, now strengthened by nuclear weapons. In 1953, few would have predicted that no new attempt would be made to cross the Parallel over the coming seventy years, but even as European communism collapsed, new states were formed around the world and the tectonic plates of global power shifted again and again, the 38th Parallel remains the border between the Koreas.

Korea changed warfare in several significant ways, proving that limited wars could be fought under MAD and introducing the United States to many of the problems that it was to face in the future in Vietnam, Iraq, Afghanistan and elsewhere. In Korea, coalition forces had to work in conjunction with an undemocratic, unpopular host government, against an ideologically motivated, largely peasant force that it badly underestimated, with airpower that found it hard to pick out worthwhile targets in mountains and forests.[101]

When MacArthur learned of the Korean armistice in July 1953, he said, "This is the death warrant for Indo-China."[102] It was true that the news that the West could be fought to a standstill electrified the communists of Indo-China (modern-day Vietnam, Cambodia and Laos). And within a year of the Korean War, they had comprehensively defeated the French at the battle of Dien Bien Phu. For once, Douglas MacArthur was not being histrionic.

Two

Wars of Decolonization

1947–1975

If the political aims are small, the motives slight and tensions low, a prudent general may look for any way to avoid major crises and decisive actions ... If his assumptions are sound and promise success, we are not entitled to criticize him. But he must never forget that he is moving on devious paths where the god of war may catch him unawares.

Carl von Clausewitz, *On War*, 1832[1]

From controlling a mere 15 percent of the world's landmass in 1450, and still only 35 percent by 1800, European nations came to control 84 percent of it by 1914. Yet between 1943 and 1975 the largest transfer of territorial control in world history took place, as all the Western European empires were swept away.[2] The rapid vanishing of European empires over one generation in the mid-twentieth century had significant ramifications for the evolution of conflict. The typical face of war since 1945 has taken the form not so much of traditional state-on-state conflict as of insurgency and guerrilla warfare – especially in the era of decolonization when the British fought in Malaya, Kenya, Cyprus and Aden, and the French in Indo-China and Algeria. British moral exhaustion, financial near bankruptcy and domestic antipathy in the wake of the Second World War resulted in a series of end-of-empire struggles. Many of these bids for post-imperial power took the form of guerrilla warfare, and Britain struggled to adapt.

Far from being a bastardization of the normal form of warfare in which armies face each other in the open field, guerrilla or insurgent warfare has in fact been the norm almost throughout history. The words often used to describe it, such as "non-traditional," "irregular" and "unconventional," are thus misleading: time and again – from prehistoric tribal warfare to the War against Terror – the side that cannot hold its own on the open field of battle must resort to guerrilla action, assuming it has enough support from the local population. Although earlier guerrilla conflicts have encompassed the Peninsular War (from where the expression originated), the Boer War, the Cuban Uprising, Orde Wingate's Chindits in Burma and American forces in Borneo, it was the Viet Minh victory over France in Indo-China which first saw a European great power humbled by guerrilla forces from the developing world.

Kashmir: long imperial afterburn

The Pashtun tribe's invasion of Kashmir in October 1947, just two months after Britain's painful partition of India, inaugurated a conflict that has lasted to the present day, three-quarters of a century later. Burma was granted its independence at the same time, and in his database of all guerrilla wars that have killed over a thousand people since 1775, the military historian Max Boot points out that Myanmar's struggle against the Kachin and Karen tribes started in August 1948 and is still ongoing. Referring to the ancient clan conflicts of medieval England and Scotland, he adds that:

> Those who imagine that long-running conflicts – say, over
> Kashmir or Palestine – can be resolved neatly and expeditiously
> through negotiations ignore the lessons of the Anglo-Scottish
> wars: conflicts of blood and soil, fought as a series of

skirmishes between guerrillas and regulars, can drag on for
centuries even among peoples far closer in religion and
outlook than the Indians and Pakistanis or the Israelis and
Palestinians.[3]

The Kashmir conflict deserves a place in a book on the evolution of
warfare because it is the exception that proves the rule. The mountainous, tribal and relatively unchanging nature of the simmering conflict
shows that progress towards "smart" warfare has been glacially slow in
some places, reflecting societal differences. A Kashmiri guerrilla fighting against Indian, Pakistani or Chinese forces today will not be using
precisely the same weaponry as his forebears – sometimes literally his
blood-relative forebears – but he will be using much the same tactics,
for much the same reasons.

These slow-burn conflicts that are measured in decades as opposed
to months or years prove the impossibility of generalizing too broadly
about the evolution of warfare, as there will always be places where it
has evolved very little, yet they are still worthy of study. "The attempt to
judge an empire would be rather like approaching an elephant with a
tape measure," wrote the historian Margery Perham in 1961, and sometimes that seems true of the history of insurgency warfare.

In her book *What Terrorists Want*, Louise Richardson has identified
the "Three Rs" that inspire terrorists, however hopeless their cause.
These are: Revenge for real or perceived wrongs, Renown that might
give meaning to the terrorists' lives and the desire to provoke a Reaction
from enemies, thus energizing their compatriots.[4] That these inspira-
tions for terrorist activity have remained powerful intoxicants over
millennia is testament to how effective they are – and how immensely
difficult they are to eradicate. All three Rs remain evident in Kashmir,
as of course does a fourth: Religion.

"History seems sometimes to move with the infinite slowness of a
glacier," Lord Mountbatten, the last Viceroy of India, once observed,
"and sometimes to rush forward in a torrent."[5] Ironically enough, it was
Mountbatten himself who precipitated the torrent that led to Kashmir's
disputed destiny. His partition of India was swift, sudden and

simplistic: it would be divided into two. Kashmir, the northernmost and mountainous region, would, at the behest of its Hindu Maharajah, belong to India rather than Pakistan – despite what its 77 percent Muslim population and geographical position would otherwise have implied.

On 27 October 1947, Mountbatten, now India's first Governor-General and thus *parti pris* in India's favor, formally accepted the Maharajah of Kashmir's accession of his state to India, although he did state that a referendum would have to be held to confirm it. Even before the Instrument of Accession was signed, however, 300 troops of the Indian 1st Sikh Battalion were airlifted to the Kashmiri capital of Srinagar, raising fears that the arrival of more Indian troops would affect the plebiscite.[6] That same day, some 7,000 pro-Pakistani tribesmen moved into Kashmir and captured the town of Baramula. Rather than marching on Srinagar, however, they instead burned and pillaged Baramula, allowing Indian airpower to catch up with their activities. As the tribesmen tried to capture Pattan on the way to the capital, they were strafed and bombed by the Indian Air Force. Meanwhile, India used transport planes to airlift in troops who were ultimately to defeat the raiders.

In his book on the global history of airpower, Jeremy Black points out that the Kashmir example proved how important these largely Second World War planes were "as newly independent states struggled for territory and primacy in the aftermath of the retreat of the imperial powers."[7] This would be the case not just in South Asia: Israel's War of Independence relied on Spitfires smuggled from an old Luftwaffe runway in Yugoslavia, and old P-47 Thunderbolt and C-47 transport planes joined the Imperial Iranian Air Force.[8]

Although the Pakistani raiders captured Pattan on 31 October, they could not take Srinagar. There were skirmishes on the outskirts of the capital in early November, but the Indian Air Force had so depleted the Pakistanis' vehicles and ammunition that it proved impossible to capture the city, and by 8 November they had retreated to the small town of Uri. The immediate prospect of Kashmir falling into Pakistan's orbit by *force majeure* had disappeared, although the fighting widened and continued until New Year's Day 1949, when the United Nations brokered the first of many transient ceasefires.

A de facto 464-mile "Line of Control" was drawn, by which India controlled roughly two-thirds (the Kashmir Valley, Jammu and Ladakh) and Pakistan controlled the other third (Azad Kashmir and Gilgit-Baltistan). Neither side was happy with the UN-forced compromise, and neither recognized the Line of Control, either formally or in practice.

India, meanwhile, consistently blocked the political means of deciding Kashmir's fate, knowing that the majority Muslim population of the region would vote for it to join Pakistan. (In the Indian-administered Kashmir Valley, for example, around 95 percent of the inhabitants were – and still are – Muslim.) Two wars were fought over Kashmir in 1965 and 1999, and today it is de facto partitioned between India, Pakistan and China, though the former two states do not recognize this. Human rights abuses by all three states are frequently noted by the United Nations, but go largely unpunished. The endless cycle of violence – including killings, disappearances and sexual assault – has left some 8 percent of the population suffering from Post-Traumatic Stress Disorder.[9]

The low-intensity conflict in Kashmir has not yet risen to the status of a guerrilla war, but has been characterized by communal riots, individual acts of terrorism, the occasional assassination of public officials and other such bursts of violence, both organized and spontaneous. These tend to be repressed by the authorities in a manner that engenders further resentment, anger and violence, but generally without provoking outright warfare. As well as Kashmir, such long-term, low-intensity bloodletting has been seen in Myanmar, Gaza and the West Bank, Sudan, Northern Ireland and several other places around the world where the British had left their empire and resorted to hasty partitions to try to achieve peace.

The long-term failure of Britain's partition-and-scuttle policy could be seen as recently as 1999 in the Kargil War between India and Pakistan of May to July that year, half a century after the creation of the Line of Control. This saw a fierce fight in the barren mountainous region along a stretch of the Line, which by 1972 had become one of the world's most heavily militarized borders – as it remains to this day, alongside the 38th Parallel in Korea. Tensions regarding control of Kashmir are not limited to India and Pakistan – the Line of Actual Control in Ladakh also exists between India and China, and it has seen lethal fighting as recently as June 2020.

The historian Sumantra Bose wrote in 2021 that the Kashmir conflict remains "rife with incendiary ingredients, and represents a clear and present danger to international peace and security in the early twenty-first century."[10] With China's growing ascendancy in the early twenty-first century, its desire to control key strategic points – including in South Asia – has grown in tandem. Given these dynamics, the "Kashmir question" will not soon disappear.

The Israeli War of Independence: an existential struggle for survival

Fighting broke out in Palestine before the post-war UN-promulgated British Mandate had even expired on 15 May 1948. Egyptian planes bombed Tel Aviv on 14 May, the same day that David Ben-Gurion proclaimed the State of Israel at the Museum of Art at 5 p.m., although there were no deaths. American recognition of Israel came only eleven minutes after Ben-Gurion's proclamation, and Russia recognized the new country too, but Israel immediately suffered invasion from five armies comprising over 20,000 well-equipped Arabs from Egypt, Iraq, Syria, Transjordan (later Jordan) and Saudi Arabia.

The 1948 war goes down as one of several conflicts of the late 1940s that shaped the future of the world right up to the present day. "Israel was born in battle," wrote Chaim Herzog, the head of military intelligence for the Israeli Defense Forces (IDF) in the War of Independence, and a future president of Israel. "Its army was forged in the fires of conflict and the ongoing struggle for existence that has characterized the State of Israel since its foundation."[11] Yet, in comparison to the highly sophisticated military force for which the country became famous, this war was, in the words of Field Marshal Michael Carver, "an unsophisticated affair" because it was primarily fought against the backdrop of the demand by the UN General Assembly for ceasefires, which forced each side to try to take what territory it could as quickly as possible, for use as bargaining chips once the ceasefire was imposed.[12]

On 15 May, Iraqi troops crossed the River Jordan at Gesher, and that night Syrian forces overran two Jewish settlements in the plain below the Golan Heights. Ben-Gurion, the new state's founder and first Prime Minister, ordered Moshe Dayan, commander of the paramilitary Haganah – which was soon to become part of the IDF – to "Hold the Jordan Valley."[13] Meanwhile, in the south, the Egyptian Army marched towards Tel Aviv as other units landed from the sea at Majdal; Lebanese troops invaded from the north, and Iraqi and Jordanian forces from the east. Saudi Arabia also sent a contingent, which fought under the Egyptians.

Although the strategic situation seemed desperate for the Israelis, they had two great advantages. Firstly, they were fighting with organized interior lines of communication, supply and reinforcement, as well as having a central command, whereas the Arab attacks were largely disjointed and uncoordinated. "What Israel had lacked in resources we had made up for in ingenuity and organization," wrote Shimon Peres, the head of the Israeli Navy at the time. "And what our enemies had bountifully possessed they thankfully had squandered in the chaos."[14] Secondly, the defenders knew that defeat would certainly result in the extinguishing of the Jewish state and possibly the annihilation of the Jews of Palestine. Only three years after the Holocaust, they were under no illusions about what that would mean for Judaism.

Israeli morale remained extremely high during the war, despite several moments when the nascent country seemed doomed. As the history of conflict demonstrates over the millennia, while morale is impossible to quantify, it is essential to victory. Although many of the Egyptian, Syrian and Jordanian soldiers fought bravely to reclaim the territory of their Arab brothers, their quest was not as existential, as biblical and as all-embracingly essential as that of the Israelis, who would go to incredible lengths in an attempt to cement the legitimacy of Zionism.

In David Ben-Gurion, the Israelis had a leader who totally grasped the big ideas, having been thinking profoundly about them for many years. It was Ben-Gurion who in the autumn of 1947 reorganized the Haganah and set up the National Command representing all the political parties of Jewry in Palestine. It was he who, as minister of defense as well as prime minister, had taken the excruciatingly hard decisions over where to send Israel's 65mm guns; who showed personal bravery as a stretcher-bearer for his own sentry wounded by Egyptian shelling; who spotted that Jerusalem was central to the struggle, whatever happened in the Negev desert region; who accepted Yitzhak Rabin's arguments to change the focus of the relief of Jerusalem; who put down the Irgun uprising during one of the short truces; and who expelled the Arabs of Lydda and Ramle.

Chaim Herzog was to describe Ben-Gurion as the person who "led the Jewish population of Palestine – besieged, cut off, subject to a British naval blockade, deprived of basic weapons necessary for self-defense on many occasions, and fighting on all fronts against heavy odds."[15] Ben-Gurion's oratorical powers, inspired choice of commanders such as Moshe Dayan and Yigael Yadin, Israel's Head of Operations, and his drive in the implementation of the strategy, as well as his capacity relentlessly to refine the strategy in response to fast-moving events, render him the personification of a great strategic leader.

Israel called up every able-bodied male resident in the country, and by October the IDF, which had been only 20,000 strong five months earlier, numbered some 90,000.[16] By contrast, the invading armies' numbers were only modestly reinforced and also remained fairly static.[17] Another aspect of the Israeli War of Independence was the presence of foreign fighters drawn to the conflict by a belief in the cause, as had already been seen in the International Brigades in the Spanish Civil War in the 1930s (and would be seen again in Ukraine in 2022). There was a stream of volunteers to fight for Israel from all over the world in 1948, primarily but not exclusively from the Jewish diaspora. The historian Sir Martin Gilbert has written of "men and women who left their work and families in order to be of service on the battlefield ... who did not want to see the Jewish State destroyed."[18] They were known as

Mahal* and probably numbered as many as 5,000. Unlike in some later conflicts, world opinion was solidly behind Israel.

Syria sent a brigade of tanks and armored vehicles across the Sea of Galilee to attack Jewish settlements at 4:30 a.m. on 20 May, assaulting the Degania kibbutz, which Israeli strategists recognized as the key to holding the northern part of the country. Sometimes the fall or survival of crucial points can depend on relatively small amounts of armaments and the sheer tenacity of those wielding them; in this case two 65mm Israeli guns and some Molotov cocktails and hand grenades saved the kibbutz from being overrun. Meanwhile in the south, the Israelis used PIAT[†] anti-tank weapons to defend the Kfar Darom kibbutz 7 miles south of Gaza. Simultaneously, a battle for the kibbutz of Yad Mordechai on the Gaza–Majdal road lasted five days, saving Tel Aviv. To the south of Jerusalem, the Ramat Rahel kibbutz changed hands four times before the Israelis finally were able to retake and hold it for good. "Outnumbered and outgunned," Shimon Peres wrote seven decades later, "we refused to be outmatched, and our forces used whatever they had to defend their positions."[19]

On 26 May, the battle for the Jewish Quarter of the Old City of Jerusalem came to a climax when the British-officered Jordanian Arab Legion artillery attacks forced the defenders to surrender on 28 May. They were taken to Jordan as POWs, and the Jordanians were to control the most important spiritual home of Jews for the next two decades. Elsewhere in the city, however, resistance continued. In late May, the Israeli Air Force deployed the four Messerschmitts they had purchased from Czechoslovakia, managing to delay an Egyptian armored column. Although they could not achieve air superiority over the Arab air forces, the Israeli Air Force was able to deny it to the enemy in turn – a crucial tactical victory in itself.

The Israelis used misinformation extremely effectively. Winston Churchill once remarked that "In wartime, truth is so precious that she should always be attended by a bodyguard of lies." While Yigael Yadin

---

* The Hebrew acronym for "Mitnavdei Hutz la-Aretz" (volunteers from abroad).

† Projector, Infantry, Anti-Tank.

told journalists that his troops had completely encircled Egyptian forces at Isdud, after the war he admitted, "This was a complete bluff. There was not a single word of truth to this at the time I gave this statement."[20] Yet when reported in the media it convinced the Egyptian high command in Cairo, which reported it as a statement of fact to the field commander at Isdud. While outright, bold-faced lies addressed to the world's media come with obvious moral and political risks, Yadin's simple, shameless lie could not have resulted in greater success. The combination of sympathetic journalism and a nation fighting for its very life had created the perfect opportunity to deploy an effective "bodyguard of lies."

Although the IDF entered Jenin on the West Bank on 4 June, the city was recaptured by a spirited Iraqi counter-attack later that day. Nonetheless, by the end of the first week in June, more than 200,000 Arabs had fled 155 villages in areas under Israeli control, and 73,000 more from Jaffa and Acre, with 40,000 leaving Jerusalem alone.[21] This was a mass population shift that was to have profound implications for the State of Israel and the surrounding territories for decades to come.

The Jordanian Arab Legion's control of Latrun had, since 25 May, closed the main road to Jerusalem and caused extreme rationing there, with soldiers and civilians alike afforded just two pieces of bread per day. On 11 June, after four weeks of conflict, a four-week UN truce came into effect. Israel by then controlled Galilee, the Jezreel Valley from Haifa to the Jordan River, a pocket of the central Negev and the coast down to Isdud, and had kept open an alternative route into Jerusalem.[22] The four weeks of the truce were spent building up resources rather than genuinely seeking a partition settlement at the United Nations. Weapons were brought in from Czechoslovakia, and Jewish immigrants from British detention camps in Cyprus made their way into IDF units.

Fighting resumed at midnight on 9 July, whereupon the Israelis extended the area of Jerusalem that they controlled and tried to push the Syrians back across the River Jordan, which they failed to do after ten days of heavy fighting in the face of Syrian air and artillery superiority. Ramle fell to the Israelis on 12 July, and some 50,000 Arabs from the area fled into the Samarian hills, leading to the deaths of some 355 refugees.[23] As had been seen in the Fall of France in 1940 and was often to be repeated afterwards, the movement of refugees blocks roads and can slow advances; the Arab Legion discovered exactly that with those on the road from Ramle and Lydda.

Another modern aspect of warfare was to be seen on 12 July when the Israelis captured Ras al-Ai from the Iraqis, allowing the water of the River Yarkon to be diverted back to Jerusalem, whose citizens had been rationed to a gallon per individual per day for washing, cooking and drinking.

The 12th of July heralded another important tactical victory for Israel. At Negba, in the southern Negev region of Israel, Egyptian forces moving towards Tel Aviv were forced to withdraw after an Israeli settlement held out – surviving an assault in which 4,000 shells were fired at its 150 defenders. When a captured Egyptian officer asked how Negba had survived the onslaught, the settlement's newspaper explained in words that demonstrated a remarkable clarity about war itself and an essential component that has influenced the fortunes of commanders of superpowers and guerrilla fighters alike:

> Quantitatively, the balance of forces was in your favor; but
> qualitatively, it was in ours. There was one item which the
> Egyptians were lacking – a clear idea about the purpose of the
> battle. But this is precisely what we possessed in full measure:
> fearless spirit, the knowledge and conviction about the purpose
> of the battle, the realization that with our own bodies we
> barricaded the way north.[24]

Minefields were instrumental in protecting other settlements, such as Gal-On, 15 miles east of Negba. Although landmines are understandably reviled in the twenty-first century for their capacity to maim innocent civilians of all ages long after conflicts have ended, they can also, just as at the battle of El Alamein in 1942, have the invaluable effect of closing or slowing down enemy maneuverability.

When the IDF captured the Arab town of Nazareth in Galilee on 16 July, looting, a common if vile aspect of war that persists down the ages, broke out among some of the Israeli soldiers.[25] Soldiering is not a well-paid profession, and the chaos of war provides plenty of opportunity for looting the local populace, especially when officers are not fully in control of their units, or, even worse, deliberately turn a blind eye to the practice. Looting is a staple of war, but no less reprehensible for that, and it understandably leads to lasting resentment among its victims. In Israel's War of Independence, which was overall a noble struggle, the practice of looting was having a noticeably detrimental effect on defensive capabilities: Arab counter-attacks were increasing in success almost entirely due to inadequate defensive preparedness which had suffered in what the Israeli official military historian Lieutenant Colonel Netanel Lorch described as the "greed for loot."[26]

On 19 July 1948, a second UN truce came into effect that was, much like the first, on both sides honored as much in the breach as in the observance. By then, the roads from Haifa to both Nazareth and Tel Aviv were under Israeli control, although the Egyptians were on the Mediterranean coast only 20 miles south of Tel Aviv and had cut off the Negev settlements, which had to be supplied by air. The Arab Legion controlled part of the main road between Tel Aviv and Jerusalem, and Jewish settlements remained under siege in the north. Water was now reaching Jewish Jerusalem, however, and arms continued to be supplied from Czechoslovakia until Stalin ordered a sudden halt on 12 August, whereupon they were brought via Italy and Switzerland.

The ceasefire collapsed on both sides on 15 October, not least as a result of an Israeli bombing raid on the Egyptian airbase at El-Arish and on the railway line from El-Arish to Rafa, alongside the resumption of fighting both in Jerusalem and beyond. The UN intervened on 19 October demanding that both parties withdraw to their 15 October positions by 3 p.m. on the 22. This gave the Israelis only three days to open up a way to the Negev through the fortified hills around Huleikat, which thus saw some of the fiercest and most desperate fighting of the war before the Security Council ceasefire came into effect. On 4 November, Israel refused to return areas it had retaken from Egypt in the Negev. Meanwhile, Upper Galilee was captured from the Jordanians. Israelis were justifiably confident that capturing territory would bring greater security than would be achieved by abiding by the UN's demands.

Further negotiations, including threats against Israel from Britain and a warning from President Truman, ensured that all Israeli troops had left the Sinai by 2 January 1949, and on 24 February Israel signed an armistice with Egypt which was followed by further armistices agreed with Lebanon on 23 March, Transjordan and effectively Iraq on 3 April, and Syria on 20 July. By the time of the last armistice, 80 percent of the Palestine Mandate was in Israeli hands, although the Gaza Strip, West Bank and Old City of Jerusalem were not. Some 6,000 Jews had died in the war – 1 percent of the entire population of the state – but they had gained 30 percent more territory than they had originally been allotted by the United Nations.[27] The whole country remained within Arab artillery range, and the state's wasp-like waist was only 9 miles to the sea at its narrowest, but Israel existed.

For the moment at least, the State of Israel had survived the invasions of five Arab countries – a remarkable achievement for a new country populated by a group of people who had suffered so terribly in their recent history. This was not to be the last of Israel's wars. After two more defeats, Israel's antagonists learned to coordinate surprise attacks, and a de facto anti-Israel coalition surrounded the state. The first peace treaty – as opposed to a mere ceasefire – with an Arab state was not signed till 1979, three decades after Israel had been founded.

Yigael Yadin had no doubt where the credit for victory in the Israeli War of Independence lay, commending "the extraordinary qualities of Israel's youth" which, he said, "had absorbed into itself the full measure of Israel's yearning, during thousands of years of exile, to return to its soil and live in liberty and independence."[28] It was precisely this quality – an existential yearning, a metaphysical stamp upon a land – that time and time again has bonded people to land and won wars which conscripted soldiers have lost.

Israel's decision to militarize its citizenry was an essential component in its continued survival. Understandably cynical about UN-imposed armistices, and economically unable to support a large standing army, today Israel's own population remains conscripted for a minimum of two years of military service, allowing it to field the largest army in the world relative to size of population.[29]

Also vital to Israel's continued existence has been the development of a distinctive military doctrine, which as Chaim Herzog noted, was:

> based on flexibility, surprise and improvisation. Night-fighting, speed, commando-style operations, the strategy of "indirect approach" – all these became the hallmark of the philosophy evolved by the Israeli Defence Forces. Above all, emphasis was laid on the inculcation of a flexibility of thought in the officers in the field, and in particular in the junior leaders, who were trained to be able to adapt themselves in the heat of battle to the inevitable changes that occur.[30]

Such developments were not confined to Israel. The rest of the world watched the Israeli War of Independence to see what could be learned from a tiny citizen army defeating five professional ones.

The Malayan Emergency, 1948–1960: a textbook counter-insurgency campaign

Malaya – so often overlooked in military histories and geopolitical tomes alike – was home to one of the most significant evolutions of post-1945 warfare. The counter-insurgency campaign there was long, complex and hard-fought: but it was victorious. While each campaign has different elements of focus, it was in the British colony of Malaya that a well-rounded counter-insurgency approach was slowly but surely developed. Its influence would extend like a vine throughout counter-insurgency campaigns of the twentieth century and beyond.

After the Second World War, the pro-communist Malayan Races Liberation Army (MRLA) attacked the British, using much the same methods, camps and equipment that they had employed against the Japanese. After the murder of three British planters on 16 June 1948, habeas corpus was suspended and a state of emergency declared. The MRLA then numbered around 5,000, around 90 percent of whom were of Chinese origin. Although they drew some support from the 1.88 million Chinese Malays, they were opposed by the 2.43 million non-Chinese Malays, 530,000 Indians and 12,000 Europeans who lived on the peninsula.[31] The guerrillas' practice of burning unsympathetic villages that refused to supply them with food further reduced their popularity with the majority ethnic group.

The British had multiple significant advantages – the majority of the Malay population had remained loyal to their hereditary sultans, who were allies of Britain, and it was only from the minority population, the Malay Chinese, that opposition came.[32] This placed Britain in an enviable position when compared to other counter-insurgency campaigns of the twentieth century, such as the French experiences in Indo-China and Algeria.

The British responded to the MRLA terror campaign by interning 6,343 terrorist suspects and deporting 10,000 to China by October 1949.[33] Although hitherto the security forces had described their enemy as bandits, in 1952 they were called "CTs" – communist terrorists – by which time they had killed 3,000 people, despite the 30,000 Commonwealth troops and 60,000 local police who opposed them.[34] In the early days of the Emergency in 1948, British conscripts also committed war crimes, underlining a phenomenon that we shall see several times more in this book whereby recently drafted soldiers who are not volunteers behave far worse than seasoned, trained soldiers, sometimes from panic but often from sheer indiscipline.

In the Malayan jungles, pro-communist villages were routinely burned, people were ill-treated in detention, 25,000 suspects were interned for more than a month and there was one massacre of no fewer than twenty-four Chinese civilians.[35] As so often, trying to meet terror with terror was entirely counter-productive.

Lieutenant General Sir Harold Briggs understood the self-defeating nature of this kind of terror. A veteran of the Burma campaign, he enacted a plan to neutralize the communist threat in the region. This was based on providing security for the rural population while bearing down on MRLA food supplies, funding and recruitment.[36] It also involved building no fewer than 500 "New Villages" for the half-million

Chinese "squatters" who were marginalized from Malayan society and provided the perfect breeding ground for communist terrorism.[37]

Providing people with electricity, drinking water, schools, clinics and property rights proved a powerful disincentive to them to support Maoism, but the heady enticement of security had a dialectical partner: firm discipline for those who disobeyed. The inhabitants had to stay in the villages behind barbed wire and were subject to curfews and searches when they left them for work. "You can't deal with a plague of mosquitoes by swatting each individual insect," stated Briggs. "You find and disinfect their breeding grounds. Then the mosquitoes are finished."[38] This approach would be copied decades later during the Surge in Iraq when entire neighborhoods in the most embattled areas of Baghdad, Fallujah and other difficult cities would be made "gated communities" by the building of huge concrete walls that enabled control of entry to those locations using biometric ID cards and other measures.

On 5 October 1951, a gang of thirty-eight MRLA guerrillas accomplished a sensational feat when they killed Sir Henry Gurney, the British High Commissioner in Malaya, in a road ambush 40 miles north of Kuala Lumpur. For all that it was a propaganda coup, it was ultimately a tactical mistake. Gurney's replacement was Sir Gerald Templer, a veteran of the Somme, Dunkirk and Anzio, who took over as both high commissioner and director of operations in Malaya in February 1952. An Olympic hurdler and the British Army's bayonet champion, Templer was to prove an exemplary strategic leader in Malaya, performing each essential task magnificently and building on Briggs' big idea of draining the terrorist swamp. Templer's ultimate boss Winston Churchill had once written in his book *The Malakand Field Force* that "Dynamite in the hand of a child is not more dangerous than a strong policy weakly carried out."[39] Templer was determined from the moment he set foot in Malaya to carry out a strong policy strongly.

One of Templer's first acts was to reiterate the British government's promise that Malaya would be granted independence as soon as the security situation allowed, which shored up support among Malays and was honored in August 1957. He also granted full citizenship to all aliens born in Malaya, another desideratum of the Chinese Malayans. He did not underestimate the struggle ahead, declaring in 1953, "I have always said that the complete cure of it all will be a long slog."[40] He had full political support back in London, where to lose in Malaya would have been a blow to British prestige, as well as to the newly re-elected Prime Minister Winston Churchill.

Templer was right about the long slog; there were no fewer than 8,750 "contacts" between the security forces and communist guerrillas during the Emergency, with jungle ambushes and skirmishes commonplace.[41] The fighting saw Spitfires strafing jungle hideouts in Johore; sweeping operations to drive the enemy towards waiting cordons; constant patrolling (nicknamed "jungle-bashing"); dawn raids on isolated police stations and counter-raids on terrorist camps; the use of poison darts against the guerrillas by nomadic aboriginals in Perak; a deep-jungle unit called Ferret Force; paratroopers deliberately landing on treetops and then lowering themselves to the ground with ropes, and the "turning" of 180 former guerrillas to undertake operations against their erstwhile comrades, providing invaluable insights into communist military practice. Templer could occasionally be harsh – he once halved the rice ration and imposed a twenty-two-hour curfew on a village harboring terrorists who had killed twelve British soldiers – but he was generally regarded as fair, while also more than earning his popular soubriquet "the Tiger of Malaya."

Fighting in the denser parts of the jungle was made harder by the terrain, where it sometimes took four hours to cover a mile, often in single file. "In primary jungle," wrote one veteran, John Scurr,

> there would be giant, vine-covered tree-trunks with roots four feet high and ten yards long over which a patrol would have to clamber. Sometimes there would be fast-flowing rivers to wade across, or vast expanses of foul-smelling swamp. The humid heat soaked soldiers in perspiration until they were visibly steaming ... Leeches sought their vulnerable flesh and sucked out their blood: a man might find anything between five and fifty of the creatures clamped on to his body. There were also vicious, biting red ants

which showered upon them from the trees ... The night's sleep
was inevitably disturbed by swarms of whining mosquitoes ...
During the monsoon months marching soldiers were drenched at
around 4 p.m. every day by torrential rain, which could continue
unabated for up to ten hours ... Men might lie for a large part of
the night trying to sleep in streams of muddy water pouring
across their groundsheets. With so much fatigue and discomfort,
plus the additional hazards of malaria, scrub-typhus, jaundice,
dysentery, snake bites, scorpion stings, ulcerated jungle sores,
ringworm, foot-rot and prickly heat, it might be supposed that
soldiers had enough to contend with, but, of course, there was
also the enemy.[42]

Yet somehow morale remained high among the security forces, largely
thanks to their faith in ultimate victory. Scurr recalled a contact in
which a patrol of the King's Own Yorkshire Light Infantry was outnumbered five to one in a disadvantageous position, but when there was a
sudden halt in the terrorists' fire, a small but sturdy private shouted
out, "What's the matter, Johnny? Have you gone for a NAAFI-break?"[43]
The security forces were also better armed than the guerrillas; in the
early years of the Emergency they carried .30-caliber American M1 and
M2 carbines and .303 Bren light machine guns, which later made way
for 12-gauge Browning A-5 semi-automatic and Remington slide-action
shotguns that performed better in the very close-quarter fighting that
was typical of the jungle war.[44]

Templer vigorously pursued Briggs' twin-track carrot-and-stick counter-
insurgency strategy. Concomitant with the New Villages building
program came aggressive long-range jungle operations, undertaken by
special units such as Michael "Mad Mike" Calvert's Special Air Service,
who used Borneo headhunters as trackers. "The shooting side of the
business is only 25% of the trouble," Templer reported to Oliver
Lyttelton, the Colonial Secretary, in November 1952, "and the other
75% lies in getting the people of this country behind us."[45] He once
summed up his and Briggs' strategy by saying, "The answer lies not in
pouring more troops into the jungle, but in the hearts and minds of the

people."[46] The phrase "hearts and minds" was picked up by politicians and overused to such an extent that even by 1968 Templer was describing it as "nauseating." Nonetheless, it remains the most succinct explanation for how to win a counter-insurgency, and its efficacy has been demonstrated far beyond jungle warfare.

After Briggs' massive increase in security forces personnel – which was expanded fivefold in two years, while the British garrison was also enlarged – Templer's counter-attack could be intelligence-led. For fifteen months starting in 1952, Britain undertook a wholesale reform of the Malayan Police Force, with 10,000 officers removed for corruption or incompetence – the two often overlap – and new police schools set up where senior British police officers taught analysis methods and investigation techniques.[47] Four months' basic training was made compulsory for every officer, and the best were trained for a year in Britain. The number of ethnic Chinese in the Malayan Police Force more than doubled in the calendar year 1952 and, with improved discipline through intensive training, closer relations between the security forces and the population, which led to the virtuous cycle of better actionable intelligence, which in turn led to a better security situation.[48]

Insurgents, taken alive, proved valuable. Torture, ill-advised on both moral and practical grounds, was spurned in favor of interrogation methods that abided by the Geneva Convention. There were two main justifications for this logic: torture radicalized those it touched, and all too often it also produced worthless information, given simply to make the pain end.[49] Templer introduced an overall doctrine based on a new manual entitled *The Conduct of Anti-Terrorist Operations in Malaya*, which his officers were expected to learn and follow.[50] This concentrated on effective "clear and hold" operations rather than large-scale "search and destroy" ones. (The merits of this point were never embraced by the Americans in Vietnam, as we shall see in the next chapter.)

Commonwealth propaganda directed against communist morale was both relentless and imaginative. Aircraft flew low over the jungle and loudly broadcast the names of individual guerrillas and the large sums that could be claimed by anyone who brought them in dead or alive. Simultaneously, leaflets were dropped that could be exchanged for amnesty by any guerrilla. Troops or planes would destroy the rice paddies on which the guerrillas depended for sustenance. When one starving insurgent surrendered, a brigadier described his "shaggy hair, emaciated countenance, ragged khaki uniform, and eyes like those of a hunted rat."[51] By contrast to these haggard insurgents, the government was able to demonstrate that peace led to prosperity, especially after 1950 when the outbreak of the Korean War brought – as large-scale wars often do – a global increase in the value of commodities. (Tin and rubber were mainstays of the Malayan economy.)[52] For an already unpopular insurgency, the evident long-term economic and military success of the counter-insurgency began to presage a protracted, slow defeat.

In the time that Templer spent in Malaya between February 1952 and May 1954, the terrorist incident rate fell from 500 a month to fewer than a hundred. Two-thirds of the guerrillas were killed or captured or had surrendered, and the Commonwealth troops' casualty rate declined by four-fifths. In April 1952, the charismatic MRLA leader Chin Peng was forced to move his headquarters from Malayan territory to Betong in southern Thailand. Between 1948 and 1960, the Emergency saw 1,865 Commonwealth troops killed, out of more than 100,000 who served. A further 3,383 Malay civilians lost their lives, and 6,698 communists.[53]

Few other Western counter-insurgency campaigns have been as successful. Aware that it would take close to a decade to complete, in 1953 Templer had said he would "shoot the bastard who says this Emergency is over" – but by 1960 it was. His focus on low-intensity, skill-based warfare meant that Malaya could look forward to an independent, non-communist future. Counter-insurgency warfare had taken a step-change forward and future campaigns would be fought in the knowledge that guerrillas could be defeated when the government forces appealed constructively to the hearts and minds of the population.

France in Indo-China, 1945–1954: the road to Dien Bien Phu

At the end of the Second World War, France was battered, morally and financially exhausted, and knew it would be hard-pressed to hold on to its Indo-Chinese colonies, comprising modern-day Vietnam, Laos and Cambodia. As soon as the war against Japan ended, Ho Chi Minh, Vietnam's nationalist leader, proclaimed his country's independence. "For more than eighty years," he announced, the French had "violated our Fatherland and oppressed our fellow citizens … Vietnam has the right to be a free and independent country – and in fact it is so already. And thus the entire Vietnamese people are determined to mobilize all their physical and mental strength, to sacrifice their lives and property in order to safeguard their independence and liberty."[54]

Ho was wildly exaggerating the level of popular support his communists enjoyed, but by November 1946 fighting had broken out at Haiphong, northern Vietnam's major industrial city. Deciding on the harshest measures, on 23 November the French Army launched an artillery attack that resulted in the deaths of no fewer than 6,000 civilians.[55] It was the beginning of a series of French successes in major cities, bought at huge cost to the native population. One month later, Ho Chi Minh was forced to abandon Hanoi, the capital. Inspired by Chinese communist doctrine, his Viet Minh forces gave up conventional warfare and adopted guerrilla tactics, at least until Mao Zedong – after his victory in China in 1949 – started to provide him with heavy weaponry, including artillery.

General Vo Nguyen Giap ("the Red Napoleon"), the Viet Minh's brilliant commander, concentrated on destroying key French border posts in order to be supplied with heavy weaponry which could then cross the border. Although France had originally sent out General Jean de Lattre de Tassigny, its most distinguished general, who had won three victories over Giap in quick succession, he died of cancer after only eleven months in command. He was replaced by General Henri Navarre, who said that he could "see [victory] clearly – like light at the end of the tunnel."[56] For all that such hubristic comments might give a temporary fillip to morale, generals are ill advised to make them until victory is actually achieved. False dawns are all too common in warfare – as the British discovered in South Africa in 1900 when they announced victory two years before fighting finished.

Wary of the close relationship between the Viet Minh and communist China, the United States began to send aid and advisers to the French forces in Indo-China. Nonetheless, the American contribution was modest, and by 1953 the war had ossified into a miserable stalemate. The battle of Dien Bien Phu would change this entirely.

"At first I had no idea where – or even whether – the battle would take place," General Giap would later recall.[57] The French had concentrated their best troops, under General Christian de Castries, at Dien Bien Phu in a valley in north-east Vietnam surrounded by high hills. It was a strategically disastrous decision. Giap's army carried their supplies on bicycles and on their backs, dragging howitzers into the hills above the French positions. Seeing that the French were well dug in, Giap decided to delay his planned attack. "Suddenly I postponed the operation," he told an interviewer later. "My staff was confused, but no matter. I was in command, and I demanded absolute obedience – *sans discussion, sans explication*."[58] Ignoring advice from his Chinese military advisers to attack the French with human-wave assaults such as had been employed in Korea, Giap instead decided to launch an artillery attack against the French forces from above, while Viet Minh soldiers crept stealthily forward, employing trenches and tunnels to approach, yard by yard, the French lines.

Over the next two months, one by one, the forts surrounding Dien Bien Phu – named Anne-Marie, Beatrice, Claudine, Dominique and Lili – fell to the Viet Minh. (They were named, in somewhat Gallic fashion, after some of the women in General de Castries' love life.) On 7 May, he was forced to surrender his army. President Eisenhower considered using US airstrikes to help the French, but decided against it. "We would have had problems," Giap later admitted, reflecting on the possibility of US air support, "but the outcome would have been the same. The battlefield was too big for effective bombing."

As a highly successful commander, General Giap saw the big strategic principle of denying the French their advantages and concentrating on Dien Bien Phu; he communicated his intentions to his staff effectively – *sans discussion, sans explication* – relentlessly driving the implementation over the vital weeks of siege, and, as the situation later morphed into a vast war against the United States, he refined his tactics accordingly – as we shall see in the next chapter.

The French Army suffered 75,000 killed in Indo-China – of whom around 28 percent were French-born and the rest African and Indo-Chinese – with a further 64,000 or so wounded.[59] These losses helped to alienate the French population.[60] It had also left men behind, the Groupement de Commandos Mixtes Aéroportés, Vietnamese soldiers who had fought for France behind the Viet Minh lines in North Vietnam. When the ceasefire was agreed in July 1954, many of them were stranded hundreds of miles behind in the communist-occupied zone. With no ability to make their way to safety, they were hunted over the next two years by Ho Chi Minh's forces. Some French officers remained with them. One day in 1956, two years after the ceasefire, a French voice was heard shouting over the radio airwaves, "You sons-of-bitches, help us! Help us! Parachute us at least some ammunition, so that we can die fighting instead of being slaughtered like animals."[61]

The French Army had been wholly unsuited to the type of warfare that it needed to undertake. It had expected traditional, set-piece battles in which it could overwhelm an economically and technologically inferior opponent. The size of Vietnam's jungles, in which opponents could disappear for months, made France's ordinarily superior firepower almost a burden instead of an asset. Ironically, the French eventually got the decisive conventional battle they had earlier wished for, but it was Vietnamese firepower that ultimately overwhelmed them. In the words of George C. Herring, Dien Bien Phu "epitomized French mistakes ... particularly their stubborn overestimation of their own ability and their repeated underestimation of the enemy."[62]

As atypical as Dien Bien Phu was, it demonstrated how important battles still were, and it proved something else that would electrify the world, which was that a well-funded and well-equipped European force could be defeated in the field by an underfunded and underequipped non-European one. This had happened before in history, of course – principally at Isandlwana in the Zulu War in 1879, at Maiwand in Afghanistan in 1880, in Japan's great naval victory over Russia at Tsushima in 1905 and in 1942 when the Japanese had forced the British to surrender in Singapore. But in three of those four cases the imperial power had returned in greater strength to exact retribution. By contrast, France fled Indo-China after Dien Bien Phu, never to return.

The way that conflict evolved at Dien Bien Phu was by proving beyond doubt that nationalist guerrilla warfare could humiliate European – and by extension American – power, when carried out on the guerrillas' own turf. In later years, Colin Powell would, like so many others, reflect that the United States should have learned more from the French failure in Vietnam: "We should have realized that it was a war as much about nationalism and self-determination within this one country as it was about the ideology of Communism or the worldwide Communist conspiracy."[63]

Algeria: a textbook case of how not to fight an insurgency campaign

The end was coming for France's northern African territories by the middle of the twentieth century, but a country humiliated by its defeat and occupation by Nazi Germany was determined to re-establish itself as a conquering military power. Instead of withdrawing from its empire with as much dignity and honor as possible, as Britain had been doing since the late 1940s, France tried to fight it out, with catastrophic results in Indo-China and Algeria. As we have seen so often, the Second World War cast a long shadow.

Unlike the British in Malaya, the French government in Algeria could not promise independence to the local population without causing immense domestic political upheaval. Furthermore, while in Malaya the European population was minuscule, in Algeria it numbered over a million (the so-called *pieds noirs* or *colons*) – out of a total population of 9.5 million. The *pieds noirs* were largely French, Spanish and Italian by descent but were born in Algeria and intended to spend the rest of their lives there. Many were as poor as the native Algerians, but they were utterly opposed to an independent Algeria. France had accepted the independence of Morocco and Tunisia in March 1956 since neither had large French minorities, but it balked at Algeria going the same way.

The significance of the Algerian conflict in the evolution of warfare lies in the way that France attempted to fight it in ways that were fundamentally opposed to the Republic's founding principles, as set out in the "Declaration of the Rights of Man and the Citizen" in 1789, the most humane legacy of the French Revolution. It was the widespread recourse to torture that convinced most French people that the cause of Algérie française was fundamentally flawed, because it was unFrench.[64] Torture confronted the French culturally and emotionally with a crisis of identity and conscience, and, more legalistically, it also of course contravened the French Penal Code, the Hague Convention and Article 3 of the 1949 Geneva Convention.

In the autumn of 1954, the Front de Libération Nationale (FLN) under Ahmed Ben Bella launched an armed uprising of nationalists, which started in the east of the country but within two years had spread across the country's Mediterranean littoral. Algeria (not including the Sahara) extended 650 miles along the coast and 350 miles inland, yet in 1954 the entire country had only 50,000 policemen. Although the police warned of the growing FLN insurgency, they were ignored until blood had been shed. Ultimately, France was to send no fewer than half a million men to fight in Algeria, including conscripts. The sending of non-volunteers to fight, which had not happened in Indo-China, polarized French society and radicalized opposition to the war in a way that the United States would soon experience during the Vietnam War.

One of the most incisive analyses of the FLN insurgency campaign was written by Colonel David Galula. Raised in Casablanca, Galula graduated from the French military academy in Saint-Cyr in 1939, before fighting in North Africa, Italy and France during the Second World War. His expertise in irregular warfare was subsequently honed by observation of the wars in China, Greece and Indo-China, and then service in Algeria as a company commander from 1956 to 1958. By

1964, he had risen to the rank of lieutenant colonel and wrote *Counterinsurgency Warfare: Theory and Practice*, a classic of military writing with important insights into the motivations and tactics of anti-colonial insurgents.

Galula recorded that the FLN's initial attacks, aimed at terrorizing the local population into compliance, were initially focused upon fellow Muslims who served in the Algerian police.[65] Obtaining the passive complicity of the local population was, as Galula points out, a crucial determinate in their success:

> This is done by killing, in various parts of the country, some of
> the low-ranking government officials who work closely with the
> population, such as policemen, mailmen, mayors, councilmen,
> and teachers ... In order to involve the population further, simple
> *mots d'ordres* are circulated, such as "boycott tobacco"; a few
> violators caught smoking are executed.* These assassinations
> have value only if they serve as examples, therefore they must not
> be hidden or committed on the sly. The victims are generally
> found with a tag explaining that they have been condemned by a
> revolutionary tribunal and executed for such and such a crime.
> When all this is achieved, conditions are ripe for the insurgent
> guerrillas to operate and for the population to be mobilized
> effectively.[66]

These attacks on the agents of the state, and subsequently on French conscripts, led to the widespread introduction of a de facto secret sanction to torture suspects – *passé à la question* – the very measure that Briggs had understood to be almost wholly counter-productive in Malaya. The issue of torture in warfare, which is still intensely relevant today, was made very public during the Algerian War because it was resorted to by the French, who had not learned the central lessons about it: that the victim will tell his torturers whatever he thinks they

---

* The FLN tended to cut off people's lips and noses for smoking, as the practice was considered to profit the French economy (Howard et al. (eds), *The Laws of War* p. 207).

want to hear, and that it is a powerful propaganda tool that ought not to be handed to the enemy. It is all the more remarkable that the French, whose Resistance leaders had been tortured routinely by the Gestapo only a decade earlier, failed to understand either fact.

In March 1955 a senior French civil servant called Roger Wuillaume took the extraordinary step of recommending that since torture had become so rampant in Algeria it ought to be institutionalized, especially the method by which suspects had their bellies filled up with water from a pipe, which he concluded "involves no risk to the health of the victim ... I am inclined to think that these procedures can be accepted and that ... they are no more brutal than deprivation of food, drink and tobacco, which has always been accepted."[67] Wuillaume concluded that the only way to restore police confidence was "to recognize certain procedures and to cover them with authority."

The Wuillaume Report, commissioned by the Interior Minister (and future French President) François Mitterrand, concluded that the use of water pipes in the stomach was acceptable whereas electric shocks were not.[68] His recommendations were ultimately rejected, but that did not change much, and there is evidence of Algerians being hung naked by their feet with their hands tied behind their backs and their heads plunged into buckets of water to make them talk.* In the reports that the French state commissioned, the word *torture* was almost never used, but the euphemisms such as *sévices* (ill-treatment) and *procédures* were.[69]

The torture method used most regularly was the *gégène*, which involved an army signals magnet from which electrodes could be

---

* The similarities with the waterboarding used in Guantánamo Bay and elsewhere after 9/11 are obvious. As General David Petraeus would later note while commanding in Iraq and Afghanistan, however, not only was torture wrong and a violation of the Geneva Convention, history had shown that it also was by no means a guarantee of accurate, valuable information. Moreover, he observed, the ultimate price of torture in the court of public opinion to the country and force employing it ultimately would be much greater than the conceivable value of what might be obtained by its use. Experience in detainee operations in Iraq and Afghanistan would show that the best way to get useful information from a detainee was for an interrogator to develop a relationship with the detainee through patient, informed engagement.

fastened to various parts of the body such as the penis. It was simple to operate and left no traces. The French parachute Commander General Jacques Massu tried this one out on himself in his office – we do not know at what voltage – and denied it constituted torture, but as the military historian Alistair Horne pointed out, "What he failed to note in his 'experiment' was the cumulative effect of prolonged application of the *gégène*, as well as of all deprivation of the element of hope – the essential concomitant of any torture."[70]

On one occasion in November 1956, Paul Teitgen, the Secretary-General of the Algiers Préfecture, was presented with an appalling choice. A Catholic Resistance hero who had been tortured nine times at Dachau, Teitgen had to decide whether to have Fernand Yveton, a communist who had been caught red-handed putting a bomb in a gasworks, tortured. Although the first bomb had been discovered, a second one had not been, and thousands of lives would have been lost if it blew up the gasometers, but Yveton refused to reveal its where-abouts. In the event, Teitgen would not allow the police to resort to torturing Yveton, although he recalled how he "trembled the whole afternoon." The bomb did not go off. "If you once get into the torture business, you're lost," Teitgen later stated. "All our so-called civilisation is covered with a varnish. Scratch it, and underneath you find *fear*. The French, even the Germans, are not torturers by nature. But when you see the throats of your *copains* [comrades] slit, then the varnish disap-pears."[71]

The varnish certainly disappeared in Algeria, with the sexual assault and torture of both genders: bottles thrust into vaginas; hoses inserted into rectums; detainees suffocated en masse. Suspicious deaths such as alleged jumping from windows "to escape interrogation" were recorded; some 3,000 detainees "disappeared"; corpses were dropped into the sea by helicopter.[72] There was a standing order in Orléansville in Algeria that any corpse of an FLN guerrilla must be displayed for twenty-four hours and hung with a black placard around its neck stating, "I will not kill my brothers any longer, I will throw no more grenades."[73] In the words of one prominent intelligence historian, French activity in the region "made the CIA look like a nunnery by comparison."[74]

During the battle of Algiers, which lasted from September 1956 to September 1957, increasing recourse was had to *la torture*, and General Massu believed that it was instrumental to his victory there. After the war, Massu admitted openly that torture had been used – "I am not frightened of the word" – yet he claimed it had been neither institutionalized nor codified. He argued, as many have before and since in warfare, that there was no other option in the prevailing circumstances if much-needed information was to be obtained from prisoners. By contrast, Alistair Horne, who entitled his book A *Savage War of Peace*, recorded that torture had become "a growing canker for France, leaving behind a poison that would linger in the French system long after the war itself had ended."[75]

Horne differentiated between the brutality of roughing up a suspected terrorist immediately following his arrest and the "systematic application of physical or psychological pain expressly aimed at making a subject talk." He concluded that "Torture ends by corrupting the torturer as much as it breaks the victim," and it certainly produced a powerful backlash against the French in Algeria, in France itself and in international opinion, much as the use of waterboarding, the revelations of the way rogue elements behaved at Abu Ghraib prison and the existence of CIA "black" rendition sites were to damage America's standing during the Iraq War.[76]

Torture is a propaganda gift to the enemy, leads to corruption and cover-ups, provokes appalling escalation and retaliation – the FLN would cut off their enemies' noses, lips or penises – and ultimately damage the cause.[77] The French turned against the *sale guerre* (dirty war) largely because it *was* dirty. "All right, Massu won the Battle of Algiers," conceded Teitgen, "but that meant losing the war."[78] As the search for FLN guerrillas expanded, the resulting alienation of the population grew too. By the latter days of the war, support for the French forces among the Algerian population had been all but extinguished.[79]

Massu's failure to get the big strategic idea right – in combination with his communicative and moral failures – meant that he could not and did not perform adequately the tasks of strategic leadership. He failed to recognize that although France started with some support among native Algerians, this would quickly leach away once the FLN targeted them and it started to look like France would lose. He also failed to appreciate how quickly domestic support would evaporate once the French people recognized that the highest principles of the Republic were being traduced through torture, and that the FLN had turned from being an extremist terrorist organization into a mainstream nationalist movement capable of governing an independent Algeria.

On 13 May 1958 there was a mass demonstration of *colons* in Algiers against the incoming French premier Pierre Pflimlin, who they feared was planning a negotiated settlement which would result in de facto Algerian independence. It turned into a coup in which Massu took power with a Committee of Public Safety. Back in Paris, the Algerian situation destroyed the French Fourth Republic, and Charles de Gaulle was called upon by leaders of the National Assembly to save the country, with something approaching civil war as the alternative.[80] De Gaulle was installed in office by the National Assembly on 1 June. "France is still faced with a difficult, bloody problem: Algeria," he said in a television broadcast on 16 September 1959, adding that it must be solved not by "sterile simplistic slogans" but instead "by the free choice of the Algerians themselves."[81] It was a brave statement to have made, and it exposed him to a series of assassination attempts for years afterwards.

Algeria became independent in July 1962. In the course of the war, more than a quarter of a million Algerians died on both sides, making it one of the worst post-colonial wars.[82] French military deaths numbered 17,000.[83] What happened in Algiers fell well outside both the formal and informal laws of war and had made France's position untenable.

The Borneo Confrontation, 1962–1966: a secret victory in the jungle

The British Commonwealth evolved warfare in an unexpected and somewhat counter-intuitive way between 1962 and 1966, when it fought an almost completely secret war against Indonesia. In a way that would be inconceivable in today's hyperconnected, technology-dominated world where satellite imagery is widely available and mobile-phone connectivity almost global, some 30,000 troops from Britain, Australia, New Zealand and Malaysia (the successor state to British Malaya) fought a war deep inside the jungles of Borneo, in almost total secrecy. Whatever the shortcomings in the evolution of warfare, the fighting showed that non-native troops could excel at jungle combat.

The Indonesian President Ahmed Sukarno had wanted to strangle the Federation of Malaysia at birth, and set his army and the Indonesian Communist Party to the task. Indonesian Army officers led guerrilla cross-border raids from Kalimantan into Sarawak, posing as "volunteers," rather as the Chinese had in the Korean War. These raids became more frequent, aggressive and far-reaching, and included seaborne and airborne raids on the Malayan peninsula from August 1964 to March 1965. On 28 September 1963, some 200 Indonesian regulars attacked a British army outpost at Long Jawai. The Indonesians also supported the terrorist Chinese Communist Organization in their sabotage attacks in Sarawak and Sabah.

Major General Walter Walker, the Commonwealth Director of Operations in Borneo, was given orders to "contain Indonesian aggression without escalation to open war."[84] It was a tough assignment, and one that rested on a deniable, covert form of warfare. He recognized the impossibility of patrolling the thousand-mile land frontier – not to mention the even longer coastline – with only five regular battalions. As he recalled, "It would have been madness to put them in penny packets along the border. I therefore held them back and relied on small surveillance groups on the most likely incursion routes."[85]

As a former director of the Malaya Jungle Warfare School and veteran of the Malayan Emergency, Walker was an expert in counter-insurgency, and the Army manual he wrote emphasized the need to go on the offensive, patrol constantly at platoon and company level and set ambushes. As he put it in his forthright autobiography *Fighting On*, "Offensive action is the very essence of successful military operations when faced with guerrilla or terrorist forces ... A policy of containment is a passport to failure."[86]

London's clear reluctance to engage in war with Indonesia was manifold: the cost of open warfare, combined with the success of the Malayan strategy and a geopolitical focus on withdrawing forces from the entire area east of Suez had all culminated in a request for a more covert form of action. As the former US Secretary of State Dean Rusk was to put it incredulously, in reference to the establishment of the National Health Service, "Free aspirin and false teeth were more important than Britain's role in the world."[87] In the face of attacks upon Malaysian Borneo being launched from Kalimantan in Indonesia, Walker decided to counter-attack vigorously. Yet this was to remain covert and deniable to the general public; Walker took a very reasonable bet that the Indonesian government would be too embarrassed by any losses suffered to publicize the secret British counter-attack.[88] He was right. No public response was forthcoming, and the tone was set: the entire war would be kept "unofficial and bilateral" by both sides. Sukarno never admitted that his troops had even crossed the border, and neither did British ministers. By not publicizing the conflict, the Commonwealth gave Sukarno what today would be called an "off-ramp" to end the conflict without public humiliation. The press was not allowed to stray from military bases and was not informed about operations.

The splendidly entitled Operation Claret started in April 1964 and saw battalions conducting one operation a fortnight in 1964, and two a week by mid-1965, a fourfold increase intended to dominate their Indonesian antagonists. Walker was determined to fight a jungle war, rather than a war that merely happened to occur in the jungle. He was thoroughly versed in the importance of winning the hearts and minds of the local (mainly Iban) native population, which he did by emphasizing that the Commonwealth was protecting them from Indonesian incursions.

Jungle warfare, unlike many contemporary forms, involves close combat operations – indeed, the average contact distance between combatants during the Confrontation was a mere five to ten yards. "Guerrillas could appear unseen and unheard," notes one historian of the campaign, "and then vanish into the rainforest."[89] Yet, with their extensive experience of jungle operations in Burma and Malaya, their dominance in numbers and their readiness to do most of the fighting inside Indonesia, the Commonwealth won the Confrontation with relative ease.

Walker's seven "Golden Rules" for every operation were summarized by the soldier-historian Emile Simpson. Walker himself:

> would personally authorise every operation; only trained and tested troops were to be used; the penetration depth would be limited, attacks being only to thwart enemy offensive action, never in retribution for one's casualties, and civilian casualties were never to be risked; there was to be no air support except in extreme emergency; operations were to be planned and rehearsed for at least two weeks; every operation was to be planned and executed with maximum security, cover plans made, codenames for each operation used, and soldiers sworn to secrecy, with no details to be discussed over radio or telephone, no ID discs worn and no identifiable material to be left in Kalimantan; and no soldiers were to be captured alive or dead.[90]

Walker stands as a prime example of someone who performed all four of the tasks of a strategic leader, especially the last, in which the strategy is refined and adapted to enable conduct of the first three tasks again and again.

Commonwealth troops did not always return to base after operations, but often stayed in the jungle waiting to ambush the enemy. (One Gurkha unit waited in ambush for forty days before the terrorists fell into their trap.) "He had to be played at his own game by living out in the jungle for weeks on end," Walker wrote of the Indonesian guerrillas. "The jungle has got to belong to you; you must own it; you must control and dominate it."[91] You must, as he put it in his autobiography, "out-guerrilla the guerrillas."[92]

The Dhofar Rebellion, 1965–1975: perfecting civil–military relations

The evolution of warfare took a leap forward during the ten-year Dhofar Rebellion, where the lessons learned during the Malayan counter-insurgency were put into operation alongside important new developments, principally the employment of special forces and wholesale social, political and economic reform in the host country. This combination had not been seen before, and its success provided a template for counter-insurgency successes of the future. Indeed, if hosts such as President Diem of South Vietnam and Hamid Karzai of Afghanistan had been more reform-minded, the tragedies that overcame their countries might not have happened. The differences between the Dhofar campaign and the Iraq War are many and obvious – not least in their size – but the lessons learned in the former were later to be put to use in parts of the *Counterinsurgency Field Manual* FM 3-24 of 2006 which laid the ground for the Surge in the Iraq War.

The Sultanate of Oman lies on the south-eastern coast of the Arabian Peninsula, in a key strategic position dominating the mouth of the Persian Gulf and abutting the 25-mile-wide Straits of Hormuz. Two-thirds of the West's oil passed freely through the Gulf in the 1960s, rendering continued access to the Straits and good relations with the Sultan of Oman vital. By the mid-1960s, however, the perpetually expanding reach of the Cold War had begun to threaten this relationship. Russian-backed communist guerrillas put themselves in a position to endanger the stability of the sultanate, and therefore the West's access to the Straits. Yemen – Oman's neighbor, and a state with its own designs upon the Straits – had nurtured a growing rebellion, also backed by China and the Soviet Union.

Sultan Said of Oman had been on the throne for three decades and was fiercely opposed to the modernization of his country; he banned trousers, transistor radios, dancing, cameras, cigarettes, dolls and gas cookers, in favor of maintaining the traditionally Bedouin culture.[93] He also opposed all but the most rudimentary economic, medical, educational and agricultural progress.

In June 1965 the Dhofar Liberation Front of pro-development dissidents was formally set up under Musalim bin Nufl. The British exodus from Aden in November 1967, under the new policy to evacuate forces east of Suez, both aided and radicalized the DLF, especially once Yemen provided it with weapons, food, equipment and a headquarters in Hauf, on the other side of the Omani border. DLF guerrillas were taught Marxist-Leninist ideology and trained in guerrilla fighting in Russia and China.

The north-east of the country, which remained loyal to the Sultan, was separated by a 600-mile near-trackless desert from Dhofar, a province around the size of Wales and comprising around 50,000 Omanis, mostly living in tribal coastal villages abutting the country's extensive oil fields. In Dhofar "ferment had been a traditional pastime for hundreds of years."[94]

June to September brings monsoons to Oman's coast, creating "a muddy, cold, insect-ridden, murky gloom" that reached, in certain places, 30 miles inland. It provided the perfect conditions for guerrilla fighting.[95] By 1970, the Dhofar Liberation Front, which changed its name to the People's Front for the Liberation of the Arabian Gulf (PFLOAG), had captured large parts of Dhofar, and its communist-led rebellion had spread across much of Oman. Its methods were vicious: opponents were thrown from clifftops or tortured by fire applied either to the back, the eyes or the genitals.[96]

On 23 July 1970, Sultan Said's 29-year-old only son, Qaboos bin Said, a graduate of the Royal Military Academy Sandhurst, overthrew his father in a palace coup. The usurped Sultan swiftly flew off to live out the rest of his life in the Dorchester Hotel in London. (He himself had deposed his own father in a palace coup, so he had limited cause for complaint.) On coming to power, Sultan Qaboos announced an amnesty for rebels, offered a ceasefire and publicized plans for development and the introduction of up-to-date education and medicine, all with the aim of modernizing Oman while simultaneously undercutting the PFLOAG.

A minority – the original Dhofar Liberation Front supporters who had been more development-minded than Marxist – left PFLOAG, mollified by the reforms and alienated by the communists' virulent rejection of Islam. Some 200 went so far as to change sides entirely; known as "Firqats," they proved an invaluable asset to the Sultan's Armed Forces (SAF). Such was the strategic importance of keeping Oman anti-communist that the SAF was eventually to include servicemen from Britain, Jordan, Iran, India and Pakistan.

Qaboos devoted over £300 million of Oman's annual oil revenues to the SAF, but building and agricultural projects proved soft targets for PFLOAG, which at the height of its strength had around 2,000 fighters supported by 4,000 militia and several thousand sympathetic locals. Although Russians, Chinese, Libyans and Cubans helped train the PFLOAG guerrillas in Hauf, they rarely made the mistake of crossing the border into Oman.

In 1973, Colonel (later General Sir) John Akehurst, a veteran of the Malayan Emergency, was appointed to command the SAF's Dhofar Brigade. Most of the fighting was done by more than 10,000 Omani, Baluchi and Iranian soldiers, but there were British officers command-ing all three services of Dhofar Brigade, whose contributions were out of proportion to their numbers in planning, logistical and combat roles. The local inhabitants, however, were essential. As Akehurst wrote in his autobiography *We Won a War*, the Dhofari people "had in common a most remarkable innate natural intelligence. To a man they were alert, quick-witted and possessed of keen powers of argument."[97] Of the PFLOAG enemy, he wrote that they were "fierce, aggressive, courageous, independent people born of a long tradition of fighting" who moreover "seemed to find small-arms fire immensely stimulating and with anything like fair odds would usually come off best." He added that the SAF's air support and superior firepower in both artillery and mortars were the most successful components in the campaign. "Every one of [the PFLOAG fighters] seemed to have a well-developed sense of impor-tant tactical ground and the use of cover," he recorded.[98]

The SAF's greatest problem was that the enemy operated in familiar home territory, supported in the form of food, money, information and concealment by local civilians who were sometimes family relations of the fighters. "It was often possible for the enemy, on the approach of the SAF, to drop his weapon and become an instant civilian; he would never be given away."[99]

Qaboos and Akehurst ultimately defeated the insurgency by employing a number of methods, which included, in no particular order: the use of barbed wire and mines to limit terrorists' movements; the maintenance of a coalition that included 1,500 Persian troops sent by the Shah of Iran; Strikemaster air attacks on the PFLOAG's daytime camel-loaded supply lines;* surprise attacks during the monsoon season; aggressive patrolling; the employment of 1,400 Firqats; and the drilling of water boreholes and innumerable other civic action projects as incentives for loyalty. If clean water could be provided at a central point, civilians could be relied upon to go there and be persuaded that the government was offering them a better deal.[100] As irritating as Templer had eventually found the phrase, the SAF had learned valuable lessons on winning over hearts and minds in unsympathetic areas and in retaining loyalty. There was also the stick: local tribesmen were warned that enemy activity in their area would result in the water being cut off.[101]

Additionally, a Civil Aid Department set up clinics and schools, sold subsidized rations and built mosques. Doctors were deployed by plane to tend to the civilian population. Amnesties and money were offered to Firqats on the Sultan's birthday. Military engineers built several miles of rail tracks per day in order to bring reinforcements to threatened areas, as well as airstrips that could take the enormous C-130 supply aircraft. The Sultan ordered that Dhofaris should have priority for employment in the development projects, keeping the population as prosperous and as busy as possible. A Dhofar radio station was instituted to counter the enemy propaganda emanating from Radio Aden. Cash offers were made to anyone who revealed where arms caches were, with a Katyusha 122mm RCL (recoilless) worth the vast sum of £2,000

---

* "Although it caused distress to many of the pilots, the camels were important targets and in two days [in September 1974] over two hundred were killed by air attack" (Akehurst, We Won a War p. 68).

to an informant, while a light machine gun commanded a £450 bounty. Officers were appointed on merit rather than for their conformity; one important post went to a Lieutenant Colonel Graham Sherwell "who paid scant regard to the finer points of uniform dress, and who kept a pet piglet called Percy," but who Akehurst recalled was "the man of the hour" as a troubleshooting administrator at Thumrait.[102]

Qaboos and Akehurst together proved to be exceptional strategic leaders. They invariably got the big ideas right, communicated them effectively, oversaw their implementation with skill and determination and demonstrated a willingness to refine the big ideas and do it all again and again. Theirs was a textbook counter-insurgency campaign. "There was no doubt in my mind who was in charge," Akehurst wrote perhaps overmodestly, referring to the Sultan's role. But he was empowered by the Sultan, and ensured that "Every military operation begins with a mission so that none of its recipients is in doubt about what the commander intends to achieve."

For the Dhofar Brigade, Akehurst put the mission, in truth a powerful big idea, in capital letters: "TO SECURE DHOFAR FOR CIVIL DEVELOPMENT."[103] In August 1974, he made the following notes for his future campaign:

1. SAF is nearer to winning this war than it thinks. There is a tendency to over-carefulness and we must take risks and be arrogant over our possessions.
2. Where the enemy is adopting a low profile our policy must be to trample over the area and dominate it. This may be best done with Firqats.
3. Initially I thought a reverse would be disastrous. Now I do not ...
4. The keys to victory are:
   a. Cutting enemy supplies as far west as possible.
   b. Civil development, especially roads and encouraging normal commerce.[104]

Much as in the Borneo Confrontation, it was against the interests of the Commonwealth and Omani governments to publicize foreign involvement in the war. Journalists were denied access to Dhofar, which, in Akehurst's words, "spared combatants the complications, then being endured in Vietnam, of conducting operations in the full and impersonal glare of televised publicity."[105] So few Britons even knew where Oman was that it was assumed by Akehurst's comrades that he had been sent to Amman in Jordan. Akehurst recalled "taking endless trouble" over the Firqats, not least because, as he put it, "tribesmen disbanded, disarmed or dismissed were potentially recruits for the enemy."[106] It was an important lesson, but one that coalition forces forgot regarding the defeated Iraqi Army in 2003.

The Special Air Service (SAS) distinguished itself in the campaign with novel approaches demonstrating valor and panache. One such example took place in July 1973, when ten SAS soldiers, along with some Omani gendarmes and artillerymen, managed to hold off 200 PFLOAG warriors attacking under the cover of a monsoon with vicious hand-to-hand combat.[107] "The enemy suffered heavy casualties and never attempted such an operation again," Akehurst recalled.[108] The role of special forces was of crucial importance to both the success and the secrecy of the Oman campaign, and has continued to be in counterinsurgency operations.

One further notable feature of the campaign was the weapon commonly found in the hands of the rebels. The Russian-made Kalashnikov AK-47 was fast on its way to becoming the most ubiquitous weapon of revolutionaries across the globe. From its genesis, the cheap, competent Kalashnikov came to symbolize guerrilla struggle in the latter half of the twentieth century and beyond. It even appears on the flag of Mozambique. It is estimated that no fewer than 75 million AK-47s have been produced, with a further 25 million other types from the Kalashnikov family of weapons.[109] Akehurst recalled that those tribesmen who defected tended to prefer to keep their AKs rather than switch to British semi-automatics or American Armalite M-16 rifles.

The PFLOAG guerrillas were also skilled in using other Soviet or Chinese weaponry, such as the light hand-held anti-tank RPG-7, recoilless rocket launchers of 75mm and 82mm that were accurate over 6,000 yards, Russian 81mm and Chinese 82mm mortars and Shapagin heavy machine guns. The SAF was also occasionally faced with the 122mm Katyusha rocket that could be effective over 6 miles. The Russian plastic PMN anti-personnel mine had only one tiny piece of metal, and was thus very hard to detect. Fortunately PFLOAG did not have any artillery in Dhofar, but they were never able to hold ground permanently and artillery would have been easy to detect from the air. Nor did they have as effective radio communications as the SAF.

Akehurst was tremendously fortunate in having in Sultan Qaboos an unparalleled host for the counter-insurgency campaign. The new Sultan was utterly supportive of what Akehurst was trying to do, and brought statesmanlike qualities to the fight that saved his country from communist takeover. He compares very favorably with the host-nation leaders of the early twenty-first-century conflicts, such as Iraq's Nouri al-Malaki or Afghan presidents Karzai and Ashraf Ghani. The only times that Qaboos frustrated Akehurst were during his occasional bouts of wariness of bad omens: the Sultan would refuse to embark on journeys or projects if his astrologers deemed them unpropitious. Nevertheless, Akehurst concluded that "It would have been folly to overrule the Commander-in-Chief's premonitions."[110]

Sir John Akehurst and Sultan Qaboos join that short but distinguished list of soldiers and statesmen who exemplified the keys of strategic leadership in action. "In the many small conflicts throughout the world in the post-World War II era," concluded one history of the art of war, "leadership and training, together with skill gained in conflict, still made the same contribution to victory that they had always made."[111]

Three

America's War in Vietnam

1964–1975

The first, the supreme, the most far-reaching act of judgment
that the statesman and commander have to make is to
establish ... the kind of war on which they are embarking;
neither mistaking it for, nor trying to make it into, something
that is alien to its nature. This is the first of all strategic
questions and the most comprehensive.

Carl von Clausewitz, *On War*, 1832[1]

On 29 April 1975, the day before Saigon, the capital of South
Vietnam, fell to advancing North Vietnamese troops, a CIA officer
was photographed on the roof of the elevator shaft of 22 Gia Long
Street, an apartment building half a mile from the US Embassy, helping
US government employees up a ladder to board an Air America Huey
helicopter. The number of people on the ladder, the precariousness of
the helicopter and the evident chaos from which the evacuees were fleeing created a powerful image of desperation and defeat and prompted
an enormous question: how had the greatest, richest and most militarily powerful democracy in human history been so humiliated by a small,
economically backward dictatorship?

America's experience in Vietnam was costly and frustrating and, ultimately, unsustainable and unsuccessful. As early as 1965 then-Harvard
Professor Henry Kissinger described it as trying to "build a nation in a
divided society in the middle of a civil war."[2] Two decades later,
American soldier-scholar Andrew Krepinevich, paraphrasing General
Omar Bradley's reflection on the Korean War, would conclude that
Vietnam had been "the wrong war – at the wrong place, at the wrong
time, with the wrong army."[3] The way the US fought the war added
considerably to its difficulties.

Vietnam consumed America throughout most of the 1960s and well
into the 1970s. It broke the presidency of Lyndon Johnson and preoccupied that of Richard Nixon. It led to massive anti-war protests that,
following the 1968 assassinations of civil rights leader Dr. Martin
Luther King Jr and Senator Robert F. Kennedy, evolved into general
protests against the American establishment that included destructive
riots in American cities, on university campuses and in Washington,
DC. The very fabric of American society was badly torn in those years,
driven in large part by opposition to the Vietnam War.

The defeat in Vietnam would be a costly setback for the United
States and produce a "No More Vietnams" sentiment that would haunt
US military thinking on the use of force for nearly three decades. That
sentiment – and a flawed belief that, if the Army avoided developing the
capability to conduct counter-insurgency operations, it would not be
asked to conduct them – would also undermine the US military's ability
to conduct such campaigns for decades following the American withdrawal from Vietnam, well into the early years of America's involvement
in Iraq and Afghanistan.[4] However, the exhaustive examinations of
every aspect of the US conduct of the Vietnam War would also lead to
substantial reforms of many aspects of the US military.

At repeated junctures, Vietnam came to be seen by American policymakers as a war that could not be won yet must not be lost. Additional
US forces were therefore committed each time the situation looked dire,
until finally the White House recognized not only that the war was
unwinnable but also that the US commitment was not sustainable,
particularly given the enormous opposition to it in the United States.

Yet withdrawal was not conceivable until 1973, when Henry Kissinger, President Nixon's brilliant and tireless National Security Advisor (and, later, Secretary of State), found a way for the United States to extricate itself from South Vietnam.

Two years after the agreement, Kissinger would despair at the collapse of the South in the face of a North Vietnamese invasion, but American forces had been withdrawn by then and were precluded by Congressional restrictions from helping the South on the battlefield. As a result, when the end came, the US could only evacuate its Embassy personnel and a modest number of the innumerable Vietnamese whose lives were endangered by their service for the US or for the South Vietnamese government. Many hundreds of thousands of Vietnamese were left behind – an outcome that would presage a similarly chaotic and terribly incomplete withdrawal from Afghanistan in August 2021.

As we shall see, it is entirely possible that the shortcomings of the South Vietnamese leaders, government and military forces, as well as cultural divides and pervasive corruption in the South's institutions and society, precluded victory. The enemy's skill and determination, its ability to access the South through neighboring countries and the South's challenging terrain also played very significant roles in the ultimate defeat. But in addition, significant shortcomings in the US strategy and conduct of the war dramatically undermined the prospects for success, perhaps as much as the inadequacies of the South Vietnamese and the tenacity of the North.

The American military leadership failed for far too long to perform what Clausewitz described as "the first, the supreme, the most far-reaching act of judgment that the statesman and commander have to make," that of establishing "the kind of war on which they are embarking; neither mistaking it for, nor trying to make it into, something that is alien to its nature." As Kissinger observed in early 1967, two years before Vietnam became part of his responsibilities, "Our military tend to expect to fight this war as they have studied war at Fort Leavenworth" – home of the US Army Command and General Staff College – "while the other side is not doing anything of the kind. The

Vietcong is using political and psychological criteria where we are applying some very traditional military criteria."

Presciently, Kissinger feared that "we were being lured into the role of a bull in a bullfight, who always forces the other side to give way, but in the process slowly has his strength drained."[5] The US, he judged, "lacked any overall concept for the conduct of military operations against the guerrillas and for the building of a nation." Beyond that, he noted, South Vietnam was also deeply "fragmented and disorganized." His observations would be more than validated when he became the national security advisor in early 1969, and the realities they captured would continue to present enormous challenges as he and President Nixon sought to bring US involvement in the war to the best possible conclusion, albeit one that would ultimately prove unsustainable.

Setting the stage

Although it is now commonplace to regard Afghanistan as America's longest war, America's *overall* involvement in Vietnam actually eclipsed by several years the nearly two decades that the American military served in Afghanistan.[6]

The first members of the US Military Advisory Assistance Group (MAAG) were deployed to French Indo-China in late 1950, as the French struggled against the communist insurgents known as the Viet Minh. The Viet Minh were led by the intellectual communist nationalist Ho Chi Minh and were viewed with increasing alarm in the West. This apprehension was fuelled by three developments in Asia: Mao Zedong's victory in the Chinese Civil War; the Korean War launched in June 1950 by North Korean communist forces; and the recognition of Ho Chi Minh's regime in Vietnam by both Beijing and Moscow. In the wake of these events, leaders in Washington came to regard the Viet Minh campaign in Indo-China as another manifestation of Soviet-supported communist expansion in Asia, with Indo-China seen as the key to South-east Asia.

The Viet Minh were, in fact, as much nationalist as communist, but that distinction was not appreciated as governments became increasingly committed to containing communist expansion in Asia and elsewhere in the world. Given that the Viet Minh drew substantial support from both China and the USSR, by the mid-1950s President Eisenhower and other American leaders started to subscribe to the "domino" theory, which posited that governments vulnerable to communist takeovers might be toppled like a row of dominoes if not stopped. Vietnam was seen as a critical domino, one of the principal countries in which the spread of communism had to be halted.

The nature of the war

The campaign pursued by Ho Chi Minh and his formidable military leader, General Vo Nguyen Giap, was adapted from Mao Zedong's theory of protracted war which had guided the Chinese communists to victory in China's civil war. Mao's concepts posited a long-drawn-out struggle conducted methodically by an insurgent movement to overthrow an existing government. This effort is typically described as comprising three phases from the insurgent's perspective. The first, the Contention Phase, consists of insurgent agitation, proselytizing and limited guerrilla and terrorist attacks, because the government has greater forces and the insurgents have to concentrate on survival and building popular support. The second, the Equilibrium Phase, concentrates more on significant guerrilla warfare, overt violence and the establishment of insurgent bases. And the third, the Counter-offensive Phase, commences when the insurgents achieve superior strength to the government forces and are able to topple the existing regime.[7]

Of course not all insurgencies progress smoothly from one phase to the next throughout a country or region; rather, the progress and growth of the insurgent movement, moving from one phase to the next, are often beaten back by the government, with its forces countering the insurgents' efforts and forcing them to regroup and start again. There are successes by the insurgents in some areas even as there are setbacks in others, with different phases present in different areas. The insurgents may conduct mostly terrorist activities in some locales, more organized guerrilla unit efforts in others and larger regular force and guerrilla operations elsewhere.

The same is true of the levels of the political progress in the different phases of an insurgent campaign. This was certainly the case in the course of the French and US wars in Indo-China and Vietnam (and it would be later in Iraq and Afghanistan as well). The result of both the French and US wars in Vietnam was the same, with substantial, organized communist ground forces – including North Vietnamese Army units joining the guerrillas and insurgents in the South – prevailing, and the French and later American forces withdrawing, ultimately leading to the unification of Vietnam under a communist regime.

Most insurgencies consist of elements that can be described as: movement leaders, combatants (comprising local, regional and main-force elements), political cadre members, auxiliaries (supporters who provide various services and forms of assistance) and a mass base (the bulk of the popular membership), though different insurgencies will exhibit different proportions of these elements and unique contextual features. All of those elements were present in Vietnam.

Defeating an insurgency requires a comprehensive civil–military counter-insurgency campaign, an effort that typically entails tasks well beyond the conventional offensive and defensive military operations for which most military forces train and equip. Moreover, counter-insurgency campaigns are normally carried out in populated areas, among the people. The people's allegiance may be gained through intimidation (more common with insurgents) or persuasion (ideally by the government, though not always). As was often emphasized during the Surge in Iraq in 2007–8, "the human terrain is the decisive terrain," and counter-insurgents have to develop a deep understanding of it.

Although geographic terrain also matters considerably in counter-insurgency campaigns – as was obvious in the challenging jungles, rice paddies, mountains, river delta and cities of Vietnam – the people are the prize, at least until the later stages of an insurgent campaign when large conventional forces become the focus.

Military operations in a counter-insurgency campaign must strive, first and foremost, to achieve security in the areas of focus. Security is the foundation that makes all else possible; without it, nothing else is attainable. But it is the performance of other tasks that solidifies the security foundation and builds on the gains, with the ultimate goal being achievement of local support for the government and rejection of the insurgents. Having achieved progress in a specific geographic area, the effort is then expanded to increase the size of the areas in which security, governance, restoration of basic services and so forth have been achieved.

This process has been described as the "oil spot" approach – clearing, holding and rebuilding one spot or area and then expanding it, as an oil spot expands, by doing the same in a contiguous area and so on, until a larger and larger area has been secured, held and rebuilt. Over time, military forces gradually thin out as local police and other local security elements augment, and then replace, the military forces, so that the military elements can push out further and focus on additional areas, while still keeping a close eye on the areas that have been cleared to ensure that they remain "held." If security is not sustained, needless to say, the counter-insurgent forces have to begin all over again, as American forces experienced repeatedly in Vietnam and later in 2006 and 2007 during some of the toughest days in Iraq.

While all combat operations require detailed intelligence about the adversary, the terrain, the various elements that comprise the population and how a country is governed, a counter-insurgency campaign requires especially granular understanding of these elements. Proper conduct of such a campaign requires particularly deep appreciation of the "human terrain," including the culture, traditions, religions, societal organizing structures, economy, governance, legal system, security forces and so on. If one is to win the "hearts and minds" of the people, that deep understanding of the people is obviously essential, as are relationships built with the people themselves. As we will see, this element of a counter-insurgency campaign was often lacking during the American effort in Vietnam.

Achieving legitimacy is particularly crucial in the conduct of counter-insurgency operations. Indicators of the progress made in this respect include: the ability to provide security for the population; selection of leaders in a manner considered just and fair; a high level of popular participation in political processes; a culturally acceptable level of corruption; adequate political, economic and social development; and a high level of central government acceptance by major public institutions. It is rare for a counter-insurgency effort to achieve lasting success without the host-nation government achieving such legitimacy.[8] And in South Vietnam, with corruption going right to the top of the government, that was always a challenge.

While noting that every counter-insurgency campaign is unique, achieving success generally involves a number of common elements, including, as already highlighted, the detailed understanding of the nature of the enemy and the local political entities, society, values and culture. Another requirement is unity of effort of all civil, military and security elements engaged in the campaign. Additionally, these elements must accept the primacy of politics and political entities, develop very solid intelligence and use it to drive the conduct of operations and achieve security that is based on the rule of law and recognizes the need for long-term commitment.

Many of these key elements were missing in South Vietnam. This was in large part because, as John Nagl observed in his superb book *Learning to Eat Soup with a Knife*, "The United States Army entered the Vietnam War with a doctrine well suited to fighting conventional war in Europe, but worse than useless for the insurgency it was about to combat."[9]

From the outset, then, there was a significant disconnect between the operational concepts, organizational structure, training, leader development and materiel proposed by the Americans for the South Vietnamese forces and what was actually needed given the challenges at the time.

Notably, the American approach was also disconnected from what the leaders of the new Army of the Republic of Vietnam (correctly) thought they needed. Although the French had been defeated by a campaign that had exhibited each of the three stages of the classical insurgency model, the Americans sought to structure the new Army more to repel a conventional Korea-style invasion by North Vietnamese forces along the eastern coast of Vietnam than to counter the insurgent threat that was already manifesting itself in the South.

The opening phase of US involvement

After the French forces' withdrawal following their defeat at Dien Bien Phu, the US Military Assistance Advisory Group (MAAG) worked hard in the late 1950s to develop a plan to overhaul the South Vietnamese force structure to meet what Washington saw as the major threat – aggression from North Vietnam. Not surprisingly, the resulting force would be based largely on the structure of US Army divisions at the time, with seven being the final number of Army of the Republic of Vietnam (ARVN) divisions to be manned, trained and equipped.

Meanwhile, the new South Vietnamese government set about trying to deal with what it viewed as the most pressing security threats at that time, those stemming from the growing insurgency being carried out by Vietcong (VC) forces that would be the guerrilla elements of the National Liberation Front (the political arm of the VC that was based in Hanoi and closely aligned with the communist government of North Vietnam). In contrast to the American emphasis, the South's approach included all the key elements of a civil–military counter-insurgency strategy, namely clear-and-hold operations, political indoctrination, proactive civic action, paramilitary security forces and economic reforms.[10] Unfortunately, numerous factors undermined the South's strategy, among them poor leadership, corruption and half-hearted execution, not to mention lack of US support for the South's approach.[11] Nor would much change in the next decade and a half.

The South's counter-insurgency strategy was substantially under-mined by the MAAG's near-exclusive focus on the development and exercising of the new ARVN divisions. Though those efforts were inter-rupted from time to time by the need for the new regular forces to come to the aid of beleaguered paramilitary forces countering the actions of Vietcong in the countryside, the MAAG largely dismissed the activities of the guerrillas. It persisted in establishing conventional forces capable of repelling an invasion by the North, even when, in late 1959, ARVN units began increasingly to be pulled into operations against the insur-gents, operations for which they were not suitably trained, equipped or employed.

Even when US special forces began training South Vietnamese personnel for special operations, the focus was not on counter-insurgency operations but on organizing guerrilla and partisan forces. A more appropriate effort would have been to counter the growing number of enemy insurgents by focusing on local security and the other elements needed to gain the support of the people.[12]

The officers who ran the MAAG were not incompetent; it was just that the principles of counter-insurgency warfare were alien to them. Thus, once the seven standard divisions had been established, American military advisers felt they had accomplished their mission. Their assumption was that South Vietnam would now be able to halt a future North Vietnamese conventional invasion until South-East Asia Treaty Organization forces could come to its aid. However, as Andrew Krepinevich points out, "The problem was that North Vietnam had no intention of conducting an overt invasion or of waging war according to the preferences of American generals."[13]

It was not until March 1960, by which time the long-neglected South Vietnamese local security forces were being "chewed up" by their communist adversaries, that the US began drawing up a plan to help the Vietnamese deal with the kind of war they were actually fighting. US special forces were deployed to set up initiatives to train and equip Vietnamese special forces, Ranger companies and Civil Guard. Yet even those initiatives would never receive the priority or the resources needed.[14]

1961–1963: the Kennedy administration ramps up the advisory effort

The inauguration of President John F. Kennedy in January 1961 brought not just a new, more activist American leader but also a new national strategy, with "Flexible Response" replacing the Eisenhower-era strategy of "Massive Retaliation." This new approach held that the threatened use of nuclear weapons underpinning the concept of Massive Retaliation had not prevented various communist actions in the 1950s and was no longer credible when it came to deterring or addressing threats. The US needed more flexible military options to establish a credible deterrent to communist expansion, with forces capable of responding to actions from insurrection, insurgency and irregular warfare through mid-intensity conventional conflict to all-out warfare including nuclear weapons. This effort gained additional urgency when Soviet President Nikita Khrushchev offered the USSR's support for wars of national liberation shortly before Kennedy was inaugurated. But Kennedy and the so-called "best and brightest" had not come into office to stand idly by while America's interests were challenged.[15] As the President had announced in his inaugural address on 20 January 1961, "We shall pay any price, bear any burden, meet any hardship, support any friend, oppose any foe to assure the survival and success of liberty."[16]

Kennedy and his advisers were far more willing to get involved in limited wars than the Eisenhower administration had been, and this was communicated in no uncertain terms to the military within the first month of the new presidency.[17] Despite this, there was not the kind of change to the key institutional elements – doctrine, organizational structure, training, leader development, preparation of forces for deployment, personnel policies and so on – that was needed to prepare the Army, in particular, for the conduct of true counter-insurgency operations. In the Vietnamese context, even when the term "counter-insurgency" was used in various documents and plans, the resulting actions were very much subordinated to the development of *conventional* Vietnamese forces and the conduct of operations by US units in large "search and destroy" operations, rather than in "clear, hold and build" operations and the development of local security forces more appropriate for a counter-insurgency campaign.

As a result, US forces in Vietnam initially focused on largely conventional operations. Even though some metrics did track critical elements of a counter-insurgency approach, such as the number of hamlets secured, they were undermined over time by lack of integrity in reporting. In large measure, then, the US military during the Kennedy and then Johnson administrations continued what they had been doing during the Eisenhower years. They largely structured the South

Vietnamese forces along the lines of conventional American units and trained and equipped them the same way as well. With some exceptions, this continued at least until the advent of the Vietnamization effort and the shift to a "One War" approach after General Creighton Abrams assumed command in 1968.

As Fredrik Logevall observed in his brilliant book *Embers of War: The Fall of an Empire and the Making of America's Vietnam*, few in Washington understood the essence of the central issue in Vietnam, that "military prowess meant only so much – the war had to be won politically if it was to be won at all."[18] American leaders did not grasp that results could be measured only over a period of several years, that ultimately the host government must carry the burden rather than the US taxpayer and that the employment of massive firepower would inevitably result in the alienation of civilians.

President Kennedy quickly became seized with Vietnam, especially after his aide Walt Rostow shared a memo by Brigadier General Edward Lansdale, who had just returned from a trip there and described the situation as very alarming and deteriorating.[19] Kennedy came to see Vietnam as the place to demonstrate the credibility of American power, and was encouraged by Robert Komer of the National Security Council, who recommended a vast increase in US involvement there. "I believe it is very important that this government [has] a major anti-Communist victory to its credit in the six months before the Berlin crisis is likely to get really hot," Komer wrote, arguing that Vietnam was a "better place than Laos to achieve the desired result."[20]

Four months into his presidency, on 11 May 1961, President Kennedy approved a National Security Action Memorandum stating that the US objective in Vietnam was "to prevent Communist domination of South Vietnam." He also approved the deployment of 400 special forces soldiers for training purposes.[21] With the situation continuing to deteriorate in the autumn of 1961, he dispatched to Vietnam General Maxwell Taylor, a former Army chief of staff who had been recalled to active duty and appointed the "Military Representative to the President." As Taylor later recalled, "I was not asked to review the objectives of this policy but the means being pursued for their attainment. The question was how to change a losing game and begin to win, not how to call it off."[22]

Taylor found in Vietnam a "country suffering from a collapse of national morale" and grappling with Vietcong strength that had grown from an estimated 10,000 at the beginning of 1961 to some 17,000 by October. There were also deteriorating situations in both Laos and Cambodia to the west and the worst flooding in Vietnam in decades. And a well-known Vietnamese colonel who had liaised with an international commission had been kidnapped and brutally murdered by the Vietcong. The situation was dire.[23]

Taylor and his team engaged in a whirlwind of briefings, meetings and visits with all key Vietnamese and US leaders and organizations, including two days in the field. In a critical meeting with Vietnam's President Ngo Dinh Diem, Diem explained the inability of his 150,000-man Army to contain the enemy forces in his country and also the deficiencies in the paramilitary and police forces for local security. He noted that those forces had not been supported in the past by the Americans. Taylor encouraged the development of "a national plan that would pull together all the resources in an integrated program to defeat the Vietcong" – something Washington had pressed Diem for in the past without success – and Diem "spoke confidently about his ability to produce and implement such a plan."[24]

One of Taylor's principal tasks was also to evaluate the political durability of President Diem, and he heard plenty of grievances about him and his seemingly dysfunctional leadership practices. Diem was a micromanager whose actions had frustrated efforts to achieve unity of effort and eroded the authority of the Commanding General of Vietnam's Field Command. There were also widespread suspicions that Diem's family and a select inner circle were engaged in extensive corruption. In the end, however, the consensus was to stick with him.

Following Taylor's return, President Kennedy approved most, but not all, of what Taylor recommended. In particular he did not agree to the recommendation to send 8,000 US troops disguised as a logistical task force to support flood relief and base security. Nonetheless, he did increase military and other forms of assistance, including advisory groups established throughout the South's bureaucracy, efforts to improve training for the Civil Guard and Village Self-Defense Corps and more equipment such as helicopters to improve the mobility of the ARVN.

Kennedy made the new aid program contingent on specific promises from Diem to reform the government, with the United States being given a voice in the decision-making process through its Ambassador. When those demands created a crisis in Saigon, however, the US backed down and a more innocuous agreement was reached in which both countries agreed not to take actions without first consulting the other. This deference to Diem probably encouraged his future intransigence and led to a missed opportunity to achieve the kind of reforms desperately needed.[25]

In 1962, the South Vietnamese, supported by the increase in American equipment and advisers, went on the offensive against the Vietcong. Simultaneously, they launched the strategic-hamlet program that was intended to isolate the VC from their principal source of support in the South. The idea was to bring peasants in scattered villages together into hamlets surrounded by moats and bamboo-stake fences, which could be protected by the Vietnamese military. It was hoped that the reinstitution of village elections, the establishment of land reform and the creation of schools and medical services would persuade the people that life under the government offered more than could be expected under the insurgents. These were seemingly textbook counter-insurgency initiatives designed to deny the VC access to the peasantry. Their execution, however, was anything but textbook.

The hamlet plan was overseen by President Diem's brother and it was projected that no fewer than 7,000 strategic hamlets would be built by the end of 1962 and 12,000 by the end of 1963, thus consolidating nearly all the rural population of South Vietnam.[26] That turned out to be a wildly overambitious plan, which was soon undermined by corruption and heavy-handedness. In late 1963, it was discovered that only 20 percent of the hamlets reported as complete had even been constructed.[27]

Around the same time, as part of the expansion of the US effort, the MAAG was increased significantly in size, renamed Military Assistance

Command, Vietnam (MACV) and given a new four-star commander, General Paul Harkins. While subsequent assessments of Harkins' tenure as head of MACV range from highly critical to assertions that he was unfairly maligned, even his greatest defender conceded that he was not a "creative or brilliant strategist."[28] Certainly, his assessment that the war would be won by 1963 proved to be wildly overoptimistic.

The substantial increase in American advisers, weapons, armored personnel carriers, helicopters and close air support aircraft did enable the ARVN to go on the offensive in the spring and summer of 1962 and to wrest the initiative from the VC, inflicting heavy casualties on them and seizing a number of strongholds. Over time, however, the VC adjusted their tactics, using the dense forests, jungle and rice paddies to their advantage. This made locating their elements much more diffi-cult. Moreover, the "hold" element of "clear, hold and build" was too often inadequate; when the South's forces withdrew from areas they had seized, the VC typically moved back into them.

By late 1962, the VC had regained the initiative. While the ARVN and their American advisers were chasing enemy units, the movement's political arm, the National Liberation Front, infiltrated the villages, mobilizing the peasants with a combination of sophisticated political indoctrination and selective violence. Consequently, the VC had gained 300,000 active members and perhaps over a million passive supporters by the end of 1962. The age-old challenges of determining friend from foe in a guerrilla war had grown enormously, and the strafing of villages and heavy use of bombing simply increased the opposition the govern-ment and the US faced.[29]

The emergency led to South Vietnam limiting the civil liberties it had previously permitted, imposing censorship and banning some public meetings. This made it harder for supporters of the war to frame it as a necessary struggle between democracy and dictatorship. Nonetheless, the US Embassy and MACV remained optimistic about the progress of the new effort, although this was all too often based on impressive statistics furnished by the South Vietnamese government that would later be revealed as wild overstatements of reality on the ground.

After a very pessimistic assessment by Senator Mike Mansfield, who visited Vietnam in late 1962, Kennedy decided to send Assistant Secretary of State Roger Hilsman to conduct a fact-finding mission.[30] Hilsman returned with serious concerns about the effectiveness of the ARVN military operations, about the strategic-hamlet program and about President Diem's increasing isolation from his people. He nonetheless concluded that the United States and Vietnam were "probably winning," but added that the war would "probably last longer than we would like" and would "cost more in terms of both lives and money than we had anticipated."[31] Despite this, General Harkins told a gathering of officials in Honolulu "that the war might be over by Christmas," an unfounded hope of the type uttered by optimists in many previous wars.[32]

In retrospect, two incidents involving ARVN forces should have been more concerning to the American and Vietnamese leadership than they were at the time. In the first, in early October 1962, the 7th ARVN Division was on a routine sweep operation in the delta region when a band of VC chose to stand and fight, rather than slip away as was usual. The guerrillas inflicted substantial casualties on the division and wiped out an entire platoon of Vietnamese Rangers.[33] Then in early January 1963 the ARVN deployed an infantry battalion, two Civil Guard battalions and a company of armored personnel carriers, supported by artillery and close air support, to engage a battalion of VC in the village of Ap Bac, only 40 miles south-west of Saigon in the delta. Here too the VC held their positions tenaciously, before slipping away during the night, leaving eighty of the government forces killed and over a hundred wounded. Three American advisers were also killed and eight wounded, and five helicopters were shot down in an engagement that showed that the VC had devised tactics to deal with American helicopters and armored vehicles.

The defeat at Ap Bac was widely reported in the press, which portrayed it as indicative of a lack of progress in the war.[34] Yet General Harkins tried to claim Ap Bac as a victory, on the grounds that the ARVN had eventually taken the village. He further accused critical members of the press of "doing a disservice to the thousands of gallant and courageous men who are fighting so well in the defense of their country."[35]

In his State of the Union Address in mid-January 1963, President Kennedy asserted that "The spear point of aggression has been blunted in South Vietnam."[36] A few days later, General Harkins presented his comprehensive plan for the war, which foresaw that US military personnel there would be reduced from more than 16,000 to 12,200 by mid-1965 and to only 1,500 in mid-1968, with MACV having been disestablished by July 1966. He also projected that, during the same period, the expansion of South Vietnamese forces would peak at 458,500 men (an oddly specific number) by mid-1964, including a nine-division army numbering 240,000.[37] Similarly, in late January, the US Army Chief of Staff, General Earle Wheeler, reported that the situation had moved "from a circumstance of near desperation to a condition where victory is now a hopeful prospect."[38] These assessments underscore the need for generals to be very careful about accepting their own side's propaganda, as fatal overoptimism often follows.

Tensions between the South Vietnamese leadership and the Americans grew throughout the spring of 1963, with President Kennedy becoming increasingly uneasy. In May, a new challenge materialized when a confrontation arose between Vietnamese troops and Buddhists at an event that violated orders forbidding the display of flags celebrating the Buddha's birthday. A photograph of the subsequent self-immolation of a Buddhist monk was published in newspapers around the world, and numerous further demonstrations and confrontations between Buddhists and government forces took place. The Buddhists drew support from other disaffected elements of Vietnamese society and quickly became a huge problem for the regime of Diem, who was Catholic, especially when harsh treatment of protesters incited further opposition and the spiral of violence escalated.

By the summer of 1963, American confidence in Diem had been seriously shaken, and secret discussions were opened between South Vietnamese generals and American officials about what to do. Following instructions in an ambiguous, inadequately cleared cable sent from

Washington on a summer weekend when many top officials were not in the city, Henry Cabot Lodge Jr., the US Ambassador to Saigon, made it clear to the Vietnamese generals that the United States would not continue to support Diem if he refused to cooperate with proposed reforms. Although there was grave concern the following Monday in Washington, Kennedy reaffirmed the instructions.

Four months later, the Vietnamese high command launched a military coup, seizing key military installations, securing the surrender of the loyalist special forces and demanding the resignation of Diem. The coup leaders assured the Americans they would spare Diem's life, but he was captured along with his younger brother and murdered in the back of an armored personnel carrier.[39]

Three weeks later, John F. Kennedy was assassinated in Dallas, and Vice President Lyndon Johnson was sworn in as president. Kennedy's defenders would subsequently argue that he had recognized the futility of the situation in Vietnam and had been intent on drawing down American involvement. His detractors would argue the exact opposite, pointing out that the American presence in Vietnam had grown substantially during his time in office and that there were no plans to reduce the number of US advisers, which was 16,000 and growing. Regardless, as George Herring observed, Kennedy "bequeathed to his successor a problem eminently more dangerous than the one he inherited from Eisenhower."[40]

Contrary to General Harkins' optimistic assessment in April 1963 that "the war might be over by Christmas," the situation on the ground had grown much more serious by the end of the year. On 21 December, John McCone, the CIA Director, reported to President Johnson that "It is abundantly clear that statistics received over the past year or more from [Vietnamese government] officials and reported by the U.S. Mission on which we gauged the trend of the war were grossly in error," and that "The future of the war remains in doubt."[41]

Unbeknown to the Americans and South Vietnamese, the North Vietnamese political leadership had already agreed on an all-out effort to "seize a favorable opportunity by massing our forces to resolutely seek to win decisive victories during the next few years." The mission of their forces in the South would be to "shatter the puppet army, the primary tool of the enemy regime, in order to create conditions that will allow us to carry out the general offensive/general uprising to overthrow the reactionary government in South Vietnam."[42] As always in war, the enemy gets a vote, and the North's decision presaged tough years ahead.

1964–1965: Johnson decides, Westmoreland takes command and the US commitment escalates

The assassinations of Presidents Diem and Kennedy were key turning points in the struggle for South Vietnam, and by the spring of 1964 that country was suffering the chaos of a society torn apart by insurgency. President Johnson could have withdrawn US forces, but the country would almost certainly have fallen to communism if he had. Instead, he decided to escalate the conflict, just when the North was deciding to do so as well.[43] With the removal of Diem-imposed restrictions on information, it became clear that the VC insurgents controlled much more of the South than the US had imagined. The strategic-hamlet program was a shambles and, in the words of Secretary of Defense Robert McNamara, the situation was "very disturbing." Unless it could be reversed, South Vietnam might be lost.[44]

The South Vietnam into which the US was now about to deploy hundreds of thousands of soldiers was an elongated country nearly the same length as California but only half the width, separated from North Vietnam by a fortified demilitarized zone along the 17th Parallel that extended 3 miles on either side of the line of demarcation, established in the years after the division of the country into North and South. In 1964, some 80 percent of the 16 million people of the South lived on 40 percent of the land, with another 40 percent largely uninhabited and covered with dense forest and jungle, scrub brush, elephant grass and swamps – ideal for guerrilla warfare. Saigon, the capital, was located in the south near the coast. The topography in the country varied greatly, with mountains in the north and low-lying areas along the coastline, and with the heavily vegetated Mekong Delta region, which also featured numerous rivers, in the south-west. Rice paddies were particularly numerous in the inland areas.

The country was made up of numerous ethnicities, including Vietnamese, Chinese, Cambodians and French. The predominant religion practiced was Buddhism, although there were significant numbers of Catholics and members of other religious groups. The elite Vietnamese who led the country included a substantial number of wealthy French-speaking Catholics who did not connect effectively with the Buddhist peasants in the countryside.

At the opening of 1964, the political situation in South Vietnam was in turmoil, just as the gravity of the security situation was becoming alarmingly apparent. The country's economic condition was also fairly precarious, with the bulk of the population, many of whom lived in poverty, engaged in agriculture. Economic inequality between the small elite class and the peasants was striking, and the country was not only heavily dependent on foreign aid, but grappling with high inflation and a large trade deficit. Beyond that, as has been noted, corruption pervaded virtually every institution and aspect of life.

In early 1964, a group of military officers led by General Nguyen Khanh overthrew the weak and divided junta that had ruled since Diem's murder. With the capital in turmoil and cities in anarchy, and with workers on strike, students demonstrating and the press highly critical of the new government, the situation was very fragile. Amid all this, VC units sought to take advantage of the situation with increasingly bold and frequent actions.[45]

President Johnson recognized the need to shore up the new Khanh government, and he publicly committed to supporting it. After a policy review in March, he approved a significant increase in the South Vietnamese armed forces the US would support. He also expanded economic assistance, increased the number of advisers from 16,300 to 23,300 and selected the well-regarded Second World War and Korea veteran and paratroop commander General William Westmoreland to replace General Harkins as the commander of MACV, a change that took place on 20 June. Shortly afterwards, General Taylor left his position as chairman of the Joint Chiefs and replaced Lodge as ambassador in Saigon.[46]

As the year progressed, the situation deteriorated further. Consequently, 900 additional advisers were approved and the US military was directed to conduct joint planning with the Vietnamese General Staff for cross-border options into Laos, where the Ho Chi Minh Trail was becoming of great concern as the logistical supply line for Northern and Vietcong forces in the South. Plans to take the war to North Vietnam were also under examination, and were given impetus when on 2 August a US destroyer was attacked by North Vietnamese torpedo boats in the Tonkin Gulf off the coast of North Vietnam. In response, Johnson authorized US aircraft to attack North Vietnamese patrol-boat bases three days later. On 7 August, the US Congress passed, with near unanimity, the Tonkin Gulf Resolution, which authorized the President to "take all necessary measures to repel any armed attack against the forces of the United States and to prevent further aggression."[47]

Meanwhile, General Khanh's imposition of severe restrictions on civil liberties led to turmoil in Saigon. With angry mobs rampaging through the streets and protesting against his dictatorial actions, he was forced to resign. Once again, the government was severely weakened, and an interim civilian government was organized to lead the country. The disorder made it very difficult to pursue some of the initiatives Washington supported until a new government was established.

Surprisingly, the military situation remained relatively static, as America's focus turned to the US presidential elections to be held on 3 November. On 1 November, however, the VC launched a mortar attack on the airfield at Bien Hoa that killed four American servicemen and damaged a number of aircraft. With a new National Security Council working group established to examine courses of action, the Joint Chiefs of Staff took a tough stance, concentrating on America's loss of prestige if it was seen to act indecisively and minimizing the problems that would arise from its being drawn into a full-scale war.[48] Crucially, the Joint Chiefs (and the command in Saigon) failed to appreciate that the major source of the problems stemmed not from infiltration from the North so much as from VC successes in the South. That was not a trivial misunderstanding.

The Joint Chiefs recommended a two-phased bombing program that focused more on the North than on the South. Phase one consisted of limited strikes against the infiltration routes in Laos and reprisal strikes against North Vietnam; phase two would be conducted once the Saigon government achieved acceptable stability and consisted of a large air offensive against the North lasting between two and six months.[49] In early December, Johnson approved phase-one bombing, but held off on phase two out of concern for possible Northern reprisals. But following additional VC attacks and the assessment of National Security Advisor McGeorge Bundy, who had just returned from Vietnam, that "Without new U.S. action defeat appears inevitable," Johnson ordered execution of Operation Rolling Thunder, a policy of gradually intensified air attacks.[50] In April 1965, some 3,500 sorties were flown against North Vietnamese targets. Subsequently, and in anticipation of likely attacks on the bases from which the sorties were flown, Westmoreland requested Marine units to protect the major base at Danang. Despite Ambassador Taylor's concerns about the readiness of the forces for counter-guerrilla operations in the jungle, as well as the possibility that such a deployment might lead the ARVN to pass security responsibilities on to the US, the request was approved.[51]

Taylor was concerned that once US combat forces had been deployed, it would be very difficult to avoid deployment of more. That worry was validated when in mid-March Westmoreland requested two US Army divisions of some 20,000 soldiers each, one to be sent to the Central Highlands, the other to the Saigon area. The Joint Chiefs not only supported Westmoreland, they pressed for the deployment of additional divisions for the conduct of offensive operations.

The administration did not explicitly approve Westmoreland's request, but in April it did approve the deployment of some 40,000 additional ground combat forces to be employed not in the Highlands but in enclaves around major US bases, and to conduct authorized operations within 50 miles of them. This marked another major step

towards large-scale ground involvement of US forces. It also precipitated the first domestic criticism in America, prompting Johnson to
state that the US was willing to enter into "unconditional discussions"
with North Vietnam, albeit ones that did not compromise the integrity
of the South.[52]

Meanwhile, a new South Vietnamese government was established,
the fifth since Diem's assassination. It featured the flamboyant Air
Marshal Nguyen Cao Ky as prime minister and the respected General
Nguyen Van Thieu as commander-in-chief of the armed forces, but was
viewed critically by most, given Ky's reputation for drinking, gambling
and womanizing. Those concerns notwithstanding, the new government would, over time, prove much more durable and bring a degree of
stability that had been missing since the coup against Diem.[53]

In July 1965, following McNamara's return from another trip to
Vietnam and his recommendation of an additional 100,000 combat
troops, President Johnson approved an intensification of the bombing
campaign against the North, although not the all-out campaign that
Westmoreland had advocated, and the deployment of 50,000 troops
immediately and another 50,000 by the end of the year, as well as whatever additional troops might be needed later. Among the units deployed
would be the 1st Cavalry Division (Airmobile) and the 101st Airborne
Division, which Westmoreland had commanded some years previously.
Johnson also authorized Westmoreland to employ US forces in combat
independent of Vietnamese forces, which was to bring about a fundamental change in the nature of the war, with fateful results over the
next seven years. From merely assisting South Vietnam, the United
States was now to take on the burden of fighting the war.[54]

This decision was an implicit admission that Vietnam had entered
the third phase of insurgency warfare, the escalation to larger unit operations from guerrilla warfare. The focus of US units would be to locate
and engage enemy units, rather than VC guerrillas, leaving the latter
task, and the all-important security of the people, to local Regional and
Popular Forces (at the provincial and district levels, respectively), even
though they had limited capability, along with ARVN units not properly
trained or employed.

Defense Secretary McNamara saw the ensuing campaign as one in which prolonged, high-intensity operations would drain the VC and North Vietnamese capacities, while B-52 strategic bombers and other air assets destroyed the enemy's chances of resupply.[55] Yet this overlooked the immense challenge of precisely locating the VC and North Vietnamese Army forces in densely wooded and jungle terrain, and of bringing to bear, without damage to innocent civilians and infrastructure, substantial numbers of large bombs dropped from high altitude. It also overlooked the possibility that the enemy forces might, in response to US operations, disperse, return to phase-two counter-insurgency operations and resume guerrilla attacks and other insurgent activities that would still generate US and Vietnamese casualties.

The American soldiers deployed in 1965 had trained and prepared together and were sent to Vietnam as units – a huge advantage over the situation in the years ahead, when the units would remain but their soldiers would be replaced individually, with their commanders rotating every six months (or frequently more often due to casualties). The units' training, however, had mostly been on large unit operations, rather than on the tasks required to combat an insurgency and guerrillas. This would not have been an issue if the North Vietnamese and VC had accommodated the US and fought primarily as large units. But the American conventional units were not prepared – nor properly employed – to conduct the kind of local security operations and other tasks that should be the centerpiece of a comprehensive counter-insurgency campaign. Nor was that their focus.

American civilian and military leaders alike had routinely talked about the importance of conducting counter-insurgency operations ever since President Kennedy had taken office. Over time, there would be numerous initiatives intended to establish the various components of a comprehensive civil–military counter-insurgency campaign. The strategic-hamlet program had been one such example, as had various local security forces initiatives, and more would follow. None of them, however, could make up for the fact that the US forces were prepared for conventional, not counter-insurgency, operations and focused on the former, not the latter.

The essence of the ultimate American strategy was that of attrition, where the overwhelming weight of American materiel and resources, especially American firepower, would defeat the enemy. Because that had worked against the Axis powers in the Second World War and also against the North Koreans and Chinese in Korea, it was assumed that it would work again.[56] Little or no consideration had been given to the reality that warfare had evolved since the Korean War had ended in stalemate in 1953, and that the experiences of the French in Dien Bien Phu and Algeria had more relevance to American strategists than what had happened in earlier, large conventional wars. Vietnam was where the United States forces ought to have learned from recent history, but instead it was where they were condemned to repeat it.

In employing the additional forces and substantial munitions, materiel and technology he had been given, General Westmoreland drew up a plan that envisaged the destruction of the VC and their bases by the end of 1967. The US military would employ forty-four battalions by the end of 1965, and twenty-four additional ones in 1966, before the insurgency was eliminated.[57] He explicitly rejected breaking down US units into smaller elements that would concentrate on pacification and the local security of the population, in favor of focusing on the enemy's big units. Much of his approach counted on American air mobility, with the introduction of large fleets of helicopters, as well as substantial ground and air firepower, all of which required very considerable logistical support and massive quantities of ammunition, fuel and spare parts – as well as the facilities in which to perform maintenance functions and store supplies.

Westmoreland's approach overlooked the possibility that the enemy would seek to avoid battle with American units. A greater focus on securing the bulk of the population – 90 percent of which lived along the coastal plain and in the delta – might have forced the enemy to engage US units rather than seeking to draw them into remote areas. The emphasis on search-and-destroy operations as the principal tactic also overlooked the fact that the conduct of such operations alongside ARVN units in the past had not yielded impressive results.

In mid-November 1965, elements of the newly deployed 1st Cavalry Division located and engaged regimental-size North Vietnamese Army units in the Ia Drang Valley. The ensuing battle was both bloody and fierce. The NVA lost over 1,200 soldiers and the US over 200. Westmoreland saw the battle as a validation of search-and-destroy operations and a strategy of attrition, as well as of the importance of air mobility.[58] However, as retired Lieutenant General Hal Moore, who had commanded the principal battalion in the battle, later observed, "General Westmoreland thought he had found the answer to the question of how to win the war: He would trade one American life for ten or eleven or twelve North Vietnamese lives, day after day, until Ho Chi Minh cried uncle. Westmoreland would learn, too late, that he was wrong; that the American people didn't see a kill ratio of ten-to-one or even twenty-to-one as any kind of bargain."[59]

Moore also noted that, following a later operation, in which he commanded a brigade (of several battalions) to liberate the village of Bong Son, and in which another eighty-two of his men were killed and 318 wounded, the North Vietnamese Army and VC main-force units returned to the area within a week of his unit turning over control to the Vietnamese. It was clear, following subsequent operations with similar outcomes, that the US command had failed to coordinate American and South Vietnamese operations with follow-on Vietnamese government programs to establish enduring control in the newly cleared areas, to hold them after they were cleared. "If they couldn't make it work in Bong Son," Moore later asked rhetorically, "where the most powerful American division available had cleared enemy forces from the countryside, how could they possibly hope to re-establish South Vietnamese control in other contested regions where the American presence was much weaker?"[60] Clearly, the "hold" aspect of clear-hold-build was lacking.

Moore also observed that, when operating in heavily populated areas, the Americans' tremendous firepower – the artillery, airstrikes and helicopter gunships that had saved his men's lives in the unpopulated Ia Drang Valley – were, despite his troops' best efforts, "taking a toll of innocent civilians killed and maimed, villages destroyed, and farm animals slain."[61] His reflections captured many of the shortcomings of the strategy that would guide operations for the ensuing four years at the least.

Somewhat ironically, more than a decade later, General Westmoreland would reflect on the battle in the Ia Drang Valley, noting that "An enemy document captured later admitted heavy losses but rationalized them on the basis that the lessons learned were worth the cost of a hundred thousand men."[62] Whether he recognized it or not, that observation captured another reality that would invalidate his strategy: the North Vietnamese leadership did not regard casualties in the same light as did the Americans. The North Vietnamese and their supporters in the South felt they were engaged in their war of independence, and they were willing to sacrifice vast numbers of soldiers to achieve victory in it.

It is undoubtedly harder for a general from a democratic country, who constantly needs to consider the domestic political effect of high death tolls, to fight an attritional war. On the other hand, ideological autocrats such as Ho Chi Minh and Mao Zedong had a cynical disregard for human life (from our perspective at least), caring only for the eventual outcome of "the people's struggle" and the victory of the dictatorship of the proletariat. By contrast, Western generals, understandably, have needed to fight in such a way that casualties are carefully kept to an absolute minimum, both on obvious humanitarian grounds and also so as not to inflame opposition to the war in Congress, in the press and among the public.

The Marine approach

Even as the US Army forces were engaging in search-and-destroy operations, typically conducted from sizeable bases, US Marine Corps units operated very differently in the northern part of South Vietnam. The Marines based their operational concepts on a superb doctrinal manual on small wars dating back to 1940, which stated, "In small wars, the goal is to gain decisive results with the least application of force and the consequent minimum loss of life. The end aim is the social, economic, and political development of the people subsequent to the military defeat of the enemy insurgent. In small wars, tolerance, sympathy, and kindness should be the keynote of our relationship with the mass of the population."[63]

Based on the guidelines in that manual, Lieutenant General Lew Walt, the Marine Commander in Vietnam from June 1965 to June 1967, pursued an important initiative, the Combined Action Program (CAP), in which platoons of Marines and Navy medical corpsmen worked together with Vietnamese Popular Force platoons that were deployed in villages and hamlets to deny the VC access to the people. As a *Life* magazine article explained, they were ordered to "help protect the villages, get to know the people, find the local Communist infrastructure and put it out of business ... If these people could be located and won over, the Communists would be hit where it hurts." In essence, the Marines were operationalizing the concept of "living with the people to secure them," one of the most important of the big ideas that would guide the Surge in Iraq more than forty years later.[64]

The CAP platoons in the villages and hamlets, together with the Popular Force platoons, routinely engaged in night patrols and ambushes to disrupt the Vietcong actions. Rather than travelling by helicopter, they used the same roads as the civilian population. While language barriers presented issues, the results were impressive. With implementation of CAP, the number of "secure" villages under General Walt's protection rose from 87 to 197 between 1965 and 1967, while the number of Vietnamese living in "secure" areas rose from 413,000 to 1.1 million, all from the work of just 2,000 Marines. Particularly noteworthy was the saturation of coastal farming areas with CAP platoons to enable the farmers to harvest their crops without Vietcong "taxation," thus denying the insurgents a key source of revenue. Moreover, the per capita casualties were half those of US Army or Marine infantry units conducting large-scale operations.[65]

Yet, despite the impressive results of the CAP initiative, General Westmoreland did not believe they were in line with the concept of operations he had drawn up, and he wanted the Marines instead to create larger helicopter-transported forces which could attack large enemy units and disrupt their bases and supply caches. While he observed in his memoirs that the CAP initiative had "achieved some noteworthy results," he also noted that he didn't have enough soldiers to put a "squad of Americans in every village and hamlet."[66] In fact, a Department of Defense study showed that the CAP approach needed only 167,000 troops, far fewer than the 550,000 troops ultimately deployed in Vietnam under Westmoreland.[67]

The war enters 1966

By late November 1965, MACV was recognizing that, due to the build-up of North Vietnamese and VC forces, US forces could not achieve their campaign objectives by the end of 1966. Westmoreland realized that his attrition strategy would take time, and that the infusion of large numbers of US troops would be required to reach the key point where the enemy's losses exceeded his capability to replace them. Following a high-level conference in Honolulu in February 1966, Secretary McNamara approved a major plan for American troop levels to be increased from 184,300 at the end of 1965 to 429,000 by the end of 1966, with the number of maneuver battalions more than doubling from thirty-five to seventy-nine by June 1967.[68]

General Westmoreland and the senior military also pushed for an increase in the geographical scope of the war, to allow MACV to take the war to the sanctuaries used by the enemy in Laos and Cambodia. This desire was born, in part, of frustration about the ability of the North to use the Ho Chi Minh Trail through Laos and Cambodia to infiltrate NVA units and supplies – although the VC took the majority of their resources from the South itself. There was also frustration at not being able to bring the enemy to battle, because the North Vietnamese and VC forces were highly elusive and were generally able to avoid contact when it suited them, fighting only at times and in places of their own choosing. Ultimately, however, the plan to interdict the sanctuaries proved unfeasible and had to be shelved.[69]

Thus the attrition strategy continued, with approximately 95 percent of American combat battalions engaged in search-and-destroy rather than counter-insurgency operations. While US forces inflicted substantial casualties on the NVA, they made relatively little headway against the VC, who actually grew in number during the first year of US combat forces being deployed.[70] North Vietnam was essentially matching the US escalation of the war with 200,000 North Vietnamese reaching draft age each year, its leaders recognizing that the higher the casualty rate the US suffered, the sooner Americans would turn against the war.[71]

In March 1966, a study commissioned by General Harold K. Johnson, the US Army Chief of Staff, confirmed the problems involved in focusing largely on search-and-destroy operations. Entitled "Program for the Pacification and Long-Term Development of Vietnam" (PROVN), it had been drawn up by a team of some of the best and brightest Army staff officers in the Pentagon and conveyed the magnitude of the task in Vietnam, enumerating the shortcomings of the ongoing campaign of big-unit operations and providing an alternative strategy.[72] The main effort needed to be in the hamlets, villages and districts where most South Vietnamese lived, it argued, rather than in pursuing North Vietnamese and VC units in the more remote areas of the country – though the latter actions were also required in certain cases. Local efforts needed to receive not just much greater emphasis, resourcing and focus, they also needed to be unified, with civil and military elements working together. "The critical actions are those that occur at the village, district, and provincial levels," it concluded. "This is where the war must be fought; this is where the war and the object which lies beyond it must be won."[73]

General Westmoreland effectively dismissed the PROVN Report, asserting that most of the issues had already been addressed and not even mentioning it in his memoirs. General Wheeler, the Chairman of the Joint Chiefs, did likewise, and disagreed with the study's assertions that the problems in South-east Asia were primarily political and economic rather than military. "The essence of the problem in Vietnam is military," he firmly stated.[74]

Eventually, following Westmoreland's replacement by General Abrams in 1968, the PROVN study's recommendations would be

embraced and helped achieve promising results; however, vital time had
been lost by then. Instead, Westmoreland continued with large-scale
search-and-destroy operations that achieved only temporary effects,
principally because the units did not hold what they had cleared.
Beyond that, the Westmoreland approach also employed extensive use
of unobserved artillery fire and close air support, even in populated
areas, which cost $3 billion a year. And all of that required the establish-
ment of enormous bases – with correspondingly enormous security
requirements.[75]

When General Westmoreland's force request for 1967, which would
bring the number of maneuver battalions to ninety and the overall force
to over 540,000, was forwarded to McNamara, the Secretary of Defense
requested a detailed analysis in order to determine those numbers truly
essential to carrying out the war plan. In mid-October McNamara and
the Chairman of the Joint Chiefs flew to Vietnam. McNamara concluded
that there was "no reasonable way to bring the war to an end soon [as
the enemy] has adopted a strategy of keeping us busy and waiting us
out." Pacification was "a bad disappointment," and the Rolling Thunder
bombing campaign had produced insignificant benefits. He also noted
that, despite substantial casualties having been inflicted on the enemy,
"there is no sign of an impending break in enemy morale and it appears
that he can more than replace his losses by infiltrating from North
Vietnam and recruitment in South Vietnam."[76]

Yet, despite that assessment, McNamara and the President approved
an increase to 470,000 troops and directed greater commitment to
pacification. McNamara would come to bitterly regret in later life his
optimistic, indeed hubristic, public statements about how soon the war
could be won. But that does not absolve him of having badly misled the
American people at the time, giving them hope that he himself did not
share.

As the campaign moved into 1967, it became increasingly clear that,
despite the very high enemy body-count figures that were being
reported, victory was still a long way off. Each American blow "was like
a sledgehammer on a floating cork," the journalist Malcolm Browne
observed. "Somehow the cork refused to stay down."[77]

The arrival of half a million American troops and the massive quantities of supplies and materials required to support the build-up also had a devastating effect on the Vietnamese economy, with ports backed up, inflation skyrocketing and an influx of goods that wrecked many of Vietnam's few indigenous industries. Moreover, crime and corruption flourished in the boom-town atmosphere. The wholesale Americanization of the war created distrust and tensions between the US and South Vietnamese militaries because of security leaks, which led to less sharing of information about future operations and even restrictions on South Vietnamese soldiers entering US bases.[78] Meanwhile, nation-building efforts floundered at all levels, especially in the critical hamlets, villages and districts.

Nonetheless, in early 1967 General Westmoreland requested an increase to at least 559,000 personnel and 108 battalions by mid-1968, with his "optimal package" being 676,000. This was received with understandable scepticism by President Johnson who, in a meeting with Westmoreland in Washington, asked, "When we add divisions, can't the enemy add divisions?"

Meanwhile, large-scale search-and-destroy operations continued, with decidedly mixed results. Operation Attleboro was a ten-week operation in late 1966 in which 22,000 US and ARVN troops, with B-52 bomber air support and massive artillery fire, drove into a Vietcong stronghold north-west of Saigon. Although it was declared a victory since the enemy suffered an estimated 1,100 dead and lost a considerable quantity of supplies, before long the Vietcong were back in occupation of the area. "They metered out their casualties," Major General William Depuy, Commander of the 1st Infantry Division in Attleboro, later reflected,

and when the casualties were getting too high ... they just backed off and waited. I really thought that the kind of pressure they were under would have caused them to knock off the war for a while, as a minimum, or even give up and go back north. But I was completely wrong on that. I was surprised a little bit, too ... at the difficulty we had in trying to find the VC. We hit more dry

holes than I thought we were going to hit. They were more
elusive. They controlled the battle better. They were the ones
who decided whether there would be a fight.[79]

Similar operations followed throughout 1967, with similar results.
Often, the only significant engagements were those initiated by the VC,
while massive amounts of firepower were expended by the Americans.
In Operation Junction City, for example, several tons of ordnance were
used, on average, to kill one VC soldier. Yet the US Army continued to
affirm that "multi-division operations have a place in modern counter-
insurgency warfare."[80] As captured VC documents revealed, however,
the US approach was playing into the hands of the enemy, which was
seeking to prevent US and ARVN forces from concentrating in the populated coastal plain areas.

Towards the end of 1967, with the Pentagon having approved an
increase of US troops to 525,000 and with additional units arriving,
General Westmoreland was invited to Washington as part of President
Johnson's "Success Offensive" – a public relations effort designed to
rally domestic opinion behind the war by focusing on the supposed
victories being won in Vietnam. "I am very, very encouraged,"
Westmoreland told reporters on arrival. "We are making real
progress."[81] Some seven months earlier, in an address to the US
Congress, he had enumerated the challenges in Vietnam but offered a
similarly upbeat assessment and assured representatives that "We will
prevail in Vietnam over the Communist aggressor."[82] While flying back
to Vietnam from his trip in November, Westmoreland sent a cable to
his deputy in which he noted that his presentation in Washington had
portrayed to the American people "some light at the end of the
tunnel."[83]

This episode reminds us that it is not for generals to propagandize
for the wars they are fighting; that is the job of the politicians they
serve. And in the Vietnam War both military personnel and politicians
did the American public no favors through their overoptimism, because
the disillusionment was all the greater when it turned out that they had
been wrong and, on occasion, untruthful.

The best that can be said of General Westmoreland as the war entered 1968 was that he genuinely believed what he was saying. In fact, he was so confident that he moved two brigades from populated areas just days before the enemy's biggest offensive to date, which took place during Tet, the Vietnamese lunar New Year celebration. That nationwide assault on populated areas would deliver such a shock that the Johnson administration would, finally, openly challenge Westmoreland's war strategy.[84]

The Tet Offensive

The week-long Vietnamese celebration of Tet was the biggest celebration of each year. Even during the war years, it had been a time to stop fighting and celebrate life. In 1968, however, the communists chose it to launch a surprise attack on the South with considerable stockpiled supplies. They hoped that simultaneous attacks throughout the country would prompt the Southern populace to rise up and support the insurgency.[85] Although Westmoreland feared the North might use Tet to launch an attack and decided to keep some of his American troops on alert during the holiday, much of the ARVN force was granted leave.

On the evening of 30 January 1968, nearly 85,000 North Vietnamese Army and VC soldiers struck targets throughout the South, attacking a hundred towns and cities including Saigon, where numerous military bases, government buildings and the US Embassy were assaulted. VC elements were able to breach the Embassy walls and were even reported to have taken it over, although in fact all the attackers were dead by the end of the day. All of the other bases in Saigon were under control and the city was safe within a week, but fighting continued elsewhere. Particularly stiff battles ensued at Hue, where the insurgents initially seized control of much of the city and massacred thousands of government supporters. That city was eventually liberated after a month of tough fighting. The battle at Khe San continued even longer, though with excellent American close air support and air-dropped logistical resupply the US forces held out and prevailed there as well.

The Tet Offensive was, as the historian Pierre Asselin later judged it, "an unqualified, unmitigated military disaster for Hanoi." What the North's leaders had believed would be a victory led to heavy losses for the communists. Their attempts to recoup with more attacks in March and in May only produced further casualties.[86] Despite that, as Australian journalist Denis Warner observed in 1972, the offensive "had been Washington's psychological Dien Bien Phu. Everything that happened thereafter ... represented a signal lowering of American hopes of winning a military victory in Indochina."[87] This was because the offensive significantly undermined the credibility of both the Johnson administration and General Westmoreland, especially given the effort in the latter part of 1967 to highlight the progress in Vietnam during the so-called Success Offensive.

Even though Tet had been very costly for the Northern forces and the VC and their associated political infrastructure, the latter of whose ranks were particularly decimated, the offensive paid huge political and diplomatic dividends for Hanoi. Most significantly, it stoked widespread anti-war sentiment across the United States and the West, and it showed that the communists were capable of striking far deeper into the South than expected. "From a psychological standpoint," wrote Asselin, "the offensive was a major strategic victory for Hanoi and, in light of subsequent events in the United States, a watershed moment in the history of the Vietnam War."[88] One reality was particularly clear: victory was not in view.

For Robert Kennedy, the Tet Offensive "shattered the mask of official illusion" about the war.[89] CBS News anchor Walter Cronkite's assessment following Tet was nuanced but also sobering, especially coming from one of the most respected of the three major television news anchors: "To say that we are closer to victory today is to believe, in the face of the evidence, the optimists who have been wrong in the past. To suggest we are on the edge of defeat is to yield to unreasonable pessimism. To say that we are mired in stalemate seems the only realistic, yet unsatisfactory, conclusion."[90]

March 1968–March 1973: drawdown, General Abrams assumes command, Vietnamization and negotiations

In the wake of the Tet Offensive, the anti-war movement grew significantly as Americans turned decisively against the war. Demonstrations
began to grow in number and location, and additional developments
also indicated that the war was at a turning point. Foremost among
these was a televised speech by President Johnson on 31 March 1968,
in which he announced four major decisions: that he would not
accede to General Westmoreland's request for over 200,000 extra
men; that the expansion and improvement of the South Vietnamese
armed forces would be the first priority; that the US would end the
bombing of North Vietnam in order to move towards peace; and that
he would not seek nor accept his party's nomination for a second term
as president.[91] As Herbert Y. Schandler observed, "The first steps on
the road to American disengagement in Vietnam began on March 31,
1968."[92]

Several of Johnson's decisions reflected the new consensus of the
"Wise Men" – a nine-man grouping of respected former presidential
advisers, as well as retired Generals Omar Bradley and Matthew
Ridgway. The group had met on 25 and 26 March, advised the President
against any further troop increases and recommended seeking a negotiated end to the war. Given the group's previous support for "staying the
course in Vietnam," those recommendations constituted a noteworthy
shift in their point of view.[93] The President's decisions were also influenced by analysis conducted by a task force for Clark Clifford, the new
Secretary of Defense, which laid out various options, costs and potential benefits, and also the likely public responses. The task force focused
the attention of the President on Vietnam and planted seeds of doubt
in his and Secretary Clifford's mind about the strategy and direction of
the war.[94] This would lead to further changes in direction in the months
ahead.

In mid-April, Clifford announced that he would increase the troop
ceiling to 549,000, which was considerably less than General
Westmoreland had requested and less than what the military leader-

ship in the Pentagon had recommended. Less than two weeks later, he made the first announcement of the policy that would later be known as "Vietnamization," whereby the South Vietnamese would have to take over more and more of the fighting.[95] Additionally, the White House announced that General Creighton Abrams, the Deputy Commander of MACV, would replace General Westmoreland in early June, with Westmoreland, in a seeming promotion, nominated to be the Army chief of staff in Washington.

As Lewis Sorley explained in A Better War, General Abrams would usher in a "dramatic shift in concept of the nature and conduct of the war, in the ... measures of merit [employed], and in the tactics to be applied."[96] The overriding object would become the security of the population, not destruction of the enemy, although that would, of course, still be a component of the new approach. The largest of the new big ideas was the concept of "One War," rather than separate wars of big battalions, pacification and territorial security. There would be roughly equal emphasis on military operations, improvement of the South Vietnamese armed forces and pacification, all of which would be closely interrelated. This represented a very significant shift, but one that would also prove challenging to execute. Abrams' approach was essentially that which had been advocated several years earlier by the authors of the PROVN study. In fact, Abrams quickly had the Army assign one of its authors to his staff.[97]

"Instead of thrashing around in the deep jungle," writes Sorley, "seeking to bring the enemy to battle at times and in places of his own choosing ... allied forces now set up positions sited to protect populated areas from invading forces. This put friendly forces in more advantageous situations and forced the enemy to come through them to gain access to the population." As early as August 1968, for example, the entire 1st Cavalry Division was operating in company-size units, rather than in battalion or brigade formations. "Where Westmoreland was a search-and-destroy and count-the-bodies man," one journalist observed, "Abrams proved to be an interdict-and-weigh-the-rice man." This reflected Abrams' emphasis on locating critical enemy logistics caches and resupply means, especially once their access to the population was reduced over time. And the key metric would not be enemy body count but population security.[98]

Killing or capturing the enemy would continue to be a major component of the effort, of course, for longer than Abrams intended, in part due to sheer inertia. Nonetheless, the new approach would be a comprehensive, integrated civil–military approach much more in line with traditional counter-insurgency operations than what had been employed previously. As MACV's deputy later described it: "Abrams ... made it 'clear and hold' instead of 'search and destroy.'"[99] Also included in the new approach would be an emphasis on restraint in the employment of air and artillery support, so as to reduce civilian casualties and property destruction, with revised rules of engagement for the employment of fire to ensure that approach.

Abrams strengthened his emphasis on the One War approach when MACV's Long Range Planning Task Group briefed him on 20 November 1968 and called for significant change in its operational approach. "All of our U.S. combat accomplishments have made no significant, positive difference to the rural Vietnamese – for there is still no real security in the countryside," it concluded, damningly.

Our large-scale operations have attempted to enable the development of a protective shield, by driving the NVA and the Vietcong main force units out of South Vietnam – or at least into the remote mountain and jungle areas where they would not pose a threat to the population. In pressing this objective, however, we have tended to lose sight of *why* we were driving the enemy back and destroying his combat capability. Destruction of NVA and VC units and individuals – that is, the "kill VC" syndrome, has become an end in itself – an end that at times has become self-defeating. To accomplish the most difficult task of the war – *and really the functional reason for the US to be here* – that of providing security to the Vietnamese people – we have relied on the numerous, but only marginally effective, ill-equipped, and indifferently led Vietnamese paramilitary and police units. The Vietcong thrive in an environment of insecurity. It is essential for them to demonstrate that the [government of Vietnam] is *not* capable of providing security to its citizens. And *they have succeeded.*[100]

Abrams strongly supported the measures recommended by the Task Force to strengthen the One War concept; however, truly "operationalizing" the shift took time and was never fully achieved. As always with a large organization, it is one thing for the direction at the top to be changed, but quite another to implement that change throughout the breadth and depth of the organization. Despite the greater emphasis on the elements of the pacification effort and security of the population, many of the major US units continued throughout 1968 and 1969 to operate as they had before, albeit with a greater emphasis on smaller-unit operations. Kill ratios were still carefully tracked in the tactical units, and firepower was still liberally applied. Moreover, search-and-destroy operations were still conducted by many units – with results that were as temporary as they had been before the shift at the helm of MACV.

One area that saw significant new emphasis and resources was the Civil Operations and Revolutionary Development Support (CORDS) program that had been established by President Johnson in May 1967. It was initially driven by Bob "Blowtorch Bob" Komer, a CIA official who became a special adviser to President Johnson in Washington. Sent to Vietnam in 1967, Komer had both the understanding of the situation in Vietnam and the determination to establish and guide CORDS in Vietnam, at least until General Abrams sent him home in 1968.

Given the rank of ambassador and appointed as a deputy to General Westmoreland, with a one-star military deputy and a rapidly growing number of military and civilian personnel under him,[101] Komer took an approach very much in line with classical counter-insurgency concepts. He established three key tasks for CORDS and the pacification effort: the first was security of the population, which entailed separating the population, even in rural areas, from the VC and their political elements. The second was to solidify the security foundation and earn the support of the population through programs that would show that people's lives would be better if they supported the Southern government and its US partners. The third task involved expanding the areas in which security had been achieved and the people's support had been earned sufficiently to help turn around a war that had been indecisive at best and deteriorating at worst.[102]

This was, in essence, the classic counter-insurgency strategy of clear-hold-build, and then to expand the "oil spot." With Komer solidly established in Saigon and the VC largely decimated after Tet, Abrams ensured when he arrived that this civil–military effort received additional priority and increased resources, and also was more fully integrated into the One War approach. He worked closely with the US Ambassador to Vietnam, ensured unity of command for all US and allied efforts and much more deliberately coordinated with South Vietnamese forces, whose development also became much higher priority, in contrast to the previous approach of essentially shoving them aside.

Ultimately, CORDS oversaw all elements of the pacification effort, including those carried out by the State Department, the CIA, the military's civic action groups and Vietnamese local security forces. There were CORDS deputies for the US military's three-star generals who oversaw the operations in each of the four corps areas. There were also advisory teams in each of Vietnam's forty-four provinces, focused on helping with every aspect of local security, community development, agriculture, public health, education, psychological operations, etc. Similar elements were established in each of Vietnam's 243 districts. Over time, the number of advisers grew from the thousand engaged in such efforts in 1966 to over 7,600 at the height of the CORDS program in late 1969 – with over 6,400 of them from the military. Financial support for the efforts also increased several-fold, and funding and equipping the Vietnamese National Police and local security forces improved dramatically as well. The increased resources and greater emphasis accorded to the pacification effort significantly improved security of Vietnamese citizens in the years after Tet.[103]

A particularly important element of the CORDS initiative was the Phoenix Program, an outgrowth of an earlier, relatively unsuccessful effort to identify and then detain or kill members of the VC's political infrastructure in the South. Established in December 1967, Phoenix was given additional impetus in the wake of Tet, during which the importance of the VC's political elements was demonstrated and many of the individuals engaged in such activities were identified and killed. As part of the post-Tet effort in July 1968 to increase emphasis on pacification, Phoenix focused on four tasks: first, establishing intelligence gathering and interrogation centers in each province and district; second, developing information on individuals suspected of activities with the VC's political infrastructure in order to capture, convert or kill them; third, establishing rules for bringing such individuals to trial and, if found guilty, prison; and fourth, employing local security and police forces (including specially trained Provincial Reconnaissance Units recruited, trained and overseen by the CIA) to carry out detention operations.[104]

The Phoenix Program has come to be seen as a controversial campaign of assassination, but careful historical analysis indicates otherwise, although there were undoubtedly shortcomings.[105] Historians now tend to see Phoenix not as a rogue operation, but rather as one that operated within a system of special laws that permitted the arrest and prosecution within the legal system of suspected communists. To ensure that the program was not pursuing personal vendettas, there was a requirement for no fewer than three separate sources of evidence to convict any individual. Between 1968 and 1972, more than two-thirds of the nearly 82,000 VC who were neutralized were detained rather than killed. Those carrying out Phoenix operations recognized that it was by capturing rather than killing VC that they could develop the intelligence necessary to capture more VC. There were abuses, including torture, to be sure, but the program, by and large, contributed importantly to the overall pacification effort pursued by CORDS.[106]

As the US presidential campaign intensified in the summer and autumn of 1968, President Johnson, no longer a candidate for re-election, launched significant efforts to pursue negotiations with North Vietnam to achieve a resolution to the war. Although they produced no tangible results despite lengthy talks and periodic halts to the US bombing of North Vietnam, the process had begun that would ultimately be concluded in Paris in 1973.

President Richard Nixon was elected in November 1968 and swiftly installed a new national security team, with Dr. Henry Kissinger as the national security advisor and Melvin Laird as the secretary of defense. Laird was a proponent of drawing down US forces and accelerating the Vietnamization process, while Kissinger would drive negotiations with the North.

Nixon took office knowing that the war had to end, since it was by then such a divisive force and one that frustrated any constructive approach to domestic and foreign policy. "Nixon clearly perceived," wrote George Herring, "that his ability to extricate the nation from Vietnam would decisively affect his political future and his place in history." However, both Nixon and Kissinger insisted that the war had to be ended honorably, meaning that any American withdrawal had to avoid the appearance of defeat and any settlement would ideally give South Vietnam a reasonable chance of survival.[107] Moreover, Kissinger believed that a "fourth-rate power like North Vietnam" had to have a "breaking point."[108] And unlike their predecessors, Nixon and Kissinger were both prepared to use maximum force to achieve the peace they sought.

Meanwhile in Vietnam, General Abrams sought to further the One War initiative. The effort to increase the capabilities of the Vietnamese military, police and local security forces gained new urgency as an American drawdown loomed larger, even as Nixon approved a massive bombing campaign of North Vietnamese sanctuaries in Cambodia. Following a meeting in June 1969 at Midway Island, President Nixon and South Vietnam's President Thieu announced that 25,000 US troops would depart Vietnam in July and August, despite private objections raised by Thieu and Abrams.[109]

The challenges of the withdrawal were compounded when, instead of redeploying entire units back to the US, as Abrams advised, the Pentagon chose to send home individuals from units throughout the country based on their length of service in Vietnam, per the recommendation of General Westmoreland, now the Army Chief of Staff. This proved to be highly disruptive, creating enormous turbulence and, ultimately, issues of indiscipline that would plague American units in Vietnam in the final years.[110]

Abrams' approach, especially his tough-minded emphasis on intelligence-driven operations, began to generate results as the effort to interdict the logistical lines of communication into the South from the Ho Chi Minh Trail achieved greater success. In fact, these achievements and other security improvements, as well as the enduring effect of the losses sustained during the Tet Offensive in 1968 and those sustained during a smaller Tet Offensive in 1969, forced the North in the summer of 1969 to decide to preserve its strength. This stemmed from a decision in Hanoi, conveyed in an order captured in October 1969, to move back to earlier-stage insurgency operations, and for large Northern units to break down into smaller ones and revert to guerrilla operations, with the emphasis shifting from urban to rural areas.[111]

These relatively positive developments in Vietnam did not, however, reduce the increasing challenges in Washington, where opposition to the war, questions about its conduct and concerns over the continued cost in American blood and treasure were stoking a deepening conviction that it was unsustainable for the United States. In response, Nixon announced diplomatic proposals that had been made to North Vietnam and a hope that all foreign troops might be removed from South Vietnam within a year of a peace agreement. Neither the diplomatic initiatives, however, nor the intensified bombing campaign, nor privately conveyed threats of greater force produced any progress with the North, which appeared ready to wait Nixon out, rightly confident that US public opinion would ultimately compel an American withdrawal.[112]

Though Nixon's ultimatum to the North had little substantive effect, Hanoi did agree to secret meetings that Henry Kissinger would conduct, but from which he returned believing that the North Vietnamese were being, in the words of one historian, "not only intransigent but also deliberatively provocative." Thus, by the autumn of 1969, Nixon was in a very difficult position, forced to choose between increasing the bombing or a humiliating retreat. He was also advised by a Kissinger study group that even escalation of airstrikes and a blockade might not force the North to make concessions. Even a heartening assessment from the British counter-insurgency expert Sir Robert Thompson that South Vietnam was daily growing stronger and with continued US security and economic assistance "might be strong enough within two years to resist a Communist takeover without external help" could not ease the difficulties of the situation.[113]

Nixon and Kissinger proceeded to embark on a complex, multi-pronged strategy that encompassed massive bombing attacks on North Vietnamese targets, including an intense campaign against North Vietnamese sanctuaries in Cambodia that had been denied by the Johnson administration. Simultaneously, there were phased withdrawals of US forces and an increased emphasis on developing South Vietnamese capabilities so that they could steadily assume more of the security responsibilities in their country. Enormous quantities of weapons systems, aircraft, vehicles, arms, ammunition and other materiel were provided to the South Vietnamese security forces to help them succeed in an "Accelerated Pacification Campaign." There were also efforts to sustain as much support at home as possible by a public campaign emphasizing force drawdowns, Vietnamization and progress in efforts to achieve a negotiated settlement.

The realities on the ground and Northern foot-dragging during the negotiations, however, made the overall effort excruciatingly difficult. In response, Nixon repeatedly demonstrated a willingness to take actions that he knew would generate enormous opposition on Capitol Hill and among American citizens, to show the North Vietnamese – and the people at home – just how tough he was willing to be. A particularly combative and controversial action was his decision to send US and Vietnamese troops into Cambodia in the spring of 1970 to destroy the sanctuaries and infrastructure the North had established there to support its forces and the VC in the South. This did, in fact, set back the North's capabilities for a period of time, though it would also contribute to a chain of events in Cambodia that would prove tragic.

The dramatic expansion of a war that Nixon had been drawing down inflamed its critics at home and sparked demonstrations and confrontations on numerous college campuses in which at least six students were killed. These deadly events led to massive protests in Washington and prompted Congressional outrage and efforts to constrain the President. All of this hardened the American public's opposition to the war. And the Cambodia incursion led Hanoi to withdraw its delegation from the Paris talks until the troops were withdrawn from there as well.[114]

In 1971 the timetable of American troop withdrawals was speeded up, with 100,000 going home by the end of the year, leaving only 175,000 personnel in Vietnam, of whom only 75,000 were combat forces. To emphasize his continued determination to secure a "just" peace and to counter the threat posed by increased North Vietnamese infiltration, Nixon stepped up the pressure against North Vietnam, mounting heavy bombing strikes on supply lines and staging areas in Laos and Cambodia. He also approved an offensive into Laos to disrupt the North's sanctuaries, infrastructure and lines of communication there, as had been done in Cambodia. In this case the ARVN, with US air support, conducted the bulk of the fighting on the ground.[115]

The year 1972 brought dramatic developments on the battlefield and at the negotiating table – as well as Nixon's successful re-election campaign. These would ultimately lead to the possibility of compromise and the agreement reached the following year. In March 1972, however, the North gambled that it could conduct an offensive into the South that would take advantage of the dramatically reduced number of US troops. It also sought to set back the Vietnamization effort and allow the VC to strengthen their positions throughout the country prior to the final peace negotiations.

The offensive from the North – launched across the demilitarized zone as well as from Cambodia and Laos – achieved early surprise and success. And with the ARVN forced to defend against 120,000 North Vietnamese troops armed with Soviet tanks, opportunities emerged for the VC to mount supporting attacks in the delta and around Saigon. In response, considerable American airpower surged to augment the South's Air Force and support the ARVN, and Nixon swiftly approved massive airstrikes across the demilitarized zone and against infrastructure in the areas around Hanoi and Haiphong, targeting fuel depots in particular. Meanwhile, Kissinger engaged Soviet President Brezhnev, warning that continuation of the Vietnam War would seriously undermine Soviet–American relations and have grave consequences for North Vietnam too. He also expressed American willingness for the first time to allow Northern forces to remain in the South after a ceasefire agreement.[116]

The North, however, continued defiant, even after Soviet officials shared Kissinger's messages with their leaders. Nixon responded forcefully, and on 8 May he announced the most dramatic escalation since 1968: the mining of Haiphong Harbor, a naval blockade of North Vietnam and large-scale bombing attacks. The last were particularly punishing and, in focusing on the North's logistics elements, made resupply of the North's units increasingly difficult, thus helping the ARVN to establish defensive lines outside Saigon and Hue. Nixon's dramatic response averted disaster and even enabled Southern forces to mount a modest counter-offensive. With the American public seeing the North's invasion as justifying Nixon's robust response, the action even resulted in broad support at home and no opposition from Congress.

With both sides having expended considerable resources and sustained substantial losses, and with Nixon increasingly likely to be re-elected in a landslide, the North returned to negotiations. Moreover, given the US concession on the issue of North Vietnamese forces remaining in the South after a ceasefire, and with the North battered by the American air campaign and exhausted by its own offensive, the prospect of an agreement and US withdrawal looked more possible. Serious discussions proceeded in the autumn and Kissinger thought he was close to achieving an agreement – until he spent five days in Saigon explaining each item to President Thieu. It was then that he discovered that he had significantly underestimated Thieu's unwillingness to accept the terms, as well as Nixon's willingness to back Thieu. Despite announcing at the end of October that "peace is at hand," Kissinger therefore found it necessary to renegotiate some issues with North Vietnam's Le Duc Tho, an effort that proved particularly difficult but was ultimately successful.[117]

The end: December 1972–1975

Finally, at the end of 1972, following more than four years of negotiations between Kissinger and Le Duc Tho, the parties were able to resolve
their differences. Contributing factors were Nixon's landslide re-election
in November and additional major deliveries of arms, ammunition and
supplies to the South. There was also another huge American bombing
operation in North Vietnam in December after Hanoi tried to demand
last-minute changes to the agreement. Nixon then insisted on Thieu's
acquiescence, and US combat operations in Vietnam were suspended on
15 January 1973 with the Paris Peace Accords signed twelve days later.
The signatories included representatives of the governments of North
and South Vietnam and the United States, as well as a representative of
the South Vietnamese communists – with South Vietnam having been
heavily pressured to sign the agreement, which its leaders, with good
reason, feared might seal their eventual demise.

President Nixon described the agreement as achieving "peace with
honor." In practice, however, the agreement provided only what some
historians have contended was the final objective sought by Nixon and
Kissinger – a "decent interval" between the departure of US forces and
North Vietnam's ultimate victory over the South.[118]

The major elements of the Paris Accords established a ceasefire in
South Vietnam; allowed 200,000 communist troops to remain in the
South; agreed on the establishment of a National Council of communist, government and neutral representatives to organize elections in
the South; called for the reunification of Vietnam through peaceful
means; established a ban on the provision of war materials to the South
(except as replacements for existing materials) and also on the introduction of military personnel into the South; provided for a sixty-day
period for withdrawal of US and allied forces and for an exchange of
prisoners of war; and agreed on a withdrawal of foreign forces from Laos
and Cambodia.[119]

In the end, very few of the Paris Peace provisions were fully implemented, though US and allied forces did depart and prisoners were
exchanged. The ceasefire broke down less than twenty-four hours after

the signing and major fighting resumed in March 1973. With no significant response from the US to violations of the agreement, the communists were able to enlarge the area they controlled in the South throughout 1973. Meanwhile, North Vietnamese forces, no longer threatened by American airpower, used infiltration routes through Laos to rebuild the military infrastructure in the areas of the South controlled by the communists.[120]

Over the course of 1973 and 1974, the US Congress cut funding for South Vietnam by more than half. In addition, it required that any spending on US military activity in Indo-China after 15 August 1973 should be approved by Congress, thus making any further air support for Vietnam highly unlikely.[121] Nonetheless, despite Nixon's growing political difficulties at home stemming from the Watergate crisis and his increasingly hostile working relationship with Congress over Vietnam, the South Vietnam leadership in Saigon remained confident of US support, given Nixon's commitments to President Thieu.

In late 1973, Thieu directed offensives that regained much of the territory captured by communist forces during their earlier campaign that year, in what was to be the last major ARVN offensive. While the South's operations were successful in the short term, though, the cost in manpower and resources was high. The North Vietnamese counterattacked, ultimately taking back much of their territory that had been lost and some additional areas as well. Thus, as 1974 unfolded, the military balance shifted in favor of the North.[122]

The war continued with devastating results for the South Vietnamese: massive unemployment; a significant decline in morale of the security forces; increasing political challenges for President Thieu; and a growing spirit of defeatism in Saigon. In August, Nixon's resignation over Watergate removed the individual who had promised Thieu support, while Kissinger's pleas to Congress for an increase in military aid fell on deaf ears. Senator Edward Kennedy captured the sentiment of many on Capitol Hill when he declared that it was time to terminate America's "endless support for an endless war." The aid approved in September 1974 for the subsequent year was only $700 million, down from $2.3 billion in 1973, and half of that was for shipping costs.[123]

This meant that the South Vietnamese forces could not afford to operate as they had been trained and equipped to do by the United States. Shortages of fuel, spare parts, ammunition and other supplies dramatically curtailed the ARVN's operations. The resulting blow to morale, as the North continued its operations in the South, led to well over 200,000 desertions in 1974.

Hanoi decided in early 1975 that the time had come to begin what it projected might be a two-year campaign of offensives to take complete control of the South. Given the situation in the South, however, two years would not be required.[124]

In March 1975, the North struck, achieving swift success in the Central Highlands, as Southern forces began a panicked retreat when Thieu unwisely ordered a withdrawal for which adequate plans had not been made. The North swiftly exploited the resulting opportunity and rapidly seized key coastal cities and ports, including Hue and Danang, cutting South Vietnam in half. With no assistance coming from Washington and political turmoil in Saigon, President Thieu resigned. Southern forces retreated under intense pressure and Saigon itself was threatened. In late April, President Gerald Ford declared that the Vietnam War is "finished as far as the United States is concerned." On 30 April 1975, Saigon fell. It would be renamed Ho Chi Minh City.

In the end, it was not the insurgents who had ultimately toppled the South. Instead, South Vietnam had been overrun by a renewed invasion from the North, with conventional North Vietnamese divisions achieving what the VC and the Tet Offensive in 1968, and the conventional force offensive in 1972, had failed to do.[125] Without the presence of US air and logistical support to bolster it, the ARVN lacked the ability to halt the North Vietnamese Army. Had US air support and other assistance to help the South not been prohibited by Congressional legislation, the outcome might have been different, though that is certainly arguable. Regardless, North and South Vietnam were united in 1976 as the Socialist Republic of Vietnam. With Cambodia having fallen the previous year to the Khmer Rouge guerrillas of the Cambodian Communist Party, and with the Royal Lao forces of Laos abandoned by the US and defeated by the communists there, the dominoes of South-east Asia all finally fell.[126]

Reflections on the Vietnam War

The US campaign in Vietnam was distinguished by many acts of bravery and compassion on the battlefield and by commendable civic initiatives, as well as by military prowess, technological innovation and other impressive actions. But there were also, tragically and unacceptably, many very serious shortcomings and mistakes that were exceedingly costly in terms of civilian losses. These included common practices such as the use of napalm and Agent Orange, "mad minutes" (firing all weapons simultaneously into the areas around a defensive perimeter) and unobserved artillery "harassing and interdiction" fires into populated areas. There were also atrocities committed – most tragically, the massacre of some 500 villagers at My Lai in March 1968, which was then covered up for over a year. And far too many operations created more enemies than they removed from the battlefield by their conduct, a critical failure given the importance of hearts and minds in such campaigns.

Most significant, however, was the failure to understand the true nature of the war and the enemy, and thus the failure to craft a correct strategy before war weariness in the United States undermined the ability to continue the war. Although there were many programs that could be cited as components of a traditional counter-insurgency campaign, the overriding priority, at least well into 1968, was the "big war." This included the conduct of large search-and-destroy operations by US units (with a focus on enemy body counts) and the development of South Vietnamese units in the image of US divisions, focusing on large operations. The most talented of those in uniform from the US and South Vietnam tended to serve in those large units. Meanwhile, support for the critical local security forces was a much lower priority.

It took far too long in Vietnam for the appropriate emphasis on the security and wellbeing of the population to be established. Such an approach would have featured well-trained and well-equipped local security forces whose actions were complemented by well-administered civic action programs that, all together, would convince Vietnamese citizens that life would be better if they supported the South Vietnamese government and its allies rather than the communist insurgents. As we have seen, there were numerous such programs pursued; however, they were invariably too late, poorly executed, insufficiently prioritized and corruptly administered, until the advent of CORDS and its associated programs. But even CORDS did not gain truly meaningful traction until after the Tet Offensive and General Abrams' assumption of command in 1968. By then, the pressures for drawing down had already begun to grow significantly. As the Chief of South Vietnam's Joint General Staff from 1965 to 1975 later observed, "The defense of South Vietnam was founded on an inappropriate basis at the start, focusing too much on anti-invasion and too little on internal security." The enemy, he added, waged a people's or insurgency war.[127]

Despite the emphasis on counter-insurgency during the Kennedy administration in particular, the US military never developed adequate doctrine for the conduct of counter-insurgency operations or crafted a coherent, comprehensive and integrated counter-insurgency campaign plan together with South Vietnam. CORDS did address many of the latter issues, but even Ambassador Bob Komer, its initial leader, would later assess it as "too little, too late."[128] Nor did the US military adequately train and prepare leaders and soldiers for the conduct of such operations, or adequately prioritize the tasks (adviser duty, support for local security forces and civic action initiatives) that were of enormous importance. Those tasks were, in fact, often referred to as the "Other War in Vietnam," and those engaged in such missions knew they were not part of the main effort.[129] There was improvement when General Abrams assumed command and established the One War approach, but by then many of the shortcomings were difficult to remedy.[130]

That is not to say that large operations were not sometimes necessary, especially as North Vietnamese units infiltrated into the South. However, a proper campaign would have featured from the outset a much more coordinated, unified approach that tied together all of the various activities in a coherent, comprehensive counter-insurgency campaign, with much more emphasis on sustained local security, without which little else is possible. The large conventional units should have supported this effort first and foremost, in addition to being used as a reserve in specific provinces with which they developed a familiarity and where they could be employed when major enemy units threatened populated areas.

As we have noted, in 1968, General Abrams did begin shifting the military emphasis away from large-scale search-and-destroy operations to an emphasis on shielding population centers and "pacifying" the countryside, while military responsibility was steadily passed to the South Vietnamese through Vietnamization; nonetheless, it still took time for that emphasis to be operationalized fully at unit level, and even then many of the longstanding deficiencies precluded optimum performance.[131]

Vietnam was also a reminder that foreign forces don't always get to choose their host-nation partners, although after the death of President Diem most of South Vietnam's leaders were either put in place or propped up by the United States. The shortcomings of a number of the South Vietnamese leaders from the top to district and unit levels were maddening in the course of America's involvement in the country – not that the Vietnamese were always wrong, such as when they wanted to have a more local security-focused force at the outset in the mid- to late 1950s than the one the US designed, funded, trained and equipped. Nonetheless, the distance and lack of connection between the South Vietnamese elites in Saigon and the people in the countryside, the levels of corruption, the counter-productive nature of many of the operations and the inability to generate in South Vietnamese soldiers the commitment exhibited by the North Vietnamese and Vietcong proved to be crippling deficiencies.

Beyond all these issues, there were innumerable other shortcomings. The US military, largely manned by eighteen- to twenty-four-month involuntary draftees, rather than longer-serving professional forces (as is now the case), employed an *individual* replacement system to replace losses in Vietnam, rather than rotation of *units* into and out of theatre that had been properly trained and prepared for deployment to Vietnam. This meant that all-important elements such as cohesion, trust and key relationships within small units were never remotely what they should have been once the tours of those who were members of units during their first year in country began to come to an end.

Additionally, most platoon leaders and commanders were rotated every six months. So just when they began to understand their jobs, the friendly and enemy situations, the terrain and other dynamics, they were replaced and served the remainder of their twelve-month tours in Vietnam in staff or other rear-area positions. As a result of that, and also of operating from large, heavily secured bases, relationships with local leaders were also shallow, if they existed at all. This meant that the US forces on the ground fought the war with insufficient learning, as commissioned officers "punched their tickets" with a six-month command and then rotated to another assignment for the remainder of their one-year tour. What this meant was captured best by John Paul Vann, who served early on in Vietnam as a military adviser and then in senior civilian capacities. He observed that "The United States has not been in Vietnam for nine years, but for one year nine times." Given the six-month command tours, it might have been more accurate for him to have said, "for six months eighteen times."[132]

All of this, and the metrics (the enemy body count, in particular) that were the focus of those at the top, also had a corrosive effect on the moral compass of the US forces in Vietnam and their leaders. The findings of a 1970 study on military professionalism conducted by two courageous US Army War College students were scathing.[133] As James Kitfield later described the study's conclusion:

> The Army leadership had instituted, or allowed to be instituted, a system of statistical measures that had almost nothing to do with winning a war, yet had come to mean everything in terms of getting an officer promoted to his next whirlwind assignment. Numbers had become an end in themselves and a crutch on which inexperienced officers leaned in rating themselves and their subordinates. To the rank and file, it seemed as if nothing

was too sacred to be sacrificed on the altar of numerology, whether the statistics tracked enemy killed, miles trekked, villages "pacified," or soldiers re-enlisted and away without leave (AWOL) ... The resulting pressures [of the situation in Vietnam] had created a conviction on the part of many junior officers that they had to be dishonest – to falsify AWOL rates, re-enlistments, court martial figures, body counts, even the incidence of venereal diseases – to advance themselves.[134]

Although the 1970 study was not fully embraced by General Westmoreland, by then the Army's Chief of Staff, he did take it seriously and allowed it to become a catalyst for a process that would have a transformative effect on the Army's professional development programs, ethical standards, command selection processes, tour lengths and promotion policies. The emphasis on integrity in reporting – including reporting on the operational readiness of units – proved to be of enormous importance. One effect of honest readiness assessments would, in fact, drive the establishment of the Army's training centers (e.g. the vast National Training Center in the Mojave Desert) and the conduct there of extremely demanding and realistic force-on-force training exercises. It would also spur the creation of policies for deployment of forces to future wars, with an emphasis on unit rather than individual replacement and improved preparation of leaders, staffs and units.

Vietnam was thus one of the most tragic cases of successive strategic leaders not correctly performing the four tasks of strategic leadership – to get the big ideas right, to communicate them effectively, to oversee their implementation and to determine how to refine them and do it all again – at least until General Abrams took command in 1968. Even after that, despite finally getting the big ideas largely right, the overarching organization for the campaign and the execution of the big ideas were challenged by the emphases and approaches that persisted from the previous decade. Before Abrams, successive leaders failed to develop the unified, coordinated and integrated civil–military effort that would have included all aspects of a comprehensive counter-insurgency campaign. This should have focused, first and foremost, on a well-resourced local effort to secure the population (while noting that major unit operations were required as well, but not as the main effort) and then built on that security by civic action to convince the people that they should support the South Vietnamese government rather than the insurgents and North Vietnamese and VC elements.

Moreover, for many years actions that could have been taken more aggressively against North Vietnam were not taken. The Ho Chi Minh Trail, the key route through Laos and Cambodia that the North used to infiltrate forces, supplies and weaponry into the South, was never adequately interdicted (though this resulted from constraints imposed by decision-makers in Washington, not by the leaders in Vietnam). Letting it be known in 1965 that the US did not intend to invade North Vietnam also communicated to Hanoi the limitations the US would put on its operations. While the wisdom of an invasion was debatable, the US did not have to tell its adversaries how it would limit itself. As we will see in Chapter 7, President Obama repeated this mistake in 2009 when he announced the deployment of 30,000 additional US forces in Afghanistan while simultaneously stating his intention to begin a drawdown of those forces eighteen months later.

Most significant, when it came to the overarching Vietnam strategy, in the words of General Bruce Palmer, Westmoreland's Deputy Commander in Vietnam, the American generals became so "engrossed in U.S. operations, [that] we paid insufficient attention to our number one military job, which was to develop South Vietnamese armed forces that could successfully pacify and defend their own country." He noted that the adverse effects of the "conscious decision to give first priority to the defeat of enemy regular forces in the field, using American forces almost exclusively" included, most significantly, the diversion of "U.S. attention, priority of effort, and precious resources from the primary task of developing South Vietnamese forces capable of defending their country from subversion or overt invasion from the North."[135]

Having failed to get the overarching big ideas right meant that successive senior leaders could not properly perform the second task of a strategic leader – to communicate what the appropriate big ideas should have been. And given that, oversight of the execution of the big ideas, the third task of a strategic leader, was inevitably flawed, despite extraordinary courage on the battlefield and many cases of impressive conduct of various aspects of the overall campaign.

Finally, it also took far too long to perform the fourth task of a strategic leader – to determine how to change and refine the big ideas to craft the right approach and to enable performance of all the tasks properly and over and over again. It took more than twelve years for the leaders to recognize the shortcomings of the big ideas being implemented and to remedy them. By the time General Abrams did so, the pressure for withdrawal of US forces had become so great that his ability to execute the new big ideas properly was increasingly constrained. Moreover, some of the deficiencies in personnel replacement policies, such as command tour lengths, preparation of leaders and soldiers for deployment and ethical deficiencies, could not be sufficiently remedied at that point. Nor could the shortcomings of America's host-nation partners be addressed easily by then either, though there were significant attempts to do so.

With the unique vantage point of observing subsequent wars, especially those in Iraq and Afghanistan, it is clear that the challenges in Vietnam were out of the ordinary. Even perfect performance of the four tasks of a strategic leader by American commanders would not have guaranteed a successful outcome. The enemy's determination, tenacity and sheer will were impressive.[136] The terrain, sanctuaries in neighboring countries, support from Russia and China, domestic Vietnamese sectarian and ethnic issues and so on presented truly formidable difficulties. And the constraints imposed by the US military's reliance on short-service draftees created increasingly significant challenges on the American home front. Nonetheless, the possibility of a successful outcome would have been vastly better had successive leaders performed the four tasks much better than they did. As soldier-scholar Andrew Krepinevich concluded, "In Vietnam, the Army ended up trying to fight the kind of conventional war [for which] it was trained, organized, and prepared (and that it wanted) to fight, instead of the counterinsurgency war it was sent to fight."[137]

A final word

The US effort in Vietnam clearly failed to achieve its objective of building a viable nation that could secure itself against North Vietnamese forces and the insurgents and political elements supported by the North. It was not, however, without enduring accomplishment. As General Bruce Palmer Jr. reflected, "Had we not committed our ground combat forces in 1965, there is no doubt that South Vietnam would have collapsed and North Vietnam would have immediately taken over at that time. Those ten years, 1965–75, bought invaluable time for the other noncommunist countries in the area exposed to the same threat." In particular, he observed, "Thailand survived in good shape, while Malaysia and Singapore fared extremely well in that period." Palmer also attributed the success of the counter-coup in Indonesia in 1965, which saved Indonesia from communism, to the commitment of US power in Vietnam. "Our stand in Vietnam," he noted, "contributed to our overall position in all of the western Pacific, as our allies, especially Japan, the Republic of Korea, and Taiwan, not to mention Australia and New Zealand, prospered under our security shield during the period."[138]

Some of those observations might be seen as an attempt to contend that the commitment to South Vietnam was not completely in vain. Nonetheless, the assessment does seem reasonable, especially if one is objective – as General Palmer was – about the failure in Vietnam itself and the enormous cost of the campaign there in US, coalition and Vietnamese lives, as well as in the considerable damage to the fabric of American society at home and to US prestige abroad.

Henry Kissinger, who negotiated the ultimate agreement with North Vietnam that led to the US withdrawal and release of the American prisoners of war, shared Palmer's assessment. Kissinger concluded in a draft memo of 12 May 1975 on the lessons of Vietnam (that was inexplicably never submitted to President Ford):

> I believe the benefits were many though they have long been
> ignored ... I have always believed ... that our decision to save
> South Vietnam in 1965 prevented Indonesia from falling to

Communism and probably preserved the American presence in
Asia. This not only means that we kept our troops. It also means
that we kept our economic presence as well as our political
influence ... I believe our efforts, militarily, diplomatically, and
politically, were not in vain. We paid a high price but we gained
ten years of time and we changed what then appeared to be an
overwhelming momentum. I do not believe our soldiers or our
people need to be ashamed.[139]

General Bruce Palmer concluded *The 25-Year War* by noting that he was
"optimistic enough to believe that we Americans can and will learn and
profit from our experience."[140] That judgment would be validated in
many respects in the decades that lay ahead as the US military engaged
in serious introspection and embarked on very significant reforms of
how it manned and organized the force, developed leaders, trained and
equipped American units, advised the President and the national security team, conducted operations around the world and partnered with
allies and like-minded nations.[141] The fruits of all this would be seen
most visibly when Iraq invaded Kuwait and the US led a coalition effort
that deployed swiftly in 1990 and liberated Kuwait in 1991.

Tragically, however, the lessons that the US military should have
learned in Vietnam about counter-insurgency operations would have to
be relearned years later in Iraq and Afghanistan.

Four

From the Sinai to Port Stanley

1967–1982

> Battle is a delicate balance between extremes of human
> behavior – selfish cowardice and selfless sacrifice,
> brutality and humanity, callousness and pity – and the
> virtuous must be made to outweigh the dishonorable,
> both in the individual and the unit, if either is to
> survive with any pride.
>
> Sergeant Graham Colbeck, 3rd Battalion the Parachute
> Regiment, Falkland Islands, 1982[1]

The Six Day War, 1967: classic pre-emption

Between May 1949 and the outbreak of the Suez War in 1956, Israel's
Arab neighbors and Palestinian refugees carried out a large number of
acts of terrorism and sabotage across the border against Israeli settle-
ments, against which Israel retaliated with raids, especially after Moshe
Dayan became Israel's Chief of Staff in 1953. Once the British left the
Suez Canal Zone in 1955, it became increasingly obvious that Egyptian
President Gamel Abdel Nasser was preparing for another war with
Israel, though the crisis over his nationalization of the Suez Canal inter-
fered with his plans.[2]

Days before the British and French attacked to secure the Canal against nationalization, Israel captured the Sinai Desert from Egypt in a six-day campaign at the cost of only 200 Israeli soldiers killed. International opposition led by the United States against the actions of Britain, France and Israel forced the last to disgorge the Sinai in March 1957. Ten years passed before Nasser felt strong enough to return to the attack, however, with his armed forces by then trained and equipped by the Soviet Union.

By May 1967 there was no doubt that Nasser was squaring up for a third Arab war against Israel in a decade – one that would be fought on several fronts. There had been attacks by Syria along Israel's northern border, with infiltration into the state via Jordan and Lebanon, and Syria had shelled Jewish farms in the demilitarized zones along the Sea of Galilee in April 1967. Israeli aircraft had attacked Syrian artillery positions on 7 April, leading to an air battle between the Israeli Air Force's French-manufactured Mystères and Syria's MiGs, in which six MiGs were shot down. Lieutenant Colonel Yitzhak Rabin, Israel's Chief of Staff, warned Syria against further provocation, to little effect.

On 17 May Nasser ordered U Thant, the UN Secretary-General, to withdraw the United Nations Emergency Force from the Sinai frontier, which Thant did without consulting the General Assembly or the Security Council. Three days later, Nasser moved 100,000 troops in seven divisions – five infantry, one armored and one lightly armored – with a thousand tanks and hundreds of artillery pieces, to Israel's south-western border, under General Abdel Murtagi. Taken together, the Arab armies fielded half a million regular troops, 900 combat aircraft and 5,000 tanks, against Israel's 275,000 citizen army, 250 planes and 1,100 tanks.[3] On 22 May, Nasser closed the Straits of Tiran to Israeli shipping, thus blockading Israel's Red Sea port of Eilat, something that Israel had already warned would be a *casus belli*. Four days later, General Nasser publicly informed the Arab Trade Union Congress that it was his intention to destroy Israel.[4]

Meanwhile, the Arab oil producers of the Organization of the Petroleum Exporting Countries (OPEC) threatened to nationalize the refineries (in Arab territory), destroy the pipelines and deny passage to the ships of any country that supported Israel. France, having hitherto been a strong supporter of Israel, immediately backed away. "Our goal is clear," stated President Aref of Iraq, "to wipe Israel off the face of the map. We shall, *inshallah*, meet in Tel Aviv and Haifa."[5] Military contingents started to arrive in the theatre from countries such as Kuwait and Algeria. King Hussein of Jordan flew to Cairo to sign a defense agreement with Nasser, who had only three weeks earlier denounced Hussein as "an agent and slave of the imperialists."[6] The Arab media announced that the destruction of Israel was imminent.

In response to these threats, Levi Eshkol, the Israeli premier, formed a government of national unity which included Menachem Begin, the leader of the opposition, and General Moshe Dayan as minister of defense. In retrospect, the only surprising thing about Israel's pre-emptive surprise attack on Egypt was that it came as a surprise to anyone, especially Egypt. The Israelis were helped by high-quality intelligence, mainly from signals, about Arab intentions, but also from agents such as President Nasser's masseur Ali Al-Alfi, and from Wolfgang Lotz, an uncircumcised German Jew who persuaded Egyptian leaders that he was a former SS officer.

At 7:10 a.m. on Monday, 5 June, the codeword Red Sheet (*Sadin Adom*) unleashed Major General Mordechai Hod's pre-emptive air attack on the Egyptian Air Force. He had chosen the time well: enemy dawn patrols had stood down, the mist in the Nile Delta had cleared and Egyptian pilots were driving from breakfast at home to their bases. "Fly, soar at the enemy," Hod ordered his pilots, "destroy him and scatter him throughout the desert so that Israel may live, secure in its land, for generations."[7]

The Israelis flew low – often at no more than 50 feet – in order to evade radar screens, and under complete radio silence. Electronic jamming equipment was used to prevent Soviet vessels detecting the planes as they flew over the Mediterranean, and the pilots were told that in the event of mechanical trouble they must crash into the sea without radioing for assistance.

Hod's pilots flew sixty-five Mirages, seventy Mystères, fifty Ouragans, twenty Vatours and forty-five Fougas, and attacked nineteen Egyptian airbases in no fewer than 500 sorties that day alone. In three hours, the Israelis were able to destroy or disable 309 out of Egypt's 340 serviceable combat aircraft (another 110 were out of service), including all thirty long-range Tu-16 bombers, twenty-seven medium-range Illyushin Il-28 bombers, twelve Sukhoi Su-7 fighter bombers, ninety MiG-21 fighters, twenty MiG-19 fighters and twenty-five MiG-17 fighters.[8] The Egyptians had failed to construct concrete hangars, and had parked their planes in easy-to-attack formations without even sandbags for protection. Only four Egyptian training flights were airborne at the time, none armed. It was, in the words of the historian Alistair Horne, "A lightning campaign of Old Testament quality."[9]

The Six Day War provides an object lesson in the dangers of overoptimistic propaganda. With total air superiority established over the Sinai Desert by 10:15 a.m. on the first day of fighting, Israel attacked on the ground in the Sinai. Not realizing the extent of the Egyptian defeat, King Hussein of Jordan entered the war promptly to the schedule he had previously agreed with Nasser.

Based on reports on Cairo radio of an Egyptian air victory, the Jordanian, Syrian and Iraqi air forces made the catastrophic decision to attack an Israeli airfield near Kfar Sirkin, some oil refineries in Haifa Bay and the town of Natanya respectively. The Israeli response was devastating, and by the evening of the first day the Jordanian Air Force had virtually ceased to exist, the Syrian counterpart had lost two-thirds of its strength, including seventy-two MiGs, and the Iraqis had suffered badly too. By the end of the second day of the war, 416 Arab aircraft had been destroyed (94 percent of them on the ground) for the loss of twenty-six Israeli planes.[10]

Egyptian forces stationed close to the Israeli border in Sinai massively outnumbered Israel's Southern Command. Lulled into a sense of security based on their numbers, they did not expect a three-pronged direct frontal attack after Israel's Air Force had already destroyed much of their defensive infrastructure there. Falling for an Israeli deception plan, they expected the main thrust to take place in the south, in central Sinai, whereas in fact it came in the north along the Mediterranean Sea via El-Arish.

The Israelis crossed sand dunes that the Arabs considered impassable by armor, bypassed Egyptian strongpoints that could be neutralized and mopped up later and outmaneuvered attempts to outflank them and cut them off. As always in rapid advances, there were complaints that too much time was wasted in consolidating the previous day's gains, but each of the six days brought such large advances that most of these critiques can be safely discarded as overactive indulgence in that most common military attribute: hindsight.

"A great part of the information obtained in war is contradictory," Carl von Clausewitz wrote in *On War*, "a still greater part is false, and by far the greatest part is uncertain." Another downside to the victorious propaganda heard on Egyptian and other Arab state radio stations on the first day of the war, which trumpeted fabricated stories about Tel Aviv being heavily bombed and great Arab victories on the battlefield, was that once it became obvious by the third day of the war, Wednesday, 7 June, that they were untrue, ordinary Egyptians became even more disappointed and angry than if they had been told the harsh truth in the first place.

So when it was known by Wednesday that the Egyptian forces were losing badly in Sinai, demoralization set in among the armies. Field Marshal Abd el Hakim Amer's conflicting orders panicked some of his commanders, a few of whom escaped back across the Suez Canal, for which they were later court-martialled. Amer himself committed suicide – or was possibly murdered – three months later.

By Wednesday, 7 June, Major General Israel Tal in the north had almost reached the Suez Canal at Kantara; General Avraham Yoffe was contesting the Mitla Pass in the center, and General Ariel Sharon had reached Nakhle in the south. Battles fought by Tal's forces at Romani and Bir Gafgafa, by Yoffe's at Bir Lahfan and Jebel Libni, and by Sharon's at Nakhle all went Israel's way, although the remaining Egyptians put up desperate resistance, especially at Um Katef, which would suffer the largest barrage of shells that Israel had ever unleashed.[11]

By the morning of Thursday, 8 June, the Israeli Army had reached the Suez Canal itself, an extraordinary achievement in only four days of war. Yoffe and Tal laid a trap at the narrow Mitla Pass, which for much of the

Egyptian armor was the only way back westwards from central Sinai. With the Egyptians funnelled into the Pass, the Israeli Air Force strafed and bombed enemy vehicles at will, leaving hundreds destroyed and burning. The Pass itself was also blocked by a small unit of Israeli tanks under Colonel Yiska Shadmi, which was able to collect fuel and ammunition from Egyptian tanks abandoned on the battlefield.

By the end of the day, the Israelis had reached the Canal and the Gulf of Suez at no fewer than six major points between Kantara in the north and Al-Tur in the south, and artillery fire was exchanged over the Canal. There remains some disagreement among historians over the number of Egyptian troops killed in the Six Day War – the Israelis claim 15,000, the Egyptians only admit to 10,000 – but 5,000 were also captured and two-thirds of Egypt's 800 tanks in the Sinai were destroyed with the rest captured. Israeli losses numbered 300 soldiers killed and a thousand wounded in that theatre.[12]

In the Jordanian theatre, the war was being fought in very different circumstances than those in the Sinai. Jerusalem and the West Bank contained densely populated cities and towns, containing Jews, Muslims and Christians. The urban setting and the hills and ridges nearby were to lead to contrasting kinds of military engagement from those on the desert dunes.

At 11 a.m. on Monday, 5 June 1967, the Jordanian Army shelled Israel, crossing the 1949 Armistice Line and demilitarized zone south of Jerusalem and capturing Government House to the south of the Old City. By the end of the day, the Israeli Air Force had attacked the Jordanian Air Force at Mafraq and Amman, destroying its twenty-two Hawker Hunters. Much like their Egyptian counterparts and allies, Jordanian forces had to struggle on in the face of complete Israeli air superiority. Although the Israeli Air Force could not attack the Jordanian Arab Legion in the vicinity of Jerusalem itself, it did prevent Arab elements at Jericho from supplying and reinforcing their forces there.

The situation by 6 June was desperate for the Arab forces. King Hussein had previously placed his forces under Egyptian control, and at midday General Abdul Munim Riad, the Egyptian Commander leading the Jordanian forces, sent an update to Cairo that read:

The situation on the West Bank is rapidly deteriorating. A concentrated attack has been launched on all axes, together with heavy fire, day and night. Jordanian, Syrian and Iraqi air forces in position H3 have been virtually destroyed. Upon consultation with King Hussein, I have been asked to convey to you the following choices:

1. A political decision to cease fighting to be imposed by a third party (the USA, the Soviet Union or the Security Council).
2. To vacate the West Bank tonight.
3. To go on fighting for one more day, resulting in the isolation and destruction of the entire Jordanian Army.[13]

The King and Riad clearly did not favor Option 3.

The battle for Jerusalem took place between 5 and 7 June. Heavy fighting occurred in relatively built-up areas such as the Mandelbaum Gate through the no man's land surrounding it on 5 June, as well as the Police School, Ammunition Hill and the Sheikh Jarrah district. It spread to the Rockefeller Museum, Damascus Gate and Herod's Gate the next day. With pressure mounting from the United Nations for a ceasefire, the Israeli Defense Forces' Colonel Mordechai Gur had little time left to take ground, and at 8:30 a.m. on 7 June his three battalions attacked across the Augusta Victoria Hill and the Mount of Olives, supported by artillery and warplanes. Gur led the entry into the Old City at St. Stephen's Gate, encountering little resistance, and by 10 a.m. had secured the Western Wall.

Fearing international outrage and sanctions, Dayan himself had previously warned against Israeli troops entering the Old City, but upon hearing that the UN was about to implement a ceasefire, he rescinded his instruction.[14] "The Temple Mount is in our hands," was Gur's response soon afterwards. For the first time in over 2,000 years, the Jewish people controlled Jerusalem. At 8 p.m. on 7 June, Israel and Jordan accepted the UN ceasefire proposal, by which time Israel's conquest of the West Bank had also been completed. The Jordanians lost over 6,000 killed and missing in the war, the Israelis 550 killed and 2,500 wounded.[15]

On the third major front of the war, between Israel and Syria, the main ground fighting took place on the Golan Heights, the strategically vital 45-mile-long plateau that dominates north-east Israel. The Syrians had fortified the Heights along the ceasefire line since 1949, from where over the years they would occasionally direct tank and artillery fire onto Israeli settlements below. When war broke out in June 1967, the Syrians increased the shelling of the settlements, but did not attack in the way they had promised Egypt and Jordan they would, beyond undertaking a few minor raids in the direction of Kibbutz Dan and Sha'ar Yashuv village. As truthful news emerged from the other two fronts, this restraint seemed wholly justified.

It was not until the morning of the fifth day of the war, Friday, 9 June, once victory was certain in the other two theatres, that Dayan, without consulting either Eshkol or Rabin, ordered Major General David Elazar of Israeli Northern Command to attack Syrian forces on the Heights, in a massive armored, paratroop and infantry assault aimed at taking as much territory as possible before another United Nations ceasefire came into effect later that day. As with the supposedly impassable deserts of the Sinai, Israeli tanks climbed steep ground that the Syrians had left lightly defended because they assumed they could not be traversed.

The IDF were motivated by the need to take the Heights as soon as possible, in order to protect their countrymen in the settlements below once the ceasefire came into effect. Their officers and NCOs led from the front; together they represented 23 percent of IDF casualties in the war.[16] Syrian resistance was powerful, and Israeli troops were soon forced to use converted bulldozers to counter their defenses. Although Syria's 250 Russian-built tanks and similar number of artillery pieces were relatively new, they had been poorly maintained. (In conflict, especially in testing climatic conditions such as the Middle East, the maintenance of weaponry is sometimes as important as its actual caliber.)

On Saturday, 10 June, Israeli reinforcements arrived on the Heights after their victory on the West Bank. Helicopters were used to capture positions that allowed further armored penetration, one of the first times that they had been used in that way. Panic set in among the Syrian forces in the course of the morning, and many fled the battlefield, abandoning tanks that would otherwise be easy targets for the Israeli Air Force. (When Israeli tanks were knocked out, any survivors were expected to join the infantry on foot.) By the time of the 6:30 p.m. UN ceasefire, the Israelis were overlooking the Damascus Plain. "For the first time in its history," wrote Chaim Herzog, a former director of Israeli military intelligence, "Israel had the benefit of defence in depth."[17] The war left Israel in control of Jerusalem, the West Bank of the Jordan River and the Golan Heights, which Israel refused to relinquish until its state had been recognized. In total, the Six Day War cost Israel 778 military and 26 civilian deaths, less than a tenth of the Egyptian losses alone.[18]

Moshe Dayan performed all of the tasks of a strategic leader brilliantly. He saw the big picture clearly and, although he was criticized during the war for not switching efforts to the West Bank and the Golan Heights until the Sinai was secured, his strategy was proven right. Israel's only significant misstep would be its own coming hubris. "The Israeli Command tended to credit itself with many achievements that were in some cases more a result of Arab negligence, lack of coordination and poorer command at the higher level, than of Israeli effectiveness," noted Chaim Herzog.[19] His son, General Michael Herzog, who fought in the Yom Kippur War of 1973, has gone further, accusing Israel of showing "conceit, complacency and stagnation" after the conclusion of the earlier war.[20] Humiliated in 1967, the Arab countries' opportunity for revenge came only six years later, when Anwar Sadat and his generals proved that they had learned several of the most important lessons of Nasser's many errors. But not all of them.

The Yom Kippur War, 1973: devastating counter-attack

"A generation that has taken a thrashing," Otto von Bismarck once said, "is always followed by one that deals out the thrashing." It seemed in the opening hours of the Yom Kippur War that the Egyptians and Syrians had not waited for a generation, but only for six years. At 2 p.m. on

Saturday, 6 October 1973 – Yom Kippur, the Jewish Day of Atonement – the Egyptians and Syrians achieved impressive tactical surprise in opening hostilities against Israel. Fine Egyptian staffwork ensured that 32,000 infantry could cross the Suez Canal into the Sinai Desert in five separate places over a 50-mile front, supported by almost 4,000 guns (including those in tanks) which fired more than 10,000 shells in the first minutes of the attack. The combined total of 3,000 mainly Soviet-made tanks of the Egyptians and Syrians equalled that of NATO itself, and outnumbered Israel's 2,000 tanks, although with their 105mm guns Israel's Centurion tanks were superior.[21] The combined population of Egypt and Syria numbered over 40 million, to Israel's 3 million.

Few conflicts have started better for the aggressor than the Yom Kippur War, the opening scenes of which have been described as "an Israeli Pearl Harbor."[22] Anwar Sadat, Egypt's ambitious President, was determined to reclaim its lost territory.[23] He had been prepared to lose 25,000 during the crossing of the Suez Canal alone, but in fact Egypt lost only 200 men in the operation. Once across on the eastern side of the Canal, the Egyptians used ladders to climb the Israeli defenses and then pushed forward half a mile beyond. Since 1967, Israel's lines on the Suez Canal, which had originally been for observation and early warning, had turned – through mission creep – into a static de facto line of defense.[24] It would have been far better to have had the capacity for mobility, and here we witness another example of how cuts in defense spending – which Israel instituted after its success in the Six Day War – seldom produce genuine savings in the long run.[25]

Yom Kippur was a reminder that deterrence works only when it threatens overwhelming punishment. But Israel's security policy was subjected to budgetary and geographical restraints. With its much smaller population, Israel could not afford to station a large standing army on its borders, was vulnerable to simultaneous attacks on several fronts and was unable to withstand long wars without American support. Its security was therefore based on a mixture of deterrence, early warning, air supremacy, rapidly deployable reserves, a commitment to taking the fight to the enemy territorially and seeking an early, decisive, Clausewitzian victory.

Of course Israel's enemies understood this, and fashioned a plan to enhance Israel's weaknesses and negate its strengths. Having licked their wounds and rebuilt their militaries, the Arabs achieved a surprise attack, split Israel's forces through multiple incursions and as in 1967 used oil to enhance Arab solidarity and dissuade any country liable to side with Israel – all with the intention of extracting the maximum price in casualties from the tiny nation, and causing the worst energy crisis since the Second World War.[26]

Egypt attacked when Israel felt most overconfident. Israeli security services had detected a week-long training exercise adjacent to the Suez Canal in the week before the attack, but had dismissed it. Syria's call-up of reserves was similarly ignored. The attack did not come as a total surprise, however: although the CIA had missed the build-up, Israeli intelligence recognized a few hours beforehand that a major assault was coming and so Israel's Air Force was made ready for a pre-emptive strike from noon on 6 October. Yet the Israelis did not unleash it, as they had so successfully in 1967. They had overestimated their defensive capabilities on the Golan Heights and at the Suez Canal, while underestimating the Arab forces' recent increase in technological capabilities.

The Egyptian and Syrian ruse of hiding genuine military build-ups within well-advertised military maneuvers fooled even their own troops; until the attack began, Egyptian and Syrian soldiers were not informed that they were going to war. Other subterfuges designed to lull Israeli intelligence had also been employed, including the demobilization of 10,000 Egyptian troops forty-eight hours before the attack, and rumors spread about problems with the war materiel supplied by the Soviets. To add to the deception, the Egyptian Foreign Minister flew to Washington in September 1973, ostensibly to reactivate peace talks. It was a world apart from Nasser's annihilationist rhetoric against Israel of 1967; Sadat's attack was professional, intelligent and initially successful. Henry Kissinger, newly appointed as secretary of state, had warned Israel that any pre-emptive aggressive strikes would cause a withdrawal of American support. He was especially concerned by certain aspects of the new conflict: not only was Israel largely armed by the Americans and the Arabs by the Russians, but a naval standoff between the United States and Soviet fleets in the Mediterranean was occurring simultaneously. Would this be revealed as Soviet proxy warfare, potentially leading to another Cold War crisis between the superpowers? Kissinger could not have known it at the time, but the Soviets were in fact not behind the attack; Brezhnev himself was informed of it only the day before.[27] Additionally, the growing Watergate scandal was destabilizing the Nixon administration, complicating the American response to the war.

When, in the early hours of 6 October, it became clearer that an assault was imminent, Israeli Chief of Staff David Elazar's recommendation for a pre-emptive attack on Syria was rejected, as was his call for fully mobilizing the reserves. These were fearfully difficult choices for the Israeli political and military decision-makers. In the event they merely warned the Egyptians and Syrians that they were aware of the planned attack and instituted only a partial mobilization. The Israelis feared that if the intelligence estimates were wrong – and they can only rarely be absolutely certain – any pre-emption might ignite precisely the conflict that it was designed to pre-empt.[28]

Major General Eli Zeira, Israel's Head of Military Intelligence in 1973, later disclosed that although he had furnished Israel's leaders with an accurate picture of the enemy's capabilities and deployments, he had failed to provide an accurate indication of their intentions – always the most difficult element of intelligence information but often the most crucial.

The Egyptian and Syrian staffs assumed that autumn would provide the best weather for the attack, that Yom Kippur, the holiest day of the year for Jews, would find the maximum number of IDF soldiers on leave and Israeli businesses closed and that the November elections would distract Israel's politicians. In fact, the empty roads helped Israel mobilize its reserves quickly when the threat became clear.[29] Though it was able to mobilize an astonishing 10 percent of its population for the war, there were almost no reserves in either the north or south during its early stages.

Determined to prevent a re-run of the Six Day War where Israeli air superiority had been decisive, the Egyptians and Syrians had invested very heavily in air defenses, especially Soviet surface-to-air missiles (SAMs) that had performed so well against the US Air Force in Vietnam. Their SAM-6 and SAM-7 missiles and ZS4-23-4 anti-aircraft guns, as well as Sagger and Swatter anti-tank missiles and RPG-7s, proved extremely effective in the opening days, as did new night-fighting and bridge-building equipment. It was the Sagger, a highly lethal Soviet anti-tank guided missile with a 2-mile range, that inflicted particularly serious damage on Israel's armored vehicles, and the dense air defenses that accompanied advancing ground troops mitigated the effectiveness of the Israeli Air Force, at least at the outset.[30]

Tactically decisive equipment afforded the Arabs numerous successes. Israel's rapid jet response to the initial attack was impeded as the low-level warplanes came under withering fire from heat-seeking anti-aircraft weaponry carried by the Egyptian infantry. Meanwhile, on the western bank of the Canal, high-level planes contended with Egyptian radar-controlled, long-range anti-aircraft missiles. The Egyptians also deployed self-propelled, four-barrelled 23mm radar-directed automatic aircraft guns. In one Egyptian unit's area, no less than 80 percent of Israeli jets suffered damage of some kind.[31] The one saving grace for the Israelis was the small warhead of the Egyptian infantry's missiles, which tended to explode in the hot gases of the jet engines' exhaust, damaging but not entirely disabling the planes.

On the evening of 6 October and throughout the next day, Israeli counter-attacks were heavily punished by Egyptian armored systems that carried powerful 100mm anti-tank guns and lightly armored vehicles that carried smaller anti-tank weaponry. One Israeli unit lost 90 percent of its tanks in ten minutes. This devastation was repeated on 8 October when a counter-attacking Israeli division lost half of its tanks, before the army in the south was forced on to the defensive having lost over 250 tanks. By the end of that day, the Israelis had also lost nearly fifty aircraft, mostly to SAMs. The fate of the country seemed to hang in the balance.

The Sinai Desert provided Israel with the necessary strategic depth to give it some breathing space after three successive days of defeats, and it subsequently provided considerable room to maneuver during the counter-attack. In the north, however, Syrian tanks were only a ten-minute drive from the River Jordan and the Sea of Galilee. They could be stopped only by the rapid deployment of reserves. The subsequent fighting in the Sinai and on the Golan Heights saw the last real tank battle in history, in which armored formations fought each other supported by artillery and air forces.'[32]

The Israelis had assumed that, because the Arabs knew that Israel had superior warplanes and tanks, they would not attack. Those attacking, however, trusted in their new technology to at least limit Israeli superiority in airpower and armor. For the first three days of the war, it looked as though they might be right.

As had happened in the Allied landings at Anzio in Italy in January 1944, however, the attacking force failed to widen the initial bridgehead fast enough fully to exploit the opportunity of the initial surprise. The Egyptians brought up nine divisions against the Israelis' three, and transported large numbers of tanks across the Canal, but remained close to the eastern side of the Canal in order to consolidate gains. It was not until 14 October that the Egyptians attacked across the whole front, which by then had well-protected and well-hidden Israeli positions with 800 tanks but relatively few infantry. Ideally, attacking forces ought to outnumber defenders by three to one and concentrate on vulnerable points, but the Egyptians managed neither and were repulsed with heavy losses, especially when Israeli tanks rallied to counterattack on the flanks.

By the end of that day, the Egyptians had lost 260 of their thousand tanks, against Israel's sixty, with many of the latter being repaired quickly.[33] The Soviet Union further worried President Sadat by providing satellite imagery that revealed an entire Israeli division moving along the West Bank of the Canal. By 19 October, after further heavy fighting, the Egyptian Chief of Staff – who had been sent to the West Bank to report back first-hand evidence of Israeli military presence in the area – recommended a withdrawal of all forces east of the Canal,

---

* Despite that battle having taken place in 1973, nearly twenty years later NATO still had over 23,000 tanks while the Warsaw Pact had over 52,000.

which the Commander-in-Chief refused, threatening to court-martial him for suggesting such a thing.[34] Ahmad Ismail Ali, a division commander, suggested that Egypt push hard in the United Nations for a ceasefire so that Israel's success could be stunted.[35]

The failure to achieve a ceasefire proved fateful, because by 21 October the Israeli Army, supported by the Air Force, was bypassing Egyptian strongpoints and had totally outmaneuvered the Egyptians, who had made the error of dispatching their anti-tank units to help elements on the West Bank that had been pinned down, rather than keeping them to contain the Israeli counter-attack across the Canal just north of the Great Bitter Lake. Those they did have were dealt with by a highly lethal combination of Israeli artillery and machine-gun fire.

In six days the Israelis drove 50 miles, crossing the Canal and pinning the Egyptian Third Army down against it all the way south to Suez, with much of the army stranded on the south-eastern side, just as the Egyptian Second Army was pinned down on the north-eastern. The Israeli punch-through and its spreading southwards still stands today as one of the greatest maneuver campaigns and strategic victories since the Second World War. Occasionally the victory had required blatant disobedience of orders. In one instance, Ariel Sharon attacked after he had been expressly ordered to defend, but it was forgiven; as Dayan later said, "I prefer his pressures and initiatives tenfold to the hesitations and excuses of other divisional commanders ... Better a noble steed you have to restrain than a lazy ox you have to beat."

In the northern theatre, the Syrians also came close to achieving a decisive breakthrough against Israel in the opening hours of the war; indeed some of their tank commanders were able to look down on the Sea of Galilee, and Israeli commanders gave the order to prepare to blow the bridges across the River Jordan. Because of their lack of reserves, the Israelis were reduced to counter-attacking in a costly and piecemeal way. In one instance, the Israeli 7th Tank Brigade was under constant fire for four days in the Valley of Tears from 6 October, until each of its seven tanks from an initial complement of one hundred was down to three or four shells. Lieutenant Colonel Yossi, returning post-haste from his honeymoon in the Himalayas, took thirteen tanks from the Barak Brigade, which had itself been badly mauled, and arrived on the battle-field just as the 7th Brigade completely ran out of tank ammunition and were arming themselves with hand grenades to carry on the fight.

With both sides about to collapse, the Syrians retreated first. The 7th Brigade and Yossi's reinforcements gave chase, only stopping at an anti-tank ditch when, in the words of one eyewitness, "The brigade had reached the limits of human exhaustion."[36] One officer slid to the floor of his armored personnel carrier fast asleep while in the middle of answering a question. Some 260 Syrian tanks and hundreds of armored personnel carriers lay destroyed or abandoned in the Valley. Across the 7th Brigade's radio network came the commendation of the divisional commander Raphael Eitan: "You have saved the people of Israel."[37]

When the UN imposed its ceasefire, in President Chaim Herzog's phrase, Israeli forces "were within artillery range of Damascus and well on the road to Cairo."[38] Although violence and conflict between Israelis and Arabs would continue after the War of Atonement, it was their last comprehensive multi-front war.[39] In the half-century since Sadat visited Jerusalem after the war, no soldiers have died on the Israel–Egypt border.[40] In one of the ironies with which history is replete, it was at a military parade to mark the eighth anniversary of his major military achievement of the crossing of the Suez Canal that President Sadat was assassinated.

Lawrence Freedman believes that Israel's victories in the Arab–Israeli wars owed as much to superior training, doctrine and morale as to Western equipment being generally better than that of the Warsaw Pact. "Israel's most potent weapon was its air force," he writes, "and this made the decisive difference in all its campaigns."[41] Israel had lost 2,500 men in the Yom Kippur War, more than three times the numbers of the Six Days War, against Egyptian and Syrian losses of around 8,000 each. Both sides saw half their tanks destroyed and Israel lost 115 warplanes out of 500, less than half the 250 lost by Egypt and Syria.[42]

Among its other legacies, the war would have a significant impact on US military doctrine. "Our interest in the Arab–Israeli War, all the analyses and all the discussions, are not just an intellectual exercise," wrote General William DePuy of US Army Training & Doctrine Command. "True, it is fascinating for soldiers, but there is a purpose to this study and the purpose is that we want our schools, our combat developers and those involved in training, to remember these lessons and to relate them to our concepts."[43] The Americans would go on to make a highly detailed analysis of the war, for, as Frederick Kagan was to put it, that conflict "rocked the American defense establishment to its foundations."[44]

One military historian estimates that "By multiple metrics – money allocated, personnel assigned, time committed, material produced – the Yom Kippur War was the most extensive investigation of a foreign war ever undertaken by US Armed Forces."[45] The story of a small country overwhelming several much larger ones was one that naturally concerned the United States, and by March 1974 the Pentagon had initiated no fewer than thirty-seven separate studies into different aspects of the Yom Kippur War, some of which – including a seven-volume data report on weapons systems – remain classified to this day. American servicemen walked the battlefields in 1974, with Israeli commanders who had fought on them as their guides.

The reason why the 1973 war merited much more scrutiny than the 1967 one was that military observers learned that the flexibility, speed and sheer intensity of conflict made the old concept of attritionally holding a line completely moribund.[46] In the future, only in wars where neither side had the capacity for flexibility and speed, such as the Iran-Iraq War and the fighting in the Donbas in Ukraine in 2022, would line-holding and attrition return to the battlefield. Another lesson was the growing lethality of combat, with weapons of ever greater destructive power and the increasing complexity and precision of warfare, in particular in the areas of electronic warfare, anti-tank guided missiles and integrated air defenses. Future wars, it now seemed, would be short, intensely chaotic affairs in which forces needed to be ready for action at the outset, and intensive training could overcome the disadvantages of being surprised and outnumbered, as the Israelis had been.[47]

The US Army and Air Force found in their detailed analysis of Yom Kippur that despite the admittedly huge advances in anti-tank guided missiles and surface-to-air missiles, the tank still had an important role to play on future battlefields. Yom Kippur also exposed the weakness of attack helicopters when flying against high-density integrated air defenses, although this was insufficiently appreciated at the time.

Much of what the US learned from the conflict was put to good use in the Gulf War eighteen years later.[48] It was because of the Yom Kippur War that the United States military introduced non-flammable hydraulic fluid; built up its chemical weapons defenses; increased stockpiles of munitions; appreciated the danger posed by ground-based air defense; introduced vastly more realistic training and simulator use; intensified emphasis on Combat Electronic Warfare Intelligence battalions and multiple launch rocket systems; and upgraded and increased the numbers of M-60 tanks, Sidewinder missiles and electronic countermeasure pods (ultimately replacing each of them).

The performance of the newer Soviet systems during the Yom Kippur War strengthened the hand of those in the US military who were calling for Patriot surface-to-air missiles, cutting-edge F-15s, new tanks and infantry fighting vehicles, more ground- and air-launched precision munitions and the Airborne Warning and Control System.[49] Yet, ultimately, the victory had come down to the soldiers rather than their kit. "The War demonstrates the extent to which *man* is the key to the outcome of war," wrote Michael Herzog. "The training and skill of the soldier, his motivation, the quality of the chain of command, initiative, courage and perseverance all underlie the War's result far more than any weapons. Even in the era of technology, man still stands at the center of the picture."[50]

The Soviet invasion of Afghanistan: military defeat as a catalyst for ideological collapse

Over Christmas 1979, the Soviet Union invaded Afghanistan with 75,000 men in support of a beleaguered pro-Russian regime in Kabul. These forces were quickly faced with a guerrilla campaign fought by

Islamic rebels, the Mujahideen, who were supplied by neighboring Iran and Pakistan, and later also by the United States. The Afghan Army "soon all but evaporated" and within weeks the Soviet forces – which, like the Guomindang in the Chinese Civil War and the French in Algeria, held all the major towns – were caught in a long-drawn-out, hard-fought war against the rebels, who controlled much of the countryside.[51]

Far from projecting strength, the Soviet invasion in fact demonstrated how weak the USSR had become. Trying and failing to contain rebellion through political means, Moscow instead resorted to a one-note military operation. Concerned that Mujahideen rebels would slip northwards into Soviet Central Asia, the Soviet leadership believed that their vastly stronger force would easily quell any disturbances. It was an invasion that was mirrored by America's invasion of Vietnam, not least in its eventual outcome.

President Jimmy Carter's administration reacted forcefully against the Russian incursion, withdrawing from the Salt II disarmament treaty, supporting a boycott of the 1980 Moscow Olympics, imposing economic sanctions, establishing a Rapid Deployment Force focused on the Persian Gulf (which would ultimately lead to the establishment of US Central Command), increasing military aid to Pakistan and, most importantly, ordering the CIA to supply the Mujahideen with lethal weaponry. It was not until 1986, during Ronald Reagan's second term as president, that the United States began delivering the devastatingly effective shoulder-fired ground-to-air Stinger missiles. These drove Soviet attack helicopters from the skies, or at least forced them up so high that their effectiveness was severely compromised. With that development, the Mujahideen were able to muster the strength to attack and occasionally overcome and massacre isolated Russian garrisons.[52]

The Russian General Staff had been skeptical about invading Afghanistan, a multi-ethnic, multi-sectarian, tribal and mountainous country; the decision to do so had been that of Leonid Brezhnev and the Politburo. The war revealed the immense complexities of Afghanistan. "There were at least nine hundred guerrilla groups," records an historian, "some under traditional leaders, others led by younger commanders from the educated middle class."[53] These groups were divided religiously, ethnically, geographically, ideologically, and by whether they were supplied by Pakistan, Iran or the United States. "Beneath the superpowers' proxy war was a kaleidoscopic local struggle," notes the historian David Reynolds.

The war, which was immensely vicious and saw a tragic dearth of pity on both sides, lasted almost a decade before the Soviet withdrawal in February 1989, by which time some 15,000 Russian lives had been lost, and an unrecorded (but far larger) number of Afghans. It was the first time since the Second World War that the Red Army had withdrawn while under attack from a nation it had occupied. When the Mujahideen ultimately captured Kabul in September 1996, they seized the pro-Russian former leader Mohammad Najibullah (who had resigned in 1992) and his brother Shahpur from a UN compound, castrated them, tortured them to death, dragged their corpses around the city from the back of a truck and then hung them from a lamppost outside the Presidential Palace.

The Soviet occupation of Afghanistan devastated both nations. Nearly one in ten Afghans died during it.[54] Unable to distinguish fighter from civilian, the Red Army killed indiscriminately, using a variety of methods: massacres and depopulation programs abounded in the region, and young women were kidnapped and sexually assaulted by Soviet forces, rendering those women unable to return home afterwards due to their perceived "dishonor." "Villages were razed in retaliation for Soviet casualties," records an historian, "and there were 'fun' shootings into passing buses or private homes."[55] Scholars agree that the Soviet Union perpetrated many gross crimes against humanity in Afghanistan, while some argue that it amounted to genocide.

By 1982, around 3 million of Afghanistan's 15 million-strong population had fled to Pakistan and Iran as refugees. Rumors abounded of chemical weapons used against the Afghans, and villages were occasionally turned to dust in a single afternoon. In some instances, Soviet soldiers, well aware of their unpopularity within Afghanistan, quietly took weapons away from their own Afghan allies at night.[56] Abused by superiors and hunted by Afghan villagers and tribesmen, bored and lonely, miserable and vengeful, many Russian soldiers took to smoking opium.

The eradication of so much of Afghanistan's population ironically left the Mujahideen militias with greater power within the region. The educated middle classes of Afghanistan quickly fled abroad as their country was ravaged around them, leaving, in the words of one Afghan historian, "the laity, the commanders, and the Islamic fundamentalist groups – or, to put it differently, bearded men, veiled women, and armed warriors" as "the principal characters of Afghan society."[57]

Afghanistan was a deeply underdeveloped tribal nation that had been deemed strategically insignificant to America for decades prior to the Soviet invasion. None, least of all the Soviets, could have predicted that it would become the arena for the bloodiest proxy war of the 1980s, and then again after the 9/11 attacks of 2001. By the time the Soviet forces withdrew, with columns of armored vehicles humiliatingly driving northwards out of the country, Afghanistan's future had changed.

Of course there were to be seismic repercussions that went comparatively unnoticed at the time, as well as deeply ironic moments: US Secretary of Defense Donald Rumsfeld was filmed in the late 1980s handing a Kalashnikov to a jihadi fighter with the words "Allahu Akbar." Zbigniew Brzezinski, Carter's National Security Advisor, justified the arming of the Mujahideen by asking, rhetorically, "What was more important in the world view of history ... A few stirred-up Muslims or the liberation of Central Europe and the end of the Cold War?"[58] Little did anyone recognize at the time that, once the Cold War ended, there would only be a short hiatus before the worldview of history was forced to concentrate on more than just "a few stirred-up Muslims."

In the meantime, the Soviet invasion led to the traditional Sufi mysticism widely practiced in Afghanistan being superseded by hardline Islamism, as Afghans turned to a different form of faith amid the harsh exigencies of wartime. The total failure of the war had also undermined the prestige of the Red Army and led to internal divisions within the state: using the Soviet military to achieve strategic goals and maintain geopolitical interests no longer seemed sensible, or perhaps even possible. For Mikhail Gorbachev, the reforming new General Secretary of the Soviet Communist Party, it was further evidence that only a certain degree of liberalization could save the Soviet Union.

The Russian invasion of Afghanistan had profound implications for how the War against Terror would be fought there two decades later. The significance of shoulder-launched anti-aircraft missiles was underlined, because once the United States had provided this key asset to the Mujahideen the Soviets were deprived of a crucial advantage, their rotary and fixed-wing close air support. The Soviet campaign and its failures also highlighted for American planners (though perhaps not strongly enough) the severe challenges of counter-insurgency operations in Afghanistan, given terrain such as the Hindu Kush Mountains, the lack of infrastructure, the very limited economy, insurgent sanctuaries in Pakistan, illiteracy and lack of human capital and the substantial support provided to the insurgents by the United States and Saudi Arabia (a dynamic that would be mirrored, twenty years later, by the support provided to the Taliban by Iran and elements of the Pakistani secret service).

The campaign also demonstrated America's ability to counter the USSR by supporting surrogates – as was done elsewhere around the world, of course, but nowhere as successfully as in Afghanistan. The Soviet-supported Afghan regime did survive for two years after the departure of the last Soviet forces, but it was the cutoff of funding that doomed that regime, closely mirroring the fate of the Southern Vietnamese government between 1973 and 1975. Above all, it became clear that small numbers of US personnel – mainly CIA – but large amounts of arms, materiel and financial support could deal a major defeat to the Soviet Union, one that propelled it towards its death throes.

The empowerment of the Mujahideen would have grave implications in subsequent years, as the Taliban (which included many of the Pashtun Mujahideen) prevailed in the subsequent Afghan Civil War over the forces of the other ethnic and sectarian elements fighting to take Kabul and control of the post-Soviet Afghanistan. And, in the late 1990s, the Taliban allowed Osama bin Laden to establish a sanctuary in eastern Afghanistan in which the 9/11 attacks were planned and the initial training of the attackers was conducted – leading to the US intervention in Afghanistan in late 2001 and the subsequent twenty-year conflict that was the longest campaign of the post-9/11 long war.

The Falklands War, 1982: a very long-distance conflict

It is perfectly possible that had any other person than Margaret Thatcher resided in Downing Street on 2 April 1982, when an Argentine Marine battalion invaded the Falkland Islands in the South Atlantic, a United Nations- or United States-brokered deal might have been agreed upon, in which Britain lost its dominion over the Islands in return for a fine-sounding agreement in which sovereignty was pooled in a condominium compromise. Mrs. Thatcher's contempt for such bromides, however, combined with her outrage at Argentina's actions, meant that she was never likely to view the prospect of a neo-fascist junta in Buenos Aires ruling over British Crown subjects as anything other than a *casus bellum*.

Many people could not see much point in going to war over 1,800 Falkland Islanders (and 600,000 sheep). The Argentine writer Jorge Luis Borges compared the war to "two bald men fighting over a comb." But the lack of raw materials in the Islands made no difference to the principle involved. When one critic of the war told the future Conservative leader Michael Howard that "there was nothing down there," Howard retorted, "Yes, there was: the Union Jack."[59] As Margaret Thatcher later put it, the Argentine attack was a "crisis of Britain's honor."[60]

In the emergency debate on the Saturday after the invasion, Thatcher told the House of Commons, "It is the Government's objective to see that the Islands are freed from occupation and are returned to British administration at the earliest possible moment." At a time when her Defence Secretary John Nott feared that it might be impossible to retake the islands, Henry Kissinger observed, "Thatcher had unequivocally conveyed her resolve by cutting off the possibility of her own retreat."[61]

The invasion caught the British armed forces and government entirely unprepared. It was one of the strangest wars in history – described by Dr. Mark Almond as "the world's first post-modern war" – as it was fought 8,000 miles away from home and with no land base from which to launch the operation.[62] It was a story of extraordinary last-minute improvisation and a test of the adaptability and flexibility of Britain's armed forces when called upon to fight a conflict of a totally different kind from the one they were expecting, against the Soviet Union.

Britain had no specific defense against the weapons of other NATO member countries, such as the Exocet, a French sea-skimming, aircraft-launched anti-ship missile that had been acquired by the Argentines. The Royal Navy expected to operate as part of NATO, complete with airborne early warning systems, air dominance and several layers of air defense weaponry, none of which would be available in the Falklands expedition.[63] (That said, the Task Force that liberated the Falklands contained more frigates, destroyers and hunter-killer submarines than exist in the whole of today's Royal Navy.)[64]

A recurring theme of this book is that money spent on deterrence is seldom wasted, especially when considered against the costs incurred when the deterrence fails. The British government's 1981 White Paper which had announced the intention to withdraw HMS *Endurance*, a South Atlantic icebreaking patrol vessel, as well as the Royal Navy's two amphibious assault vessels HMS *Intrepid* and *Fearless*, led the Argentine junta's General Leopoldo Galtieri and Admiral Jorge Ayala to believe that the British would not go to war for the Islands (which Argentina had claimed since 1833 and called Las Malvinas).[65] Here was a prime example of disarmament sending the wrong message to potential aggressors. The White Paper also suggested selling Britain's aircraft carrier HMS *Invincible* to Australia and questioned the future of the Royal Marines, sixty-eight of whom were stationed on the Falklands at the time of the invasion and were overwhelmed on the first day.

"From the military man's point of view, she was an ideal prime minister," recalled Admiral Sir Terence Lewin, Chief of the Defence Staff, of Margaret Thatcher afterwards. "One wanted a decision and she gave it."[66] The decision – supported by a small, hastily formed War Cabinet – was immediately to send a naval task force that ultimately, once it was joined by other vessels exercising off Gibraltar, comprised forty-four warships, twenty-two naval logistic ships and forty-five merchantmen, with 28,000 men.[67] These forces included four naval and one RAF squadron of helicopters, one brigade of Marines, two parachute and three infantry battalions. The ships included luxury liners, ferries, trawlers turned into minesweepers and a children's education cruise ship that was converted into a floating hospital.[68]

"Service officers conditioned by years of petty economy, niggling bureaucracy and chronic supply problems were astounded by the efficiency of the machine that now moved towards war," recorded Max Hastings, who reported on the war for the *Daily Telegraph*. "The whole of southern command was ransacked, an immense transport plan put into effect to move the stores to the docks."[69] Equipment such as arctic clothing, new radios and laser rangefinders were sent to units. "It was like Christmas, that Monday," recalled the 39-year-old Lieutenant Colonel Hew Pike, a general's son and the Commander of the 3rd Battalion Parachute Regiment (3 Para).

Brigadier Julian Thompson, the Commander of 3 Commando Brigade, increased the order of Rapier anti-aircraft missiles from a troop to a battery of twelve, and ensured that Scorpion and Scimitar armored reconnaissance vehicles were loaded for their cross-country capability. The general rule was to take "something of everything" to be ready for every circumstance, especially the air threat.[70] Nor was it just military equipment being loaded; watched by television cameras, Sergeant Graham Colbeck climbed the gangplank on to the SS *Canberra* to the sound of the battalion band and "noticed that many hundreds of cans of beer were being loaded on to the conveyor belt below."[71] In a triumph of organization and dispatch, the Task Force left port on 5 April, only three days after the Argentinian invasion.

Thanks to Caspar Weinberger, Ronald Reagan's Defense Secretary, and US Navy Secretary John Lehman, Britain was able to use the American Wideawake airbase on the British-owned Ascension Island, and the United States started to provide aviation fuel, spare parts and air-to-air Sidewinder, radar-seeking Shrike and anti-aircraft Stinger missiles.[72] Field Marshal Carver later stated that Wideawake "played an essential part as an air staging post."[73] Even excluding the fuel and Sidewinders, the US provided $60 million in aid to Britain during the war. Critical American signals intelligence on the Argentine forces was also made available to the British.[74]

Although President Reagan knew of this material support, General Alexander Haig, his Secretary of State, who was trying to negotiate a peaceful settlement based on a condominium, initially did not, which took the concept of need-to-know to an entirely different level. After the war, Lehman revealed that in the event of the loss of one of the British carriers, Reagan had agreed to lend to the Royal Navy the assault ship USS *Iwo Jima* which could accommodate Sea Harrier fighters. As Reagan told Weinberger, "Give Maggie everything she needs to get on with it."[75]

On Monday, 4 April, the day before the Task Force's departure, Brigadier Julian Thompson briefed forty commanding officers. Major Ewen Southby-Tailyour of the Royal Marines gave a presentation that emphasized the difficult, boggy terrain on the Falklands, as well as the freezing winds and chill factor, total absence of local food and drinking water, and nil cover except during the hours of darkness. "British planning for war in Europe had taught a generation of soldiers to assume that troops crossing the sea would be received by a friendly government and population, able to offer vital support facilities," point out Max Hastings and Simon Jenkins. "Yet there would be no local resources to call upon in the Falklands."[76]

Thatcher was insistent that 3 Commando should leave as soon as the ships were loaded, while she still had political support for the expedition. Julian Thompson and Hew Pike both recognized from the start that there would be fighting. In his address to his battalion in the gymnasium at their Kandahar Barracks at Tidworth immediately after the Argentines seized the Islands, Pike said that although diplomatic efforts were being made to resolve the crisis – which all sensible men must hope would succeed – they were nonetheless "very fortunate" that they would indeed probably see action.[77] One of his listeners, Graham

Colbeck, later wrote that "If there were any sensible men in the gymnasium that day, then I was not one of them – I would have been bitterly disappointed if there had been a diplomatic solution."[78] Of the mood at the time, Pike later said, "There is enormous residual self-confidence in a parachute battalion. People don't wonder whether they can do things."[79]

On his way down to the Falklands, General Jeremy Moore, Commander of British land forces, brought along a photo of General Mario Benjamin Menéndez, his Argentine counterpart, in much the same way that Montgomery had kept a photograph of Erwin Rommel in his caravan during the Desert War. It was just as well that Moore had made no assumptions about Menéndez from his physiognomy, however, as during the surrender negotiations it became clear that the photo was of the wrong General Menéndez. (It was an easy mistake to make; there were no fewer than five of them in the Argentine Army at the time.)[80]

On 28 April British forces retook South Georgia, an island 870 miles east of the Falkands that the Argentines had captured, and two days later a Total Exclusion Zone of 200 miles around the Falkland Islands was imposed by Britain, meaning that Argentina could not resupply its 13,000-strong army on the Islands. On 2 May, the Argentine cruiser *General Belgrano* was sunk some 40 miles outside the Zone by the British nuclear submarine HMS *Conqueror*, using fifty-year-old Mark 8 torpedoes, with the loss of 323 Argentine sailors from her thousand-strong crew. It is possible that the sinking might have ultimately saved Argentine lives in the long run, as it kept their navy in port for the rest of the war, whereas a naval battle against the Task Force would almost certainly have led to an even greater number of casualties.

The Islands were over 250 miles off the Argentine shoreline, which meant that Skyhawks from the carrier *Veinticinco de Mayo* were forced significantly to reduce the amount of time they could operate over the Islands.[81] If the Argentines had extended the runway at Port Stanley to take their Skyhawks, the British operation might have been endangered, but they did not, so the Argentine pilots were forced to fly from the mainland and were almost at maximum range as they arrived. In early

May, British Hawker Harriers attacked Argentine positions on the Islands, including the Port Stanley runway.

On 4 May, the Type 42 destroyer HMS *Sheffield* was hit by an AM-39 Exocet missile, fired from a French-built Super Étendard fighter, and on 10 May she sank under tow.[82] Worryingly, the ship's satellite communications had interfered with her electronic detection system and she sank without firing a shot in her own defense. The explosion and fires killed twenty of her crew. Although Exocets were to become famous during the war for what they wreaked against British ships, in fact more were sunk by low-flying aircraft dropping bombs. Of the five ships sunk and thirteen damaged, Exocets accounted for two sunk and one damaged.[83] The Royal Navy was lucky that no fewer than thirteen bombs that hit five damaged ships failed to explode.

On 19 May, Hew Pike addressed 3 Para in the Meridian Lounge of the ocean liner SS *Canberra*, urging them, among other things, not to delay operations by stopping to help the wounded before all objectives were secured.[84] (In Northern Ireland a feature of contact had been immediate aid to the wounded.) Two days later he quoted Brigadier Hill of 6 Airborne Division, who had told his troops on the eve of D-Day, "Do not be daunted if chaos reigns, because it undoubtedly will."[85]

At 2:50 a.m. on Friday, 21 May, British forces made an amphibious landing at San Carlos Water. The Royal Navy's decision to choose the bay on the opposite side of East Falkland from Port Stanley was a controversial one. In the plan's favor was the fact that the bay was far from any Argentine artillery, and was nearly impenetrable by enemy submarines and Exocets. However, it was 56 miles from Port Stanley, might possibly have been mined (it wasn't) and was overlooked by high ground that could possibly be occupied by an enemy garrison only 13 miles away. (The high ground later proved invaluable, cutting down the time that Argentine aircraft were able to choose targets in the bay.) In the event, and despite the losses of three ships to Argentine warplanes, the choice of San Carlos by Admiral John Fieldhouse, the Commander of the Task Force, Brigadier Julian Thompson and their staffs has been described as "superbly judged" and one that "contributed critically to victory."[86]

On 25 May an important merchant ship, the SS *Atlantic Conveyor*, was sunk by an Exocet, which deprived the Army of the majority of the Chinook helicopters needed to transport troops, forcing 3 Commando to march to Port Stanley from their disembarkation point, carrying everything with them. After doing that, they had to fight for three days against well-entrenched and more numerous defenders.[87] Yet, as we have seen so often in this book, high morale was the key. "The Argentinians' problem was not their tactical position, let alone their logistics," recalled Max Hastings. "It was morale on the ground. Among the troops on the mountains it was worse than the British ever dared to imagine."[88]

Hastings interviewed Argentine POWs after the war, and the reasons for demoralization were clear. "No one really told us where we were going," said Guillermo of the 7th Regiment. "We weren't prepared psychologically ... In A, B and C companies, who were the ones most in the front line, there were boys with whom I talked who didn't even know what the Malvinas were."[89] The Argentinians were also extraordinarily brutal to their own men, administering beatings and forcing soldiers who went AWOL to sit for hours with their naked feet in freezing mountain water.[90] As we saw with the Guomindang Army in the Chinese Civil War, this kind of ill-treatment of recruits is devastating for morale. As they marched across East Falkland, the British squaddies had to face cold, diarrhoea and a severe First World War-era rotting malady called "trench foot," but it did not affect their morale, partly because they had a trust in ultimate victory that many Argentine soldiers did not.

"The campaign," writes Max Hastings, "demonstrated that the Argentine command had no grasp of the principles of strategy: their air force had failed to mount effective attacks, as the British had feared. Their army had failed to defend key features, to interdict the British advance, to harass or counter-attack positions that the marines or paras had occupied ... They pinned all their hopes on holding their ground and seeking diplomatic, rather than military, deliverance."[91] At the twelve-hour battle of Goose Green on the night of 28 May, the Argentines did show that they could defend entrenched positions bravely, but 690

British troops overwhelmed the garrison of over 1,160 men, which surrendered.

On 31 May, Ronald Reagan telephoned Margaret Thatcher urging her to be "magnanimous" in victory and allow the Argentines a face-saving compromise.[92] But by then much blood had been spilt on land and at sea. "I'm not handing the islands over now," she replied. "I didn't lose some of my finest ships and some of my finest lives to leave quietly under a ceasefire without the Argentinians withdrawing."[93] Thatcher refused Reagan's request, and as her former speechwriter John O'Sullivan points out, "To make a diplomatic retreat the climax of what had become a great national cause would have confirmed and perpetuated the dismal myth of postwar British decline ... The entire Thatcher 'project' of national revival would have been derailed."[94] As a British sergeant who was fighting in the Falklands put it, "If they're worth fighting for, then they must be worth keeping."

More British reinforcements landed between 4 and 8 June, but on the last afternoon the landing ship *Sir Galahad* was bombed and set on fire at Fitzroy causing the deaths of forty-eight crew and soldiers. On the morning of 11 June, the commanding officers of 42 Commando, 45 Commando and 3 Para gave orders for a night attack on the Argentine defensive positions on Mount Longdon at 8:01 p.m., using models of the terrain fashioned from lumps of peat, pieces of canvas, rifle slings and twigs. "It is hard ... to imagine how much information commanders at all levels in an infantry battle at night have to carry in their heads," wrote Julian Thompson later.

The commanding officer orchestrating the show is not sitting in a warm well-lit operations room surrounded by staff, powerful radios and radar displays showing the position of friend and foe; or even in a relatively warm armored command vehicle with an onboard satellite navigation system giving him a constant read-out of his position. He is walking, running, crouching, lying, probably in pouring rain, under fire, trying to update his mental picture of the battle by conversations on a low-powered man-pack radio and going to see for himself if he can.[95]

Thompson took the deliberate decision for his brigade to fight at night. "The confusion of battle in the darkness," he later wrote, "would be offset by the greater skill of the marines and paratroopers, the better leadership and the intimate and flexible support made possible by our magnificent gunners."[96] It was risky, but it paid off.

As Pike had predicted before disembarking at San Carlos Water, chaos did indeed reign, and, as in every conflict, the fog of war produced tragic friendly-fire incidents. One left eight soldiers wounded, after which a British sergeant remarked that "Dangerous mistakes will occur as long as soldiers remain fallible."[97] He blamed fatigue, poor visibility and an overeagerness to engage the enemy as some of the factors that made them more likely. He noted how easily messages passed orally down long snakes of troops could get corrupted. On one occasion the phrase "Air Raid Warning Red" somehow turned into "Galtieri's dead."[98]

Because 3 Para had to fight along the spine of a ridge at Mount Longdon with only enough space to deploy one company at a time, its attack has been likened by Lawrence Freedman, the official historian of the war, to "advancing down a bowling alley in the dark being fired on from both sides."[99] As in so many of the engagements recounted in this book, raw courage had a vital part to play in the victory. In the ten-hour battle of Mount Longdon between the Argentine 7th Regiment and the 3rd Battalion of the Parachute Regiment, to take but one example from among many in that campaign, positions were captured by close-quarter "gutter fighting" with grenade, rifle and bayonet. "It was a battle in which junior officers, NCOs and private soldiers fought with courage, tenacity and aggression," recalled Thompson.[100]

Sergeant Ian McKay won a posthumous Victoria Cross trying to silence an Argentine Marine machine gun on his own with grenades after comrades from his platoon had been killed and wounded; Captain Willy McCracken called down artillery fire to within 50 yards of his own position (such was the accuracy that no Britons died from it). "Sometimes the fire was so close to us," recalled a sergeant, "that it was impossible to tell our own gun and mortar fire from that of the enemy, until the 4.5 inch gun of HMS *Avenger* joined in with its regular shell bursts."[101]

"I shall never forget the sight that morning," recalled their commander Hew Pike of 12 June 1982, "of A Company advancing through a thick mist with bayonets fixed."[102] Of the twenty-three men of 3 Para killed and forty-seven wounded in that battle for a key strategic position outside Port Stanley, Private Neil Grose died on his eighteenth birthday. As one officer wrote after the battle, in words that could have been repeated in any conflict in this book, "Such good people lost, so terribly young."[103]

By the end of the contemporaneous battles of Mount Longdon, Mount Harriet and Two Sisters fought by 3 Para, 42 Commando and 45 Commando respectively on 11 June, of the 850 Argentine defenders, 50 were killed and 420 captured for the total British losses of twenty-five dead and sixty-five wounded.[104] The capture of the heights running north–south between 6 and 7 miles from Port Stanley left the capital unprotected from direct attack, which began on 13 June, and the Argentine surrender took place on the evening of 14 June. The Islands had been reconquered only twenty-four days after General Moore's army had landed. General Galtieri and the rest of the junta resigned on 17 June.

After the fall of Port Stanley, Hew Pike wrote to his family in England: "The Duke of Wellington was indeed right about 'the melancholy nature of victory.'"[105] Some 255 British servicemen had been killed in the war and 777 wounded. Three Falkland Islanders had been killed, one after medicine was withheld. Six ships had been sunk and ten damaged.[106] The Argentines had lost 649 killed, almost exactly half on the *General Belgrano*. They lost ninety-one planes to gunfire and Harriers.[107]

Although it was a great victory for Britain, the Falklands drew attention to several important shortcomings. The Royal Navy's trust in the supremacy of new missile technology had meant that it had failed to concentrate on its elderly stock of anti-aircraft guns, some of which were of Second World War vintage and proved inadequate against brave, fast and accurate Argentine aircraft attacks. Sometimes they were merely machine guns fitted to improvised mounts, such as office swivel chairs welded on to the deck.[108] The assumption that missiles and detection technology would protect ships had – along with cost-cutting – also led to them being all too often fitted with kit that was flammable or splintered easily. Over 70 percent of those sailors admitted to the hospital ship SS *Uganda* had sustained burns, not least because their new polyester uniforms melted into the skin in a way that the previous cotton ones did not. The planned budget cuts to the Royal Navy were at least reversed in the short term.

Another lesson was found in the continuing importance of the Special Relationship between Britain and the United States. "In the Pentagon there had been a massive de facto tilt toward Britain from the very first day," recalled John Lehman.[109] This was not replicated right across the Reagan administration, as the Argentine junta had been a supporter of America's anti-communist stance towards Soviet, Cuban and Nicaraguan subversion in Central America. Yet the place that mattered most was Caspar Weinberger's Defense Department, where, as Lehman noted, "The depth of and breadth of cultural, social and historic ties between the United States and the United Kingdom were overwhelming."

The first naval war since the 1940s was studied carefully in the US Navy.[110] Even before the war was over an expert team had been put together to learn its lessons, which concluded that "Next to the quality of the personnel, the most important factor in the Falklands was intelligence."[111] By virtue of being able to use US intelligence, Britain was well informed about Argentine locations and probable intentions, in a way that Argentina could not emulate. Another lesson was that ships needed stronger armor. "If any of the sixteen successful attacks [on British ships] had instead hit the [US] battleship *New Jersey*," argues Lehman, "it could not have done sufficient damage to prevent continuing operations of the ship. The Exocet that sank *Sheffield* would not have been able to penetrate the armor anywhere on the battleship."[112]

Margaret Thatcher, Terence Lewin, John Fieldhouse, Jeremy Moore, Task Force Commander John "Sandy" Woodward, Julian Thompson and other senior commanders all performed the strategic leadership tasks superbly during the Falklands War. Sending a task force 8,000 miles to expel the Argentines from the Islands was a bold and correct decision, as was the sinking of the *Belgrano*, the siting of the landing at San Carlos Water and the plan of attack towards Port Stanley; the mission was communicated extremely effectively to the troops, whose morale stayed high despite extremely testing local conditions; the implementation was then driven through by every means at commanders' disposal, with the commanders providing admirable examples, direction and energy, and the attacks were adjusted under the changing circumstances of the fighting, both on land and at sea. The result was, as Henry Kissinger has written, that "Taken in tandem with the decisive economic reforms Thatcher had instituted at home, the Falklands victory effectively transformed Britain's standing on the world stage."[113]

As well as reiterating the old truth that money saved in piecemeal defense cuts – in this case the axing of HMS *Endurance* – often costs more in the long run, the Falklands War reminded the West that conflicts can emerge suddenly from out of what seems like a clear blue sky. No one imagined that Britain might have to fight a war 8,000 miles from home before it happened; yet, as this book has shown, surprise attacks are surprisingly common. The lesson that a country with such a far-flung set of responsibilities – the result of its imperial history – needed to provide for every eventuality was learned the hardest way. The efficacy of submarines against cruisers was underlined. Every other conflict in this book so far had essentially been a land struggle, although several have naval components. The Falklands saw conflict evolve because it was essentially a naval campaign with a land action added on. Western powers had to prove they could do both.

Five

Cold War Denouement

1979–1993

1991 saw the perfect marriage yet seen between
fire and movement.

Christon Archer et al., *The World History
of Warfare*, 2002[1]

In the four decades after 1945, the percentage of headquarters
personnel in a typical Western army increased fivefold, and the cost
of corresponding command, control and computer systems rose even
higher.[2] Armies, once simplistic if vast monoliths, had rapidly become
much more intricate and complex. Increasingly sophisticated communications systems resulted in escalating amounts of data and
information processed by a soaring number of technicians who
required ever evolving training, until, in the words of Martin van
Creveld, "staff piled upon staff, procedure upon procedure, machine
upon machine."[3]

The existing break line between formal, organized and well-funded
armies and their underfunded, scrappy guerrilla counterparts – and
often adversaries – widened further. As the gap grew between those with
billion-dollar budgets and those with pugnacious know-how, the Cold

War entered its final phase. Spectators were to learn not only that the former failed to guarantee success, but that it did not even guarantee survival of the state itself.

The wars of this chapter saw the United States recovering its military self-confidence after the disaster in Vietnam and learning the best ways to project power efficiently wherever it was needed in the world. There were teething troubles and setbacks, undoubtedly, even in successful operations that are covered here, but warfare evolved because the multifarious lessons from them were learned. The United States was fortunate, even statesmanlike, in that the operations it undertook in the 1980s provided a steep and invaluable learning curve, but without a major war.

El Salvador's Civil War: the classic proxy war

The tiny, tropical Central American nation of El Salvador had suffered instability and impoverishment throughout the late nineteenth and twentieth centuries. Long, tumultuous decades of coups and dictators culminated in the country finding itself the setting for one of the last indirect American military engagements of the 1970s and 1980s. A mere half-decade after miserably failing to, in the words of Robert McNamara, "pound a tiny, backward nation into submission" in Vietnam, US military personnel once again found themselves entangled in a jungle conflict thousands of miles from home.[4]

The mission would operate primarily as an extended, albeit dangerous, advisory and training effort for the United States, but it was also a significant, and ultimately successful commitment. The conflict would continue – with varying levels of intensity – for over twelve years before being brought to a reasonably satisfactory close.

In October 1979, Salvadoran President General Humberto Romero, fresh from suspending civil liberties and ordering massacres of students, peasants and priests, was overthrown in a military coup and replaced by a revolutionary junta that swiftly expelled all non-military personnel from its upper ranks.[5] The coup leaders, who included among them two colonels, had no democratic mandate, but were concerned that

Romero's heavy-handed repression was leading to a left-wing revolution driven by the country's persecuted. Extreme economic inequality, coupled with growing sympathy for Marxist-Leninist ideals, seemed to make the country ripe for a communist insurgency, and the new regime sought to quash burgeoning left-revolutionary feeling, with the former Mayor of San Salvador José Napoleón Duarte selected to lead a civil–military junta.

Reluctantly supporting regimes it otherwise disapproved of had, by the late 1970s, been an imperfect, flawed, domestically unpopular but well-established US foreign strategy as a means of suppressing Soviet-backed sympathies within an unstable state. Such action was no endorsement of the supported regime; President Kennedy's non-intervention after the assassination of Ngo Dinh Diem in 1963 (just twenty days before his own) was proof enough that these relationships were strictly professional.

The increasing unrest and instability in Central America was worrying even the most dovish of presidents. President Jimmy Carter, widely considered weak on foreign policy – at least until responding to the Soviet invasion of Afghanistan late in his term – nonetheless feared another Cuba-style regime taking hold in the western hemisphere. He was thus willing to support a regime in San Salvador enacting increasingly brutal recriminations against its citizenry as it became a "model counterinsurgency state."[6] The Salvadoran military forces had even taken inspiration from US psychological warfare tactics against the Vietcong.[7]

The Marxist-Leninist Farabundo Martí Front for National Liberation (FMLN) was a sufficient threat to stability in the minds of the US State Department to draw America once again into conflict. Carter authorized $5.7 million in non-lethal military aid to El Salvador, and the Pentagon later provided helicopters for its military.[8] American aid was briefly suspended in 1980 following gruesome acts of cruelty perpetrated by the Salvadoran government – including the assassination of El Salvador's Archbishop while he was celebrating mass, and the rape and murder of American missionary nuns – but it resumed six weeks later, before Carter's presidency came to a swift and undignified end at the hands of the electorate in the face of multiple other foreign-policy setbacks.

Carter's successor, President Ronald Reagan, was a more passionate Cold Warrior and dramatically increased aid to the Salvadoran government. For him, anti-communism was the most effective human rights campaign of all. In February 1981, just a month after taking office, he approved $25 million in military aid and increased the number of US personnel on the ground. More aid followed, reaching $128 million by 1985. This money funded the government of El Salvador and enabled it to take the fight to its FMLN enemies without any direct engagement by American personnel, who were officially termed "advisers" or "trainers," with the latter designation preferred by the administration to discount any untoward comparisons with Vietnam. Armed with assault rifles, pistols and grenades, yet never to be given a direct combat role, Americans could influence the tide of the war through indirect means, providing training, advice and security and economic assistance, in essence expending dollars rather than men. In the theatres of war and Congress alike, such maneuvering proved invaluable.

These US trainers operated from carefully selected regional garrisons, only going to conduct training on day trips rather than being permanently stationed there.[9] The high tensions of the post-Vietnam years had resulted in stiff restrictions that, over time, were eventually relaxed, but the order never to engage in actual combat remained firm, even if it proved unrealistic on several occasions in which the trainers found themselves in intense firefights with insurgents. Nevertheless, trainers received "hostile fire" pay, a special monthly bonus paid to soldiers in danger of being killed, and some years later, after the conclusion of their mission, there was formal recognition that they had been in combat, including decorations and the authorization of the wearing of "combat patches."

Small though it was – publicly limited to fifty-five trainers in the field and additional personnel in the Embassy, albeit with counting rules that were skirted in order to have more than that number on the ground most of each month – the American military presence in El Salvador had the potential for severe military escalation, should US advisers in the country be directly and successfully targeted by FMLN forces. This precarious situation worried policymakers in Washington and contributed to a general sense of unease about the mission in the Pentagon. While the US presence in El Salvador never remotely exploded into another Vietnam, the war did drag on for twelve years and cost America almost $1 billion in aid.[10] American intervention in El Salvador would only end in 1992 under George H. W. Bush, the third President to oversee the conflict.

American military and economic assistance did not win the war, but it did allow the El Salvador government to remain in power and to make important progress against the insurgents, in addition to making modest governmental reforms.[11] Importantly, over time, especially in the critical middle years, the US had exceptional partners. Foremost among these was Napoleón Duarte, leader of the junta from 1980 to 1982 and later elected President, who led the country for five years from 1984 into 1989 and who performed all the tasks of a strategic leader superbly. He exhibited impressive strategic judgment, was an impressive communicator, demonstrated considerable skill in overseeing the "United for Reconstruction" campaign plan (developed at the outset of the effort with American assistance and refined repeatedly over the years), was cool under fire (literally as well as figuratively) and had the flexibility of mind repeatedly to adapt and update the plan.

Duarte's senior military leader, General Carlos Vides Casanova, also proved very effective over the decade when he first led the Salvadoran National Guard (from 1979 to 1983) and then was Minister of Defense (1983–9), although his tenure did, early on, include the terrible atrocities noted above, for which he must share accountability. Nonetheless, over time, he guided the military aspects of a relatively effective comprehensive counter-insurgency campaign and also pursued reforms both in how the military operated and in its role in society and the government. Those two leaders worked exceedingly well together and proved to be the kind of partners, albeit not without shortcomings, that military leaders and diplomats hope to find in such situations but seldom do.[12]

In essence, American military and economic assistance in El Salvador enabled the Salvadoran forces to grind down a smaller and scrappier opponent, and significantly improved the security situation, solidifying the security progress with various civil initiatives. The considerable domestic popularity of President Reagan – at least once inflation was crushed early in his tenure – guaranteed continued support of the Salvadoran government and allowed America to carry on throwing an almost inexhaustible quantity of resources at a determined but struggling enemy.

The fighting in El Salvador was vicious and bloody; around 10,000 Salvadoran troops died in the conflict, as well as a far greater number of FMLN guerrillas. In spite of the advisory nature of their role, twenty American personnel lost their lives. The trainers and officials who carried out the shadow war operated in a dangerous, morally ambiguous environment that taxed their abilities and required uncomfortable choices. It was a military engagement undertaken in pursuit of the paramount foreign-policy objectives of the American government, but one that was starkly different from the more conventional conflicts to come. El Salvador provided a successful template for the United States in helping to develop a comprehensive civil–military counter-insurgency campaign plan, and to advise, assist and enable its execution with minimal numbers of boots on the ground.

The Iran–Iraq War: where conflict regressed rather than evolved

Resentments had existed between Iran and Iraq for aeons, but the centuries-old rivalry between Persian and Arab was not helped in the early 1980s by a dispute over access to the 120-mile Shatt-al-Arab waterway which provided Iraq with its only access to the Persian Gulf, let alone by calls by Ayatollah Khomeini, Iran's Supreme Leader, for an extension of his Shi'a revolution across the globe. These exhortations caused particular concern in Baghdad, given that Iraq was led by Saddam Hussein, a Sunni Arab, and that at least 55 percent of the "Land of the Two Rivers" was Shi'a. Moreover in 1980, Saddam executed an extremely important Shi'a cleric who had been a close friend of the Ayatollah.[13]

Seeking to take advantage of Khomeini's precarious situation follow-ing the overthrow of the Shah in Iran and to resist the tide of a potential Shi'a uprising in Iraq, Saddam Hussein struck at Tehran's Mehrabad Airport on 23 September 1980, and afterwards in the oil-rich Khuzestan region of Iran, which, led by its powerful Arab population, had risen up against the Islamic Republic in 1979. Such ethnic and sectarian tensions served as a justification for war: Saddam considered himself the "defender of the Arab world" and ostensibly intended to "free" the majority-Arab population of Khuzestan – although, in practice it would merely have meant shifting them from one bloodthirsty dictatorship to another.[14]

The price that both sides paid as a result of Saddam's decisions would be staggering: two lands, with rich ancient and medieval histo-ries, and alike in cruelty and pride, had set in motion the longest conventional war of the twentieth century.

Iraq's initial offensive ran out of steam within two months, and Iran's human-wave frontal assaults caused significant Iraqi losses. By the summer of 1982, the Iranians, whose ayatollahs had persuaded them they were fighting a jihad (holy war) and whose Martyrs Foundation was generous to the widows and orphans of fallen soldiers, had turned the tide. In 1986 alone, Iran had managed to recruit hundreds of thou-sands of volunteers to join its already 650,000-strong army.[15] Confident in the legitimacy, sanctity and success of its efforts, it decided to launch its own invasion of Iraq.

Because both countries fought all out, with a refusal to observe any limits to their ultimate goals or the means to which they would stoop to achieve them, the casualties on both sides mounted as the decade wore on.[16] Saddam's generals (at this point nearly mutinous against the Iraqi leader) managed to muster more than 200 missile strikes on Iran in January 1987 alone, killing more than 2,000 Iranian civilians and wounding 6,000 more. But the Iranians gave back what they received. As General H. R. McMaster later observed,

Ayatollah Khomeini sent unarmed teenagers to certain death, with instructions to pick up the rifle of the boy who fell in front of them ... The Ayatollah gave them small metal keys that he promised would gain them admission to Paradise when they were martyred. Many were bound by ropes to prevent their desertion.[17]

The war made strange bedfellows, with both the United States and the USSR supporting Iraq, or, as Lawrence Freedman notes, "more accurately, opposing Iran."[18] Further logistical support for Iraq came – at a price – from France, Saudi Arabia and Kuwait. As a pariah state, Iran could not ensure a regular supply of arms and assistance from outside the country.[19] The Iranian prospects in the war were, however, boosted by the hubris of Saddam Hussein, who was, in the words of the military historian Richard Overy, "a hopeless supreme commander with a poor understanding of his enemies, fantasies about Iraqi military might and a command style that included executing subordinate commanders who in his view had failed (three hundred alone in 1982)."[20]

Since the heyday of British early nineteenth-century Whiggery, the West has operated under the general assumption that mankind progresses as it evolves, learning from the past and perfecting techniques – advancement and improvement entwined as one. The Iran–Iraq War serves as a stark counter-argument to this theory, as it was reminiscent less of the Second World War than of the First. Poison gases, trenches, barbed wire and frontal assaults across a barren no man's land all led to horrific levels of casualties and soldiers suffering agonizing, often slow deaths, with the images of their religious leaders tucked into their shirt pockets as their bodies lay in a harsh land.

It is impossible to understand the psychology of today's Iranian mullahs without remembering that in many cases their youth was dominated by an eight-year-long existential state-on-state struggle that resulted in the death of hundreds of thousands of their brothers, fathers and countrymen. "There were no lessons for Cold War armies in the tactical conduct of the Iran–Iraq War," writes Freedman, "except that certain regimes, when they go to war, were prepared to squander manpower."[21]

It was Iraq that first undertook the use of chemical weapons against Iranian troops in 1983, and also first began to attack Iranian oil tankers in the Straits of Hormuz the following year. In 1985, Iraq unleashed bomber and missile attacks on Iranian towns and cities over a hundred miles inside Iran. For two months in early 1988, around 200 Al-Hussein missiles (an upgrade of the Soviet Scud with a slightly longer range) hit over three dozen Iranian cities, prompting what the Western media called a "Scud duel."

The year 1988 was a difficult one for Iran: Iraq's recapture of the Al-Faw Peninsula during Ramadan (aided by the use of sarin gas) was coupled with Operation Praying Mantis, in which the United States attacked Iranian surveillance platforms and ships after Iranian naval mining incidents had damaged the USS *Samuel B. Roberts*. This, along-side the threat of military sanctions being raised in the UN – where Iran had neither friends nor allies – ensured a ceasefire was agreed, though no peace treaty was signed.

The war illustrated the recurrent truth that totalitarian dictator-ships can (and usually do) wage war that is far more costly in both blood and treasure than wars waged by their counterparts in democra-cies, where leaders can be evicted when the public turns against the war. With no fear of political blowback or criticism from domestic media or elected assemblies, dictators are emboldened in launching expensive assaults almost regardless of the cost in human life. China's human-wave assaults in Korea, the brutal Khmer Rouge tactics in Cambodia in the late 1970s and Vladimir Putin's invasion of Ukraine all demonstrate that an insulated leadership, violent and intimidating to its populace, can cause massive bouts of destruction with little domestic trouble (though there may yet be repercussions in Russia in the years ahead).

Reluctant Western support of Iraq was best articulated by Henry Kissinger when he said of Iran and Iraq that "It's a pity they both can't lose." Yet in a sense both sides did indeed lose: they were both severely weakened by the conflict, which degraded their economies and popula-tions alike. Eight years after their country had been invaded, Iranians' per capita income had collapsed by 45 percent.[22] At the time of the ceasefire agreed in August 1988, a total of 262,000 Iranian and 105,000 Iraqis had died (although several estimates put the numbers far higher), as well as over 100,000 civilians.

A United States-backed, reluctantly pro-Iraq coalition had dealt severe blows to an Iran now practically on life support, while allowing Iraq to take the brunt of the counter-attack had weakened a partner few wanted to remain strong. By the late 1980s, Saddam Hussein had been left with such serious war debts that he decided to try to recoup his losses in a wild, desperate manner that we shall examine later in this chapter.

The US invasion of Grenada: protecting America's backyard

On 5 October 1983, the United States invaded Grenada, a tiny island nation a hundred miles off the coast of Venezuela with a population of just 91,000, a brief but important Cold War strike by the United States against communist infiltration of the Caribbean. A former British possession, Grenada had fallen under communist control on 13 March 1979 after a coup led by Maurice Bishop, the leader of the leftist New Jewel Movement,* who installed himself as prime minister. Four years later he was himself overthrown in a coup led by his even more extreme deputy, Bernard Coard, and was executed by firing squad on 19 October. (His body has never been found.)

Despite the diplomatic difficulties thrown up by Grenada being a member of what was then still called the British Commonwealth, and thus under the purview of his friend and ally Margaret Thatcher, President Reagan was concerned about the communist inroads being made in the Caribbean. The dangerous unrest on Grenada left him concerned for the safety of around 600 mainly American students being held hostage at St. George's University Medical School. Haunted by the seizure of over fifty American hostages in Iran just four years before – which had been one of the factors that undermined his predecessor

---

* New Joint Effort for Welfare, Education and Liberation.

Jimmy Carter's credibility – Reagan was determined to avoid a similar disaster, telling his advisers, "There are Americans there and they are in danger. We are going!"[23]

On Reagan's command, 7,300 US soldiers, sailors and Marines descended upon the island. Opposing them were around 1,500 troops of the People's Revolutionary Army, 784 Cubans (posing as "airport construction workers" yet armed to the teeth), forty-nine Russian "advisers" and small contingents from East Germany and Muammar Gadaffi's Libya. Though Grenada was essentially a client state of Cuba by 1983, Fidel Castro decided not to intervene with more substantial forces. Unwilling to give the US government an excuse to depose him once and for all, and not finding support from Yury Andropov's Kremlin, he wisely kept his distance.

Operation Urgent Fury got off to a tragic start, when four Navy SEALs died after a miscalculation of the time they were to dive into the open ocean from a C-130 cargo plane.[24] When the cargo bay door opened, the SEALs were startled to encounter pitch darkness; their original plan had been for a daylight deployment. Eleven SEALs leaped into the dark regardless, but the unfortunate four, burdened by heavy equipment, got tangled in their parachutes and drowned. Their bodies were lost.

The subsequent invasion of the island began with a heliborne assault by the Marines, who had been heading for a deployment in Lebanon until their Navy troop carriers executed a sudden southern diversion. They were initially briefed that their mission was solely to evacuate the American students, but after Reagan received a request from the Organization of Eastern Caribbean States to stabilize the political situation in Grenada – a request that was drafted with considerable US assistance and called for intervention against one of their fellow member states – he authorized an all-out invasion and regime-change.

Rough seas and coral reefs prevented the amphibious landing that was originally planned, and so the Marines were landed by twenty-one CH-46 helicopters at locations near the airport. They were the first American servicemen deployed in combat since the Vietnam War, and arrived just two days after 220 of their comrades, along with eighty-five other people, had been killed by an Iranian-backed suicide bombing of the barracks at Beirut airport. In Grenada, Army Rangers and Navy and Army special operations forces would soon follow.

After a sharp engagement, the airport was captured on the first day, but only just. "The fighting was at least as intense as any of the fights I was in in Vietnam," recalled Lieutenant Colonel Wesley B. Taylor Jr., who commanded one of the two Ranger battalions, "and I saw two years over there."[25] A Navy SEAL team landed on the beach at St. George and assaulted Government House, rescuing Paul Scoon, the Governor-General, along with his family and aides. They were unable to carry out the planned removal, however, as the house was surrounded by Grenadian forces ensconced in armored vehicles.

Lightly armed and lacking anti-tank weaponry, the SEALs and their charges lay on the floor as "oil portraits of implacable British royals and daguerreotypes of Queen Victoria" stared at them impassively from the walls.[26] The residence remained under heavy fire until the arrival of a Marine task force of 250 men who had landed at Grand Mal, several miles north, reached the house and overcame the stiff resistance. The civilians were withdrawn by helicopter to the operation's flagship USS *Guam*, and Scoon returned to the capital as a temporary head of government until elections could be held.[27]

To the surprise and initial confusion of military commanders, only around 140 of the American medical students were discovered at the campus. Rangers learned for the first time of the existence of a second, larger campus that housed the majority of students, and the fact that it had previously been unknown meant that no plans existed to capture it. The assaults from Grenadian and Cuban forces on the first day were so unexpectedly ferocious that they prompted General Edward L. Trobaugh, Commander of the 82nd Airborne Division units airlanding by then, to roar into his satellite radio for his division rear staff at Fort Bragg, North Carolina to "Send me battalions until I tell you to stop."[28]

On the second day of the invasion, US forces advanced towards Frequette, near which a series of warehouses, surrounded by barbed-wire-covered fences, housed enough Soviet- and Cuban-supplied weaponry to arm six entire battalions – far more than the Grenadian military could use. The next few days were marked by sporadic engagements with Grenadian and Cuban soldiers, with the US forces experiencing some difficulty in dealing with large numbers of civilians and prisoners. Meanwhile, the American media were clamoring for access to the island, since concerns for journalists' safety (and fears that they would get in the way) had prompted the Pentagon to keep them off the island, at least for the first few days.

Ultimately, victory was achieved within forty-eight hours, and combat operations formally ended on 2 November, the same day that the UN General Assembly by 108 votes to 9 denounced the invasion as a "flagrant violation of international law."[29] Militarily and politically, if not diplomatically, the invasion had been a resounding success, however. Enemy resistance was overwhelmed, 600 mainly American citizens had been rescued and a communist government had been toppled, with a non-communist government winning the elections held the following year. Today, the anniversary of the invasion is Grenada's national holiday.

"Our days of weakness are over," President Reagan observed after Operation Urgent Fury. "Our military forces are back on their feet and standing tall."[30] In 1832, Carl von Clausewitz had written, "We may infer, that it is very difficult ... for the most talented general to gain a victory over an enemy double his strength." The American forces had outnumbered the Grenadian defenders by more than the three-to-one ratio required for an attacker over a defender.

Yet for all its success the expedition had not been pretty. It had exposed considerable shortcomings in the US military in the areas of inter-service communication and coordination. Nineteen American servicemen had been killed and 116 wounded during the seven days of the engagement, and friendly-fire incidents and other accidents accounted for a significant number of them. The short-notice contingency aspect of Operation Urgent Fury had placed sudden and significant stress upon logistical systems, and the quality of planning and intelligence were both decidedly poor – the latter egregiously so.[31] Tactical errors had led to needless civilian casualties, including eighteen patients in a mental hospital mistakenly fired on by Navy A-7 Corsairs.

"We planned the operation in a very short period of time – in about 48 hours," General John W. Vessey Jr., Chairman of the Joint Chiefs of Staff, later told NBC's *Meet the Press*. "We planned it with insufficient intelligence for the type of operation we wanted to conduct. As a result, we probably used more force than we needed to do the job, but the operation went reasonably well ... Things did go wrong, but generally the operation was a success. The troops did very well."[32] Nevertheless, following Vessey's retirement two years later and his replacement by Admiral William J. Crowe, some much needed reforms were instituted, including a strengthening of the role of the Chairman by the Goldwater–Nichols Act of October 1986, which brought about the streamlining of the US military's command structure in order to reduce inter-service rivalry and bureaucracy.

The subsequent overhaul of its organizational structure would be the most sweeping restructuring of the Department of Defense since its establishment in 1947. The enhancement of the Chairman's office led to better coordination across the services with the service chiefs now reporting to the Chairman rather than directly to the Secretary of Defense. Crowe and his successors as chairman were officially designated "the principal military advisor to the president, the National Security Council, and the Secretary of Defense."[33]

Beyond that, the reforms also significantly empowered the so-called combatant commanders, who were responsible for the various regions of the world, and they would become the "war-fighting commanders" in the decades ahead, with all forces in their areas of operation under their control, including those provided by US Special Operations Command. The US armed forces were slowly emerging from the post-Vietnam doldrums, better funded now but perhaps not quite fully ready for war on a larger scale. Small-scale operations near and far in the 1980s under Reagan helped prepare the US military for a far greater conflict to come, under his successor. Absent Grenada, it is doubtful that the United States would have woken up to the urgent need for considerable improvement in the conduct of joint operations.

Deposing General Noriega: the ousting of "Mr. Pineapple Face"

Confident in the wake of successful operations in neighboring territories but recognizing that there was work to be done in coordination between the services, the next major American military operation, in late 1989, had the goal of punishing a dictatorial Central American regime responsible for criminal activity at home and abroad. The target was in Panama, where the United States had built, controlled and guarded a canal through which much of the world's shipping passed between the Atlantic and Pacific Oceans.

Since 1983 Panama had been ruled by General Manuel "Pineapple Face" Noriega, who for decades had been simultaneously on the payroll of both the CIA and various Central American drug cartels. The US government, embarrassed by revelations of Noriega's double-dealing, sought to prosecute him for drug trafficking and other crimes. In response, Noriega's regime fired upon American military personnel and harassed other American civilians in the country. After a sustained legal and diplomatic effort to oust him from power, President George H. W. Bush resolved to invade Panama and capture its corrupt leader.

Operation Just Cause was launched on 20 December 1989 and saw nearly 28,000 American ground troops deployed from a constellation of C-130 Hercules and C-141 Galaxy transports. The original proposed name had been Operation Blue Spoon, but that was considered to be much too frivolous.[34] General Colin Powell, Chairman of the Joint Chiefs of Staff, later expressed his approval of the new name on the grounds that "Even our severest critics would have to utter 'Just Cause' while denouncing us."[35]

The operation was the largest and most complex for the US Air Force since Vietnam.[36] An ice storm threatened to delay deployment of transports from Pope Air Force Base in North Carolina, but de-icing equipment from other bases was swiftly flown in and employed to good effect.[37] AH-64 attack helicopters, F-117 stealth fighters and other aircraft attacked the headquarters and other facilities of the Panamanian Defence Force, 16,000 of whose troops were mobilized to resist the invasion. The High Mobility Multipurpose Wheeled Vehicle made its combat debut; before long, a commercial version of the "Humvee" would appear on American roads.

For the Army's 82nd Airborne Division, whose parachutes filled the sky over Panama, it was the first combat jump since the Second World War and the largest since Operation Market Garden in Holland in September 1944.[38] Six men of the 82nd would not return home. Navy SEALs were deployed to capture Noriega, and four of them were killed in an operation to shoot out the wheels of his plane. Noriega's capture seemed inevitable, but he was able to delay it for more than a week by strolling into the diplomatic protection of the Vatican's Apostolic Nunciature in Panama City, carrying an AK-47 in each hand. The US Army took up positions around the building and tried to pressure him into surrendering by playing hard rock music at a high volume for days.

Alas, even the sound of Guns N' Roses blasting at the windows was insufficient to dislodge Noriega, but the weary and unsympathetic Vatican Nuncio finally insisted that he leave on 3 January 1990. Noriega was taken into custody and brought back to the United States where he was tried on drug-trafficking and racketeering charges before being imprisoned in a minimum-security Florida prison for seventeen years. The United States designated him a political prisoner, so he received special privileges that included a television, a telephone, two rooms and an exercise bicycle. His cell was nicknamed "the presidential suite."[39]

By the time Noriega left Panama in handcuffs, a new president was in place there and democracy had been restored. The United States had lost twenty-three military personnel, the Panamanians 314. Estimates vary as to the number of civilian casualties, ranging from 5 to 200. For the Americans, Operation Just Cause had been a brief, highly successful and valuable exercise in inter-service cooperation and rapid deployment of forces, the significance of which would become particularly evident just nine months later.

Panama was important in Pentagon thinking because it demonstrated that the United States had made considerable improvements in the conduct of joint operations since Grenada. It was a very complex operation, so much so that it would have been challenging even as a

The Soviet Union's first
nuclear test at Semipalatinsk,
29 August 1949.

Kibbutz members being briefed during Israel's War of Independence, Yad Mordechai, 1948.

US Marines in Korea, 1951. It was the first time helicopters were used to land men and supplies in combat zones.

Troops and helicopter, Dien Bien Phu, 1953.

Victorious Viet Minh troops wave a Vietnamese flag over captured French positions at Dien Bien Phu, 1954.

An injured member of the Special Air Service is carried by fellow soldiers to a helicopter for evacuation, Malayan Emergency, 1953.

British soldiers in Borneo, 11 March 1965.

A US Air Force B-52 Stratofortress plane releasing its bomb load over Vietnam, having flown from its base on Guam.

Tet Offensive: South Vietnamese PLAF (People's Liberation Armed Forces) soldiers in action in South Vietnam, 1968.

American advisers watch as Vietnamese troops run to board helicopters for an anti-Viet Cong operation in the jungles near Saigon, Vietnam, 25 October 1962.

An Israeli soldier studies an oil refinery across the Suez Canal set alight by Israeli shelling during the Six-Day War, 1967.

Israeli tanks crossing a pontoon bridge over the Suez Canal, 25 October 1973, during the Yom Kippur War.

Royal Marines wait to go on patrol from Ajax Bay during the Falklands War in 1982.

A CIA employee helps Vietnamese evacuees on to an Air America helicopter from the roof of a building half a mile from the US Embassy, 29 April 1975.

General Gromov and his son crossing into Uzbekistan during the Soviet withdrawal from Afghanistan, with the final Soviet vehicle following behind, February 1989.

training exercise, much less one in which there was a real enemy, however limited in capability.

The Gulf War, 1991: victory after a forty-three-day air campaign and a 100-hour ground war

In the spring of 1990, after eleven years of vicious dictatorial rule, President Saddam Hussein of Iraq was a worried man. The debts from his protracted eight-year war with Iran were enormous, and economic pressures were tempting him into maverick solutions. Iraq's economic lifeblood was rooted in oil, the high price of which he hoped would help pay off the country's debts and secure his hold on power. Yet just across his southern border the sovereign nation of Kuwait was exporting oil in amounts vast enough to depress the global price. Saddam met the Emir of Kuwait in May 1990 and demanded he reduce production. When he was refused, an increasingly desperate Saddam settled on a drastic solution.

Despite the Iran–Iraq War having ended less than two years previously with huge loss of Iraqi life, Saddam still had a million men under arms. When in the summer of 1990 he ordered the mobilization of 30,000 of the troops in the south of the country along the Kuwaiti border, policymakers in Washington looked on with concern, but the consensus in President George H. W. Bush's administration was that despite his saber-rattling Saddam would not dare invade a sovereign neighbor without provocation.

Yet, on 2 August 1990, the Iraqi Republican Guard – the elite units of Saddam's army – did indeed cross the border of Kuwait and began to assault Kuwait City. As tanks secured the southern and eastern edges of the capital, special forces units landed on the beaches. The Iraqis met no military resistance from Kuwaiti forces, other than a brief defense of the Emir's palace. The Emir had been targeted for assassination but managed to escape to Saudi Arabia just in time, as his country was swiftly devoured by its much larger neighbor. Saddam Hussein, who the day before had possessed one-tenth of the world's oil supply, now controlled one-fifth of it.

The invasion was driven by a false confidence: Saddam was convinced that the United States would not intervene. He had previously confided to a henchman his belief that the American defeat in Vietnam had so chastened the Americans that they would accept, however reluctantly, his annexation of Kuwait. After all, they had tacitly supported his government during the Iran–Iraq War, and recent diplomatic representations from the US Ambassador to Baghdad apparently left him with the impression that he would have a free hand.

In fact President Bush and Margaret Thatcher were outraged and appalled by Saddam's actions, and quickly resolved that they would not stand. Bush had envisioned a "new world order" after the end of the Cold War, and did not want it characterized by aggressive action in the Middle East. His advisers were also concerned lest an emboldened Saddam, having so quickly vanquished Kuwait, should push further into foreign territory. Iraq would likely next target Saudi Arabia, a US ally and one of the world's most important oil-producing nations. Seizing the eastern province of Saudi Arabia alone would leave him in control of between 45 and 50 percent of the world's oil production.[40]

As was customary when trouble loomed overseas, the Navy was dispatched first, with the aircraft carrier USS *Eisenhower* and its task force sent to the eastern Mediterranean, and the USS *Independence* and its escorts to the Straits of Hormuz. On 6 August, President Bush signed an order to deploy 122 F-15s and F-16s to the Middle East, which arrived within a week, as did the lead elements of the 82nd Airborne Division, the US Army's rapid-deployment force, that drew a line in the sand with their bayonets and light tanks. Five weeks later, no fewer than 700 warplanes were in the theatre of future operations and additional Army units were in the process of deploying as well. Saddam Hussein, faced with a choice between standing his ground and conceding defeat, chose to recommit to his oil-thirsty invasion rather than regurgitate Kuwait and pay reparations. For dictatorship is a juggernaut that can rarely survive public humiliation.

Bush dispatched a delegation led by Secretary of Defense Dick Cheney to Saudi Arabia to discuss the impending danger with King Fahd. In his meeting with Fahd at the Saudi summer palace in Jeddah on 7 August Cheney was blunt: an Iraqi invasion of Saudi Arabia had to be deterred before Iraq seized the Saudi oil fields. Any war undertaken to expel Iraq from Saudi Arabia would be far more costly in both economic and human terms than a pre-emptive deployment of US military force. "If you ask us, we will come," Cheney told the King. "And when you ask us to go home, we will leave."[41] For any other state at any other time, a large number of American troops on sovereign soil might have been resisted and been seen as destabilizing – let alone a "Christian" army on Muslim soil – but the shadow of Saddam's vast army on the border made it an offer that the King could not refuse. The Americans were invited to deploy in Saudi Arabia.

The central military figure in the build-up was General Colin Powell, a general held in high regard and Chairman of the Joint Chiefs of Staff. Born in New York to Jamaican immigrants, he joined the Reserve Officers' Training Corps while a student at City College of New York. He served two tours of duty in Vietnam, the second in 1968 as a major, during which he earned a Soldier's Medal (for saving the life of another soldier) and a Purple Heart. After serving in a number of senior posts in Washington and in Europe, he was appointed national security advisor by President Reagan in 1987. Two years later, President Bush appointed him a four-star general and chairman of the Joint Chiefs. Powell, acutely aware of the mistakes made by America during the Vietnam War, was determined to avoid perpetrating similar ones in the Middle East.

In his memoirs, Powell responded to charges that he was a "reluctant warrior" with the words, "Guilty. War is a deadly game; and I do not believe in spending the lives of Americans lightly." Thoughtful, telegenic, personable and media-friendly, Powell proclaimed to reporters what came to be known as the Powell Doctrine, insisting that before the United States undertook military action, the following questions had to be answered by policymakers:

1. Is a vital national security interest threatened?
2. Do we have a clear and attainable objective?
3. Have the risks and costs been fully and frankly analyzed?
4. Have all other non-violent policy means been fully exhausted?

5. Is there a plausible exit strategy to avoid endless
   entanglement?
6. Have the consequences of our action been fully considered?
7. Is the action supported by the American people?
8. Do we have genuine broad international support?[42]

Most importantly to Powell, any American military force that was
finally deployed had to be massive and overwhelming. The gradual
deployments and creeping escalation that marked American involvement in Vietnam must be avoided. If a huge deployment somehow
failed to intimidate Saddam into giving up Kuwait, it would certainly
help ensure the application of maximum might and bring about a swift
victory. After the deployment of the elements of the three divisions, an
armored cavalry regiment and other units of the XVIII Airborne Corps,
Powell urged President Bush to double the proposed American troop
commitment in the region to more than half a million men, a vast
deployment far beyond that envisioned just weeks before, and one
which would be possible only with major troop commitments from a
wide coalition of allied countries.

Powell later recalled that his advice was greeted by gasps from the
President's advisers in the White House Situation Room, but that Bush
himself "had not blinked."[43] Most of the troops would be ferried over in
commercial aircraft thanks to the Civil Reserve Air Fleet, but vehicles,
major equipment, helicopters, spare parts and other military kit had to
be shipped by a huge flotilla of merchant vessels. If the American war
effort failed, it would not be due to a lack of manpower or equipment on
the ground.

Overall command fell to a far brasher character than Powell. General
H. Norman Schwarzkopf was Commander-in-Chief of Central
Command, based in Tampa, Florida. A towering bear of a man with a
ferocious temper that unnerved many subordinates, "Stormin' Norman"
brought to the battlefield decades of experience, brains, considerable
blunt competence and the scars of combat in Vietnam. More recently he
had filled a command vacuum on the USS *Guam* off the coast of
Grenada. He now presided over the build-up of allied forces, designated

Operation Desert Shield, from a headquarters established in Riyadh, the Saudi capital.

The logistical challenges in implementing Powell's vision were enormous. As Michael Gordon and General Bernard Trainor observe in their book *The Generals' War*, "The problem was that the United States needed a force that would arrive in the region fast enough to fight but heavy enough to survive."[44] Although Ronald Reagan's much needed military build-up had ensured the acquisition of new weapons, less certain was the ability to deploy them so far away. As Gordon and Trainor drily note, "The Pentagon behaved like a bachelor who owned a Porsche but forgot to pay his electric bill."[45]

As military cargo planes were not enough to transport so vast an army and its equipment over such a long distance, sealift vessels, many of them aged and creaky, were hurriedly pressed into service, manned by merchant seamen and retired sailors brought back on duty. Commercial aircraft requisitioned by the Department of Defense transported many of the troops. Saudi airfields and bases were stretched to breaking by the arrival of the huge ground and air forces.

Senior officers nonetheless believed that the large numbers of troops massing on Kuwait's southern border would make more of an impression on Saddam – whom they still hoped would give in without conflict – than the materiel. Such logic cannot work on tyrants, however, for whom backing down means undermining the supposed infallibility and strength that had been relentlessly relayed to their people through state propaganda.

As American troops continued to pour into the region and hopes that Saddam would back down began to fade, Schwarzkopf and his team pondered the best way to force him out of Kuwait. This appeared to be a formidable task, for although the Iraqis had precipitated the crisis with only 30,000 troops, they now had half a million deployed in and around Kuwait, the southern approaches to which were fortified with minefields and trenches. American planners considered a wide flanking maneuver, but decided that it would require more troops than were available at the time.[46] Schwarzkopf therefore pressed Powell to intercede with the President, who was already preoccupied with

assembling what has been described as "the most elaborate political
and military coalition since the Napoleonic Wars."[47]

Ultimately, the massive US-led coalition facing a lone Iraq comprised
more than thirty-five countries from Singapore to Sierra Leone.
Traditional adversaries and other unnatural bedfellows came together:
Argentina contributed frigates as the United Kingdom supplied sixteen
ships and fifty-eight aircraft. Afghanistan sent over 300 Mujahideen as
Saudi Arabia provided 100,000 troops.

The United Nations Security Council Resolution No. 678 adopted
uncompromising language in its condemnation of Iraq on 29 November,
and it was passed with overwhelming support. It imposed a deadline of
15 January 1991 for a complete Iraqi withdrawal from Kuwait, in the
absence of which "all necessary means" to enforce it were authorized.
An earlier resolution, No. 660, had already demanded "that Iraq withdraw immediately and unconditionally all its forces to the positions in
which they were located on 1 August 1990," but now the countdown to
war had officially begun. Saddam responded by sending additional
troops into Kuwait, where the military occupation had already led to a
gruesome policy of violence and repression.

President Bush condemned the Iraqi practices of summary executions and "routine torture." To him, Saddam was not just an amoral
dictator with a desire to raise oil prices through any means necessary,
but "Hitler revisited," and the President insisted that:

> America will not stand aside. The world will not allow the strong
> to swallow up the weak ... What is at stake is far more than a
> matter of economics or oil. What is at stake is whether the
> nations of the world can take a common stand against aggression
> or whether Iraq's aggression will go unanswered, whether we live
> in a world governed by the rule of law or by the law of the jungle.[48]

This was much the same stance as Margaret Thatcher had enunciated
during the Falklands War.

For all its various hiccups and complications, Operation Desert
Shield had been a signal success. A massive coalition force had been

transported over very long distances to a distant and hostile theatre of operations with impressive speed. Luck was no small part of it; had Saddam deployed his armies against his enemy early on during the build-up, the result might have been bloody and chaotic (though the one brief incursion into Saudi Arabia by the Iraqis was fought off by Saudi forces). The potential logistical issues were aggravated by the dreaded old bugbear of American military operations – inter-service rivalry. The Army and Marines differed regarding the direction from which an Iraqi attack might come, and General Schwarzkopf's desire for his field commanders to retain their independence and flexibility meant he was reluctant to intervene.[49]

The ultimate conduct of Operation Desert Storm – the combat phase of the Gulf War – was an unparalleled success in the air, the mastermind of which was Lieutenant General Charles A. Horner, a skilled pilot and veteran of many combat missions in Vietnam. Horner, like many of his peers, was still smarting sixteen years after the defeat in Vietnam. Pained by memories of the civilian interference that had both limited the Air Force's effectiveness and increased the danger to its pilots there, he extracted a promise from Schwarzkopf that he would have a free hand. Horner was immune to the illusion, cherished by some enthusiasts of airpower, that strategic bombing attacks might so demoralize the enemy that they would give up without the need for ground troops to engage. This was the dazzling vision conjured by other air planners working for the Chairman of the Joint Chiefs, but even Powell knew that the mission would not be accomplished so easily and he expressed concern about the campaign's focus on targets such as communications centers, power plants and oil depots. He wanted to ensure that Saddam's invading army would be destroyed as a fighting force and accordingly ordered that the air campaign be expanded in scope.[50] Horner was determined to do just that.

At 7 p.m. EST (2 a.m. local time) on Wednesday, 16 January 1991, the war began with a furious air assault on Iraq. Under Operation Instant Thunder, eight Apache helicopters from the 101st Airborne Division (Air Assault) destroyed the radar installations threatening allied forces, as precision-guided Tomahawk cruise missiles arced upwards from naval vessels and plunged from the sky, racing through the streets of Baghdad, turning left and right in a remarkable display of technical wizardry. Crucially, this pinpoint-precision targeting was enhanced by a sophisticated intelligence-gathering effort that had relied upon businessmen, journalists and other expatriates with detailed knowledge of the Iraqi capital.[51]

Ten F-117 Nighthawk stealth fighters flew almost invisibly through the night from Saudi Arabia. Their shape and composition were designed to reflect radar waves at odd angles, rendering conventional air defense systems useless. The skies over the Persian Gulf region were very far indeed from Panama, where the F-117 had first been deployed, but the fighter's effectiveness was unchanged. Having slipped through the enemy's guard, the Nighthawks attacked targets in Baghdad, including one of Saddam's many presidential palaces. Other waves of stealth fighters followed. Despite heavy anti-aircraft fire, all returned safely to base.

Some of the strongest assaults on the enemy came from much further afield: seven B-52 Stratofortresses, known affectionately as BUFFs,* a bomber type first deployed in 1954, with periodic updates, took off from Barksdale Air Force Base in Louisiana and flew more than 7,000 miles, with multiple air-to-air refuellings to deliver their huge and deadly payloads. It was an impressive display of the US Air Force's global reach, and a reminder to America's enemies that destruction from above was just a bombing run away.[52] The same type of aircraft had dropped 15,237 tons of bombs on Vietnam during a single American operation. "A young Norman Schwarzkopf witnessed the B-52's psychological impact in that war," notes the war's historian Jim Corrigan, and "he wanted Republican Guard troops to experience the same fear in this one."[53]

The air campaign began with 850 sorties and the loss of only two planes, a remarkable start.[54] Television viewers around the world were treated to an astonishing display of military might, with green tracers

---

* "Big Ugly Fat Fuckers."

and glowing flares against the night sky. To them, as Corrigan puts it, "the air campaign seemed precise, effortless, and almost risk-free ... like a high-tech romp carried out by mysterious stealth fighters and Tomahawk cruise missiles."[55] The Iraqi Air Force, though less technologically advanced than that of the United States, was sizeable and capable of inflicting considerable damage, not least because Soviet MiGs made up a large share of Saddam's aerial arsenal. Nonetheless, it proved little match for the US-led air forces.

The Canadian intellectual and sometime politician Michael Ignatieff dates the genesis of precision weaponry to the US Air Force's destruction of the Thanh Hoa Bridge in Vietnam by laser-guided bombs from eight Phantoms on 27 April 1972. By the late 1980s, America had emerged as so far in the lead in military technology that the Soviet Union could barely compete and would not last the decade. Ignatieff lists these advances as:

> Lasers to improve guidance and targeting, computers linked to satellite positioning systems which made pinpoint accuracy possible; propulsion systems which increased the range of conventional rockets; refinements in explosives technology which reduced damage to civilians, as well as unmanned and robotized surveillance drones which eliminated risk to aviators.[56]

The Iraqi military's reliance on Soviet technology that was swiftly being outdone by their US counterparts provided a clear and rapid demonstration of the Eastern Bloc's technological inferiority.

While their weapons systems were outclassed by those of their opponents, the Iraqis still had the means to create havoc on the battlefield and on the diplomatic front. Their most formidable weapon was the antiquated but deadly Scud, a Soviet-designed short-range ballistic missile. It rarely hit its intended mark, but could still wreak considerable collateral damage, especially against soft civilian targets. And Saddam had discerned a weak spot in the coalition diplomatic behemoth assembled by President Bush: the Arab nations would be unlikely to maintain their alliance were Israel to attack Iraq. American officials therefore asked Israel to stay its hand if it were attacked by Saddam. The Israeli response was that "It would depend on whether the Scud hit an empty parking lot or a crowded section of Tel Aviv and whether Israel felt that the United States had done everything possible to prevent it."[57]

In a private conversation with Powell, the Israeli Chief of the General Staff, Ehud Barak, warned, "If we don't go in and clear out the Scuds, Saddam may use them to deliver chemical warheads when you launch your ground offensive. They may fire nerve gas or a biological warhead at our cities. If that happens, you know what we must do."[58] Israel's forbearance would be sorely tested. The Iraqis fired more than forty Scud missiles at Israeli population centers during the war, leading to dozens of civilian deaths, starting on 17 January 1991 when eight fell on Tel Aviv and Haifa.[59] The relatively primitive nature of the missile rendered it no less effective against such targets.

Israeli restraint in face of the attacks, encouraged by the deft diplomacy of President Bush and Colin Powell, and assisted by Scud-hunting missions by US and UK special forces, helped maintain the coalition. The Iraqis also fired forty-six Scuds into Saudi Arabia, with one striking a barracks in Dhahran, killing twenty-eight members of the Pennsylvania National Guard and wounding several more, with others hitting various civilian sites. For reasons of diplomacy and public morale, it was vital that the coalition respond decisively to the Scud threat, so launch sites and production facilities were targeted from the air and attacked, albeit with mixed success. It was important, however, to show the public (not to mention Israel) that no effort was being spared in the hunt for Saddam's favored weapon.

Despite being less than completely effective, the response to the Scud threat that captured the imagination of the viewing public and cheered the Israeli and Saudi populations, was the American-built Patriot (Phased Array Tracking Radar to Intercept on Target) missile defense system which had initially been designed to target enemy aircraft. Fired from a truck-pulled launching station with a four-missile capacity, the Patriot was repurposed during the Gulf War primarily to attack Scud missiles in the air, a feat of tremendous complexity.

The army had been working on the challenge for decades; General Dwight D. Eisenhower had likened it to "hitting a bullet with another bullet."[60] For all the Patriot's promise, however, its effectiveness against Scud missiles was far from perfect. And while military commanders, including Schwarzkopf, hailed its performance, their praise was sometimes based on information that was not entirely validated during post-war analyses. The first declared success of a Patriot in bringing down a Scud was on 18 January, but it later emerged that no Scud had been fired that day. The speed at which Scud missiles travelled meant that it was difficult to confirm whether or not one had been intercepted.

It remains unclear exactly how many Scuds were brought down by Patriot missiles; the latter's champions claim around forty, while its critics concede only a fraction of that. But as Lawrence Freedman and Efraim Karsh record, "The psychological impact of the Patriots was more important than their actual combat performance. It meant that those potentially on the receiving end of the missile had reasonable hope that it would be intercepted which, in turn, helped to blunt the edge of the Scud as a terror weapon."[61] In the realm of psychological warfare, the Patriot was of enormous importance and it was long seen as one of the great technical successes of the war, recognition validated when it generated great applause when it appeared during the post-war victory parade through Washington, DC.

Overall, coalition air forces launched 109,500 combat sorties and dropped 88,500 tons of bombs, but there was only so much that could be accomplished from the skies.[62] Powell had earlier told President Bush, who had expressed hope that airpower alone might be sufficient, "I'd be the happiest soldier in the Army if the Iraqis turn tail when the bombs start falling."[63] But an air assault, however withering, allowed the "defender to decide whether or not he has had enough punishment." It was becoming evident that only a ground campaign could forcibly eject the enemy from Kuwait.

In his public statements Saddam claimed to be looking forward to a ground campaign. "Not a few drops of blood," he prophesied, "but rivers of blood will be shed."[64] President Bush and his team were confident that they had assembled a force in the Gulf so formidable that the Iraqi

Army would not be able to resist it for long. The coalition boasted a total ground force of about 700,000 troops; but, to the surprise and frustration of Bush, Secretary Cheney and Chairman Powell, Schwarzkopf seemed to develop what President Abraham Lincoln had once called – out of frustration with General George McClellan – "a case of the slows." This was ironic, as Schwarzkopf had previously written out a quote from General William Tecumseh Sherman's memoirs and affixed it to his desk, stating, "War is the remedy our enemies have chosen. And I say let us give them all they want."[65]

Adverse weather forecasts prompted Schwarzkopf to request a two-day delay of the launch of the ground war from 24 February 1991, which led to a tense telephone confrontation with an irate Colin Powell. "I've already told the President the twenty-fourth," Powell told Schwarzkopf. "How am I supposed to go back now and tell him the twenty-sixth? You don't appreciate the pressure I'm under." Powell added, "My President wants to get on with this thing." Not one to back down from an argument, Schwarzkopf replied, "And you're telling me that for political reasons you don't want to go in and tell the President he shouldn't do something that's militarily unsound? For Chrissakes, Colin, don't you understand? My Marine commander has come to me and said we need to wait. We're talking about Marines' lives!" "Don't patronize me with talk about human lives!" roared Powell down the line.[66]

Tempers cooled and the weather cleared. On 24 February, the President instructed Schwarzkopf "to use all forces available, including ground forces, to eject the Iraqi army from Kuwait."[67] "G-Day" had come. What Saddam had predicted would be "the mother of all battles" started with three US divisions, two Marine and one Army, crossing the Saudi border directly into Kuwait, bound for the capital. Smaller Arab units guarded the American flanks. Powell worried that this "secondary attack could cost us far more casualties than the main attack because the main attack, and people flying in helicopters at one hundred miles an hour or in tanks moving twenty or thirty miles an hour, were pretty invulnerable. These were infantrymen who were going right in."[68]

As Powell's observation made clear, for all its boldness and directness the border crossing into Kuwait was but a diversion from the main component of the coalition's plan. During the months of preparation for Desert Shield, Schwarzkopf had given his planners free rein, directing them simply to "Assume a ground attack will follow an air campaign. I want you to study the enemy dispositions and terrain and tell me the best way to drive Iraq out of Kuwait given the forces we have available."[69] The plan they devised was elegant and deceptively simple: the main effort would be a left hook in which the army would "swing like a giant scythe" and "attack from the western desert."[70] Its only apparent flaw was the lack of secrecy; according to the New York Times, "Everyone seems to know that the likely scenario calls for US and British forces to wheel around Kuwait and cut across the southern part of Iraq towards Basra."[71]

It is often said that no plan survives contact with the enemy, but this one did more than most. Its success was made more likely thanks to the enemy conforming to an earlier Army intelligence prediction "that Iraq had fixed its forces and would not reinforce or redispose them in any significant way; that the Iraqi defense would be positional, with counter-attacks likely by tactical but not operational or Republican Guard forces; that the Iraqi military was fixated on the defense of Kuwait City and Basra and would not improve defensive barriers or move forces westwardly."[72]

The rapid advance of the Marines against what was initially little resistance encouraged Schwarzkopf to accelerate the main attack. "If we moved fast," he later recalled, "we could force them to fight at a huge disadvantage; if we stayed with the original timetable, they might escape relatively intact."[73] He was exhilarated by the relative lack of coalition casualties and urged his commanders to press forward as quickly as possible, ordering them to "inflict maximum destruction, maximum destruction on the Iraqi military machine."[74]

The XVIII Airborne Corps, the outer element of the left hook, raced across the open desert into Iraq with a French armored division guarding its left flank. The 101st Airborne Division used its armada of helicopters to launch an air assault deep behind enemy lines, its mission both to prevent the Iraqis sending reinforcements into the battle area and to stop those already there from escaping along Highway 8, which ran along the Euphrates River to Kuwait. And the corps' heavy division, the 24th Infantry Division (Mechanized), attacked swiftly on the corps' right, with the 3rd Armored Cavalry Regiment screening its right flank.[75]

By contrast with the speed of XVIII Corps, Lieutenant General Frederick M. Franks' VII Corps, with its 1,500 tanks, a key part of the scythe between the forces of XVIII Airborne Corps on the left and the Marine and Arab forces on the right, was making slower progress. This frustrated Schwarzkopf all the more as Franks had the luxury of the British 1st Army Division guarding his right flank. Schwarzkopf's unwillingness to intervene in the decision-making of his subordinates was severely tested by Franks' perceived lack of receptiveness to his orders to accelerate, and Franks came close to being replaced by a more vigorous officer before speeding up the execution of his mission. In the end, the sweeping maneuver across the desert deep inside Iraq destroyed many Iraqi tanks, without losing a single coalition tank.

A large convoy of Iraqi vehicles retreating along the six-lane Highway 80, which ran north from Kuwait City to Basra in Iraq, made an irresistible target for coalition commanders on the night of 26 February. Air Force, Navy, and Marine aircraft bombed both ends of the convoy to immobilize it and then unleashed their full fury along its entire length, with around 2,000 vehicles destroyed or abandoned and anywhere between 200 and 1,000 Iraqis killed. The road soon came to be known as the "Highway of Death" and became a symbol of the utter technological dominance of coalition forces. As the military historian Sterling Michael Pavelec has written, "All war is asymmetric."[76]

What some pilots referred to as "the turkey shoot" on Highway 80 pricked the consciences of President Bush, Colin Powell and others, who began to wonder whether the continuation of such devastating assaults on a clearly defeated enemy could be justified. On 27 February, Powell told the President that soldiers on the ground were becoming unsettled by the sheer scale of the massacre on the Highway of Death.[77] Decision-makers in Washington wondered whether civilians back home might be repelled by the slaughter, though, as Rick Atkinson shrewdly observed, "The belief that Americans would recoil from the Highway of Death may have misread a culture steeped in violence ... The national contempt for the military that developed during the Vietnam War had taken years to form ... and was a consequence more of American casualties than of the enemy's."[78]

Schwarzkopf certainly had few qualms about the devastating punishment of the convoy, recalling that:

> The first reason why we bombed the highway coming north out of Kuwait is because there was a great deal of military equipment on that highway, and I had given orders to all my commanders that I wanted every piece of Iraqi equipment that we possibly could destroyed. Secondly, this was not a bunch of innocent people just trying to make their way back across the border to Iraq. This was a bunch of rapists, murderers and thugs who had raped and pillaged downtown Kuwait City and now were trying to get out of the country before they were caught.[79]

Concern over the size of enemy losses was a relatively new feature that entered the consideration of warfare in the 1990s. Of course, it had existed before – German civilian losses had been raised by vocal critics of the Combined Bomber Offensive in the Church of England and the House of Lords during the Second World War, for example – but it had not before affected operational decision-making to any great degree. Much more common historically has been a disregard for enemy losses, though few instances can beat that of General Herbert Kitchener, who after killing 11,000 Dervishes at the battle of Omdurman in 1898 rued the "dreadful waste of ammunition" involved.[80] With domestic criticism of the sinking of the *Belgrano* in 1982, the destruction of Saddam's armor on the "Highway of Death" and Serbian military and civilian losses in 1999, however, military commanders of the twentieth and twenty-first centuries – at least in democratic countries – now had to concentrate on a new factor: the acceptable number of enemies to kill.

For his part, President Bush, conscious of the slaughter on Highway 80, feared that continued attacks on a vanquished enemy might seem cruel and excessive. "We need to have an end," he said. "People want that.

They are going to want to know we won and the kids can come home. We do not want to screw this up with a sloppy, muddled ending."[81]

The Kuwaiti Army was given the honor of formally liberating its nation's capital on 26 February 1991, after Iraqi tanks had beaten a quick retreat. "Tonight, the Kuwaiti flag flies above the capital of a free and sovereign nation," declared Bush soon afterwards. The ground campaign ended at 8 a.m. on 28 February with a ceasefire, only a hundred hours after it had begun, a speed that exceeded even the most optimistic forecasts (although the roundness of the number had more to do with public relations than with military realities). Schwarzkopf and his staff felt some misgivings about ending the war so quickly, as many Republican Guard units were still intact, with VII Corps and the fleet of F-117s poised to administer a *coup de grâce*.

Lessons had been learned. "We've licked the Vietnam syndrome once and for all," President Bush observed.[82] Forty-three days after it began, with acrid smoke billowing over Kuwaiti oil fields set alight by the defeated regime before they retreated, the Gulf War concluded with a complete coalition victory. In an inhospitable region and against a larger enemy force, the coalition achieved its objectives with the loss of 292 lives, 219 of which were American. It was a fraction of the 5,000 killed and wounded that Schwarzkopf and others had feared, and nearly a quarter of all American casualties were due to friendly fire.[83] By contrast, there were between 20,000 and 30,000 Iraqi military fatalities.

In many respects, the Gulf War had been sent from central casting for the US Army, which trained its forces rigorously at the National Training Center in the Mojave Desert, against an opposing force (OPFOR) that was highly skilled, knew the ground and taught hard lessons to most of the units conducting training rotations. On the other hand, the war was waged against an enemy that had been ground down by months of being deployed in desert defensive positions and then pounded from the air for forty-three days. There were numerous stiff fights, but the US and the coalition had taken every conceivable action to ensure they did not engage in a fair fight. Iraqi forces had been completely overwhelmed.

Machiavelli wrote that it was not enough for a victor to have "exterminated the family of the [defeated] prince, because the lords that remain make themselves the heads of fresh movements against you." In 1991, America decided not only to leave Saddam and his family in power, but to leave much of the Republican Guard still intact – a decision that was controversial at the time and was to remain so in years to come.

President Bush's understandable and humane desire in the quest for peace to stem bloodshed in the Middle East by humiliating but refusing to annihilate Saddam Hussein's elite guard or his position of ultimate authority would, ironically, lead to a significant conflict years later. Saddam let his generals sign the surrender documents, saving himself from the indignity, and Dick Cheney later reflected that "We had been told repeatedly by our own intelligence services and by our Arab allies that Saddam would never survive after the blow the coalition had delivered, but he was able to turn the fact that he had stood up and survived a massive assault into a personal victory."[84]

The Gulf War had been an extremely high-tech war, justifying the huge amount of money expended and research undertaken over previous decades. "This war didn't take a hundred hours to win," General Barry McCaffrey told a post-war Senate hearing, "it took fifteen years," underscoring the enormous effort that had been required to transform the "Hollow Army" of the post-Vietnam years into the exceedingly well-trained, well-equipped and well-led force that rolled over Saddam's military.[85] The breakthroughs in technology had borne fruit in the suppression of enemy air defense and precision munitions delivery, with the US Army's Big Five weapons systems (the M1 tank, the M2 infantry fighting vehicle, the Black Hawk helicopter, the Apache attack helicopter and the Patriot air and ballistic missile defense system) and its AirLand Battle Doctrine turbocharging the extraordinarily rapid land advance.

Nevertheless, technological superiority alone wasn't enough, as America's experience in Vietnam had proven. The past decade and a half had necessitated advances in strategic and tactical thinking, as well as enormous rigor in training and investment in professional development of leaders. A desert setting and conventional combat operations, with virtually no civilians on the battlefield, had also proven much more advantageous for US forces than triple-canopy jungle and countering guerrillas and insurgents in a war among the people.

The Gulf War demonstrated that the US had learned several important lessons from Yom Kippur, the last great desert engagement. In discussing expensive new systems that had long development times, such as Joint Surveillance Target Attack Radar and the Army Tactical Missile System, the military historian and former four-star general Donn Starry noted that "All the things that flowed from that train of thought out of the 1973 War went into the front end of the Gulf and came out the back a success."[86] Lieutenant General Chuck Horner, who orchestrated the air campaign for Desert Storm, had taken part in a post-Yom Kippur War evaluation while on the Air Staff and recognized the dangers posed by dense integrated air defense systems such as the Egyptians had used in the opening stages of their attack in the Sinai. Then-Colonel Larry Henry, who headed up the electronic-warfare aspect of Desert Storm, also credited Israel's experiences in 1973 and 1982 as having inspired his concept of operations.[87]

In the wake of the four-day success in the ground campaign, America's era of long, humiliating and escalating conflicts appeared finally to be over. The Gulf War had been an obvious triumph of speed, training, superior technology, professional expertise and crushing American global dominance. Needless to say, however, future enemies were paying attention. Going head to head with the US military in open desert with conventional forces was clearly a losing prospect. So they looked to something else.

Somalia: setting a dangerous precedent

For twenty-two years, General Siad Barre had controlled much of Somalia in east Africa through dictatorial, repressive power. He had left no clear successor, and so, when he was overthrown in 1991, his land was left, like that of Julius Caesar, to the strongest. In the resulting power vacuum after Barre's ousting, Mohamed Farrah Aidid, a general in the national army, emerged as the leading warlord in a by then devastated Somalia. At the height of the famine that had resulted from the civil war, some 30,000 Somalians were dying each month. By the close of 1992, an estimated 300,000 had starved to death that year alone.

In the waning days of his administration, with dramatic scenes of starvation and humanitarian crisis playing out on CNN and other television networks, President George H. W. Bush, who had been defeated in his campaign for re-election by Governor Bill Clinton of Arkansas, deployed 25,000 American troops to a United Nations operation in Somalia. The country was in chaos and its citizens were trapped in the crossfire between brutal warlords. Food supplies were disrupted as UN troops were shot at, attacked and looted from, and people continued to starve en masse. A UN Security Council resolution of 3 December 1992 authorized a Unified Task Force to provide humanitarian relief.

The United States agreed to provide the majority of the Task Force's 38,000 troops on condition that American troops would not be under United Nations command. The mission was codenamed Operation Restore Hope and the I Marine Expeditionary Force from Camp Pendleton, California, was given the leading role. Navy SEALs and soldiers from the Army's 10th Mountain Division would follow. On 9 December, helicopters delivered 1,300 Marines to Mogadishu Airport, and Navy SEALs emerged from the surf off Somalia's capital Mogadishu.[88] Some UN units entered Mogadishu while others went directly to assigned territory elsewhere. While the conventional forces concentrated on the major cities, US special operations units moved quickly to establish a presence in the rest of the countryside.[89]

Bound by strict rules of engagement, US forces spent almost a year patrolling Somalia's cities and countryside, allowing for the distribution of food and other supplies to the population and gathering information. The situation improved significantly, and the United States, not wishing to engage in nation-building, pressed the United Nations to take on more of the burden, with only minor success as the US began to reduce its presence.[90]

Immediately after the US Marines had pulled out of Somalia, however, General Aidid began attacking the remaining UN peacekeepers, prompting the Clinton White House to implement Operation Restore Hope in August 1993. Restore Hope followed naturally on from the successful Operation Provide Comfort which had flown humanitarian aid to the Kurds from April 1991 and Operation Provide Hope that gave it to former Soviet republics in February 1992. "In those [first] five months, it worked pretty well," argued John L. Hirsch, an adviser to the US Ambassador and the Commander of the US mission. "People forget the early successes of Operation Restore Hope to feed the hungry and break the famine."[91]

On 3 October 1993, all that changed when US Army Rangers and Special Forces raided Bakara Market in Mogadishu and captured two of General Aidid's lieutenants. A Black Hawk helicopter was shot down nearby by an RPG-7, killing both pilots and leaving five survivors stranded. Two surviving special operations snipers began to defend the site. At this point, according to declassified US documents, the mission "irrevocably changed from one of capturing [Aidid's] supporters to one of safeguarding and recovering American casualties."[92] A second helicopter landed nearby to rescue the two snipers; Staff Sergeant Daniel Busch would die of his injuries, having taken four bullets while defending the site of the crash. Three others remained stranded.

Around ninety members of the assault team travelled to the crash site to perform a rescue operation. Special Forces operatives formed a defensive perimeter around the downed craft and engaged in a furious firefight with Somali militia, who brought down a second helicopter nearby. At the second crash site, snipers Master Sergeant Gary Gordon and Sergeant First Class Randy Shughart worked to keep back an attacking crowd. Both were killed by a thousand-strong mob of militiamen, who then ran to the site to beat and kill all the remaining survivors. Only the pilot, Michael Durant, survived – because Aidid's men had ordered that he be taken prisoner.[93]

The battle of Mogadishu lasted for eighteen hours. By the time it was over, eighteen Americans had been killed, with a further seventy-three wounded, and hundreds of Somali civilians had died. Durant was released by Aidid after eleven days in captivity, and Gordon and Shughart were posthumously awarded Medals of Honor – the first since the end of the Vietnam War. The streets of Mogadishu had become host to the fiercest urban combat experienced by US troops in almost two decades.

President Clinton, who had inherited the operation from his predecessor, was unnerved by the severity of events. The sight of a dead American being dragged through the streets of Mogadishu and another as a wounded prisoner of war sparked outrage in the United States, alongside questions as to why the operation was still ongoing and whether the soldiers had been adequately supplied. The criticism ultimately forced the resignation of the Secretary of Defense, Les Aspin. Soon afterwards, the President announced that the armed American intervention would end by 31 March 1994.

For a purely humanitarian mission, Operation Restore Hope was a costly endeavor. More than thirty US servicemen lost their lives and 172 were wounded. Smith Hempstone, the experienced Africa hand and US Ambassador to Kenya, had been prescient when he warned the State Department before the operation, "We ought to have learned by now that these situations are easier to get into than to get out of, that no good deed goes unpunished."[94]

After the overwhelming triumph of Desert Storm, in which a massive American force had been deployed to achieve a clear and limited objective, Operation Restore Hope represented a regression. Ironically, it was carried out while General Powell was still the Chairman of the Joint Chiefs and in seeming contrast to the conditions he had established for committing US forces. It was Somalia which had inspired one Washington columnist to coin the term "mission creep" to describe ever increasing military operations with hazy, moveable objectives and situations that could easily increase in scope. One retired general described it as "the policy equivalent of Post-Traumatic Stress Disorder."[95] Murky, muddled and unpopular, the intervention was a reminder that the United States military was designed more for fighting wars than for providing humanitarian relief.

As General Stanley McChrystal wrote:

The fight in Mogadishu was to have lingering effects on America, her special operations forces, and my experiences in the years ahead. Just as Grenada, Panama, and the first Gulf War had done much to erase the frustrations of Vietnam, Mogadishu carried a whiff of failure, a reminder that despite the progress we'd made since Eagle Claw [the attempt to rescue the US Embassy members held hostage in Teheran that failed in the desert halfway to the objective and turned around] thirteen years earlier, the possibility of death and defeat was always at hand.[96]

Six months later and 1,500 miles away, the Rwandan genocide was sparked by the downing of the plane of President Juvenal Habyarimana, an act which culminated in the slaughter of many hundreds of thousands of innocent people. Although the debacle in Somalia was never explicitly stated as the justification for American non-intervention in Rwanda, it was widely understood to have been the most important contributing factor.

Six

The New World Disorder

1991–1999

If the Cold War is truly behind us, the stability of the past
forty-five years is not likely to be seen again in the
coming decades.

John Mearsheimer, *International Security*, 1990[1]

The South Ossetian War, 1991–1992: a nasty, small nationalist conflict in the Caucasus

The collapse of the Soviet Union reverberated around the world in the
early 1990s and remolded the geopolitical order. No countries felt the
aftershocks of the disintegration more strongly than the former constituent republics of the USSR, several of which soon found themselves
riddled with internal dissent and restless nationalism. This was particularly true of Georgia, on Russia's southern border. Like its neighbors in
the mountainous Caucasus region, complex internal ethnic conflicts
and economic troubles would arise in the post-Soviet republic.

Georgia had been a possession of several empires down the centuries
and had enjoyed only a brief period of independence immediately after
the Russian Revolution, before being invaded by the Bolsheviks and
absorbed by the nascent Soviet Union in 1922. Its most infamous native

son, Iosif Besarionis Jughashvili, a trainee priest turned bank robber, would eventually rule over the USSR under his revolutionary pseudonym Joseph Stalin. Georgia's history with Russia had always been that of a reluctant stepchild disciplined by an authority to which it did not feel wholly submissive. As the Soviet empire began to totter in the late 1980s, Georgians saw their opportunity to throw off the yoke of Russian domination.

Zviad Gamsakhurdia, the Chairman of the Supreme Council of Georgia, was a poet and literary scholar, noted in the West as a brave dissident who had survived arrest and captivity during previous Soviet crackdowns. His nationalist anti-Russian rhetoric had, for a time, courted international sympathy. In a burst of opportunistic nationalist zeal, Gamsakhurdia intensified his longstanding political and rhetorical assault on South Ossetia, a semi-autonomous province in the north-central region of the country with a population of just under 100,000. More than two-thirds of the population were Ossetians, an ethnically Iranian, Persian-speaking minority with a strong cultural and historical affinity to Russia, with which the province was contiguous.[2] Gamsakhurdia and his supporters portrayed such pro-Russian minorities as fifth columnists who were being exploited by Moscow in order to sabotage Georgian independence.[3]

The government's rhetorical onslaught turned more serious with the abolition of South Ossetian autonomy, and fighting broke out on 5 January 1991. Small pockets of skirmishing would soon intensify into increasingly common instances of death, displacement and extreme violence. Georgian forces attacked the Ossetians in Tskhinvali, the provincial capital, but their poor training soon became apparent once the Ossetians resisted fiercely.[4] A dangerous moment for the Georgian Army arose when the Ossetians converted unguided air-to-surface missiles taken from the helicopter regiment stationed near Tskhinvali into improvised shoulder-launched bazookas for use in street fighting.[5]

Russian President Mikhail Gorbachev dispatched troops at the request of the South Ossetian authorities, and Georgian troops engaged them in sharp fighting. Georgian authorities accused the Russians –

probably accurately – of having previously covertly supplied arms to the Ossetians, and as the struggle progressed there was widespread devastation across Tskhinvali. The city divided into two warring zones, both of which tended to ignore ceasefire agreements – a phenomenon we have already seen in the Israeli War of Independence and will see repeatedly on the Russian side of the Russo-Ukrainian War in Chapter 9. As the fighting in South Ossetia spread, the ethnic Ossetians, supported by Russians, drove out the minority ethnic Georgian population from their homes.

Against the backdrop of the fighting, the people of Georgia voted overwhelmingly for independence from the moribund Soviet Union on 31 March 1991. Gamsakhurdia was appointed president of Georgia two weeks later and won an overwhelming electoral victory the following month. A Western darling was about to become a Russian nightmare.

Even as large nations were born around the Caucasus, small conflicts such as the one in South Ossetia continued, with terrible suffering and loss of life. The Georgians blockaded Tskhinvali with more than 17,000 troops, with tanks surrounding the city preventing food and medical aid from entering.[6] Perhaps as many as a thousand people were killed, and unchecked fires consumed many of the city's buildings.[7]

Gamsakhurdia's nationalist triumphalism helped his rise to power, but weakened his attempt to maintain his hold on it. The increasingly hysterical nationalist rhetoric and authoritarian repression of a man once imprisoned for his human rights work and political dissidence had now sparked increasing inter-ethnic conflict within the country, which resulted in a bloody coup against his government just before Christmas 1991.[8] He was forced to flee the country weeks later, and was replaced by the calming though corrupt Eduard Shevardnadze, who had been the USSR's Foreign Minister under Gorbachev. The violence and unrest within Georgia – quite separate from the South Ossetian War – continued for months until Gamsakhurdia's violent death on 31 December 1993, in circumstances that have raised suspicion. Reports at the time that he had committed suicide were somewhat undermined by the fact that two bullet holes had been found in his skull rather than the more traditional solitary one.

Some of the most violent acts of the South Ossetian War occurred just before its conclusion. Georgian forces killed three dozen South Ossetians, including women and children, as they were attempting to flee the capital city on 20 May 1992. Hardliners in Moscow used the incident to accuse Georgia of genocide, and helicopters with Russian markings bombarded Georgian-controlled villages.[9]

On 24 June, Shevardnadze and the new President of Russia, Boris Yeltsin, signed a ceasefire agreement that finally brought to an end a conflict that the new Georgian leader had described as "senseless and pointless." Although still legally part of Georgia, South Ossetia was granted de facto autonomy, but the uncomfortable balancing act would bring continued conflict in the years to come – with Russian military action in the region far from over.

The South Ossetian struggle is instructive because it showed the shape and tactics of Russian war-making to come, especially after Vladimir Putin reassumed the presidency in 2012. Several of the features that had been seen in the early 1990s in Georgia – nationalist chauvinism, ethnic cleansing, the ignoring of ceasefires, the deliberate targeting of civilians – were to recur once Putin embarked on his campaign to make Russia great again and would culminate in his invasion of Ukraine a full thirty years after the fighting in South Ossetia.

Former Yugoslavia: was ethnic hatred in the Balkans solved by aerial bombing?

The loose, discordant union of republics that came to be known as Yugoslavia was born in the aftermath of the First World War, an attempt to unify and stabilize a region that had been the genesis of the near-total destruction of Europe. It was on a street corner in Sarajevo that Gavrilo Princip, a young terrorist in the employ of shadowy elements of the Serbian government, fatally shot the heir to the Austro-Hungarian throne Archduke Franz Ferdinand on 28 June 1914, sparking a diplomatic crisis that in only five weeks fulfilled Otto von Bismarck's (possibly apocryphal) prediction that the next great European war would "likely come out of some damned foolish thing in the Balkans." It was one of history's little ironies that Ferdinand's assassination occurred on the anniversary of the battle of Kosovo in 1389, when the Serbs had battled invading Ottoman forces. The Serbs had been defeated, but 28 June remained a landmark date in the Serbian historical consciousness, a potent symbol of nationalist resistance and a legend whose intoxicating quality had only grown in the subsequent half-millennium.

The Austro-Hungarian Empire collapsed within hours of the First World War's end in November 1918, and several of its southern territories joined Serbia to create the new Yugoslavian kingdom after the Treaty of Versailles was signed the following year. Born out of violence, the new nation was torn by internal strife, including the assassination of its King Alexander I in 1934, and in 1941 Axis forces invaded. Approximately 4 million Yugoslavs died under the ensuing German occupation. A ferocious resistance led by Yugoslav partisans under the command of the communist Josip Broz Tito resulted in him being established as the dictator of the Socialist Federal Republic of Yugoslavia. Tyrannical but politically skilled, he preserved an uneasy balance among the constituent republics, maintaining relative stability for three and a half decades.

Tito's death in May 1980 marked the beginning of the end for Yugoslavia. The communist leader – cowed by neither America nor the Soviet Union – had been the one figure capable of maintaining stability in the Balkans. In the wake of his death, according to historian Tim Judah, "Communist power began to recede, or rather the fear of it receded, which amounted to the same thing."[10] The end of the Cold War and the subsequent collapse of the Soviet Union destroyed virtually overnight any ideological basis for the continued existence of a unified Yugoslavia.

Even before the Soviet Union collapsed in December 1991, therefore, Yugoslavia, a nation of almost 24 million people, had begun to break apart – at first splintering and then suddenly imploding. Amid an upsurge of nationalism and economic chaos, two of its six constituent republics, Slovenia and Croatia, declared their independence on 25 June 1991. The subsequent break-up of Yugoslavia would bring about a catastrophe that killed hundreds of thousands, led to international war crime tribunals and resulted in the displacement of 4 million people.

On 29 February 1992, the people of the Socialist Republic of Bosnia Herzegovina also voted for independence from Yugoslavia. Most Bosnians were Muslim, a legacy of Ottoman rule, but there were also substantial minorities of Orthodox Christian Serbs and Catholic Croats. The three groups had lived in relative harmony under Tito's rule, until atavistic hatreds were unleashed by the collapse of central authority and the cynical machinations of regional politicians. Ultra-nationalism was the new creed, with, in the words of the journalist George Packer, "The irresistible taste of bitterness, flavored with the sediment of ancient grievances, distilled to a dangerous potency that induced hallucinations of purification and revenge."[11]

Serbia was the dominant power in Yugoslavia, and its capital, Belgrade, was also the capital of the Federal Republic. The ostensibly communist Serbian President, Slobodan Milošević, was in practice a populist and nationalist who proclaimed himself the champion of Serbs throughout the crumbling Yugoslavia and urged them to take up arms. His ally in Bosnia, the leader of the Serbian minority Radovan Karadžić, a former psychiatrist, was his equally vicious sidekick.

Together Milošević and Karadžić would unleash carnage on an unimaginable scale, the most horrific bloodletting seen in Europe since the end of the Second World War. Their Croatian counterpart was Franjo Tudjman, President of Croatia throughout the 1990s. The historian Niall Ferguson has judged that their joint plan to partition Bosnia "was always genocidal in its intent."[12] Tudjman stated publicly that the former Yugoslavia would have "no Muslim part," despite Muslims accounting for nearly 40 percent of the population.[13]

In an ominous echo of the First World War, the Bosnian conflict began with the sound of gunfire in the streets of Sarajevo. On 6 April 1992, Serbian troops killed eleven Bosnians who were marching in support of their newly declared republic, the first victims of a conflict that would eventually claim the lives of over 100,000 of their countrymen.[14] Sarajevo itself, already battered by history, would suffer an epic siege that would impose a staggering human cost. It lasted from 3

October 1992 for 1,200 days, with an average of ten people dying per day. According to one foreign reporter embedded within the besieged city, "the two dominant sounds [were] of children playing and bullets flying."[15] Half of all children within the city were to witness at least one killing. At one point, the city authorities had to dig up soccer fields to make room to bury the corpses.[16] A particularly low point was reached when Serbian snipers fired on civilians queuing for water in the freezing winter.

In the spring of 1992, Serbian paramilitary death squads moved into Muslim-majority eastern Bosnia and began killing Muslims in horrific ways – burning alive, beheading on bridges with bodies flung into rivers – in order to terrorize the local population into leaving their homes.[17] Rape was, in the words of the UN Commission on Human Rights, "massive, organized and systematic," intended to "humiliate, shame, degrade and terrify the entire ethnic group."[18] The European Union estimated that some 20,000 Bosnian women were raped, while other sources estimate that the figure was more than twice that. "Sexual violence, in the form of rape and the seizure of women," records Jeremy Black, "was very much linked to a drive to dominate, humiliate and cause pain. Conflict in the former Yugoslavia, in Rwanda, the Congo, the Caucasus, the Sahel and Sudan all took this form in the 1990s and thereafter."[19] In 2022 sexual violence was used by the Russian Army against Ukrainian women for exactly the same purposes.

The exceptional brutality of the Serbs and Croats against their neighbors horrified those who might have hoped that Europe's time as a bloody battleground had ended in 1945. Those responsible for such acts were committing one of the most egregious examples of ethnic cleansing in the modern era. So appalling was the violence, and so racially targeted were its victims, that those ordering the attacks would be deemed guilty of genocide. The chilling phrase "ethnic cleansing" entered the lexicon to describe the Serbian campaign.

The United States recognized the new Bosnian state and airlifted supplies to Sarajevo until Bosnian Serb forces seized the airport. The US's Operation Provide Promise, under a UN banner, would average three deliveries a day made by C-130s of the 37th Airlift Squadron, flying from Germany and Italy. These relief missions expanded in early 1993 with airdrops of vital supplies to Bosnian Muslim enclaves surrounded by hostile forces, with the C-130s flying at high altitudes at night to avoid attracting enemy fire.[20] Colossal C-5 cargo aircraft were used to transport two 18-ton water-purification systems from Texas to Croatia, from where they were individually ferried to Sarajevo.[21]

The mission continued with airlifts of wounded Bosnians to safety and even the delivery of toys and other gifts before Christmas. There were no American casualties during this humanitarian mission, although several C-130s were damaged by artillery and other fire, and several pilots escaped injury or death by only narrow margins.[22] The operation formally ended on 9 January 1996 after the delivery of nearly 160,000 tons of supplies from twenty-one nations, 63,000 tons of which was carried by the US Air Force over three and a half years.[23]

Even as Operation Provide Promise was unfolding, another airborne operation was launched, this time under NATO auspices and dubbed Operation Deny Flight. Designed to enforce UN Resolution 816, which imposed a "no fly" zone over the region, Deny Flight began in April 1993, with missions flown by the 36th Air Force Wing based in Bitburg, Germany, augmented by carrier-launched Navy aircraft flying from the Mediterranean. Air Force F-16s shot down four Bosnian Serb aircraft on 28 February, and over the following months the mission destroyed numerous military targets on the ground.[24] Strict rules of engagement, however, and the overlapping and sometimes conflicting authority of UN and NATO leaders kept Operation Deny Flight from decisively changing the course of the war.

Although the Bosnian tragedy was taking place in Europe, and America had been severely dissuaded from international intervention by the disaster in Somalia, it was a sign of the United States' role as the sole remaining post-Cold War superpower that the West turned to it to provide the leadership and military might necessary to stop the genocide of the Bosnian Muslims.

As Operations Provide Promise and Deny Flight were under way, American political, diplomatic and military leaders were considering more expansive policies, including arming the Bosnians, but they were understandably cautious about entering such a well-known cauldron of seething hatreds, where no obvious American interests were at stake. With Russia historically seeing itself as the protector of the Serbs, no one wanted to see Bismarck's prediction become a reality for a second time in a century.

General Colin Powell resisted intervention in a place that did not fulfil many of his criteria as listed in the previous chapter. "My constant, unwelcome message at all the meetings on Bosnia was simply that we could not commit military forces until we had a clear political objective," he recalled in his memoirs.[25] As one historian put it more colorfully, Powell perfectly rationally "wanted to keep the military's powder dry for the next conventional war like Desert Storm – anything else would be mission creep, quagmire, another Vietnam."[26] Powell bristled when Madeleine Albright, the US Ambassador to the UN, snapped at him, "What's the point of having this superb military you're always talking about if we can't use it?"[27]

Yet, as General Wesley Clark later reflected, even the "principle of allowing the Bosnian Muslims in Sarajevo to acquire the arms to defend themselves was directly in conflict with the principles of remaining neutral, containing the conflict, and ameliorating its humanitarian impact."[28] The Western arms embargo on both sides had been sold as creating, as British Foreign Secretary Douglas Hurd put it, "a level playing field," but because the Bosnian Serbs had access to weapons whereas the Bosnian Muslims did not, the policy was instead creating a level killing field, something Hurd always refused to recognize.

The Clinton administration therefore grudgingly developed a half-hearted proposal known as "lift and strike," which involved lifting the arms embargo on Bosnia and engaging in limited airstrikes against Serb forces. The proposal was only coolly received in Europe, where leaders were increasingly exasperated by what they viewed as American timidity, despite not doing anything of note themselves.[29]

Only after the Serbs had intensified their assault on Sarajevo and captured 300 UN peacekeepers for use as "human shields" did Britain and France finally respond by forming a Rapid Reaction Force, complete with artillery and armor.[30] Air force cargo planes that had been carrying food and supplies were temporarily assigned to carrying troops as NATO and the United States edged ever closer towards a decisive intervention.

This did little to deter the genocidal campaign the Bosnian Serbs were enacting, however. The apogee of the horror unleashed by Milošević and Karadžić in the wreckage of Yugoslavia was the massacre in Srebrenica in Bosnia, which had been under siege by Bosnian Serb forces since 1992, despite being in a UN safe zone. On 11 July 1995, Bosnian Serb troops entered the town and began a murderous rampage, separating 8,000 Muslim men and boys from the rest of the town and slaughtering them over the subsequent days, as the Dutch-led UN forces stood by. Remains within mass graves are still being identified three decades later. The failure of UN peacekeepers to intervene in such a grotesque massacre was an international and moral catastrophe, described as "the darkest moment in international involvement in Bosnia" and "the ultimate in international humiliation."[31]

On 28 August 1995, the Bosnian Serbs mortared a Sarajevo marketplace, killing forty-three civilians. Footage of the grisly aftermath was beamed around the world, and Western leaders finally admitted that it was time to put a decisive stop to the Serbian killing spree. Secretary of State Warren Christopher announced that the response would not be "a pin-prick ... not just a bomb or two, not just a day or two, but as much as it took."[32] Western leaders, just four years on from the success of the Gulf War, had another clearly righteous cause to unite them.

On 30 August, only two days after the market outrage, NATO launched Operation Deliberate Force, an intensive bombing campaign employing 350 aircraft, mostly operating from Italian airbases. Three-quarters of the sorties were undertaken by the US Air Force; the rest by seven NATO allies.[33] On the first day of Deliberate Force, 300 sorties took out air defense targets, including ammunition dumps, radar installations, command posts, communication relay stations, surface-to-air missile (SAM) sites and anti-aircraft artillery units. Yet the unwieldy coalition of the United Nations and NATO began to creak after only two days, with Lieutenant General Bernard Janvier, the UN Commander in the theatre, requesting a temporary ceasefire after meeting with the leader of Bosnian Serb forces.

Assurances that the Bosnian Serbs would abandon their siege of Sarajevo were received with far greater enthusiasm (and gullibility) by the UN than by NATO. While the UN commanders wanted to give the Bosnian Serbs time to show that they would honor their promises to remove heavy weapons from the exclusion zone, NATO leaders were furious with the bombing suspension, believing that Janvier, in the words of a Brookings Institution report, "had been tricked into stopping the air campaign in exchange for more Bosnian Serb assurances, half-truths, deceptions, and outright lies."[34]

The Bosnian Serbs were at a severe disadvantage, but they were far from powerless. Recent experience had taught NATO planners that one well-placed SAM could destroy an F-16, and the enemy had many of those missiles, along with a thousand anti-aircraft artillery pieces and an unknown number of SA-7 portable air defense systems.[35] In addition, the Bosnian Serbs were well aware that NATO was walking a tightrope; one off-target bomb that struck a school or hospital and they would cease to look like the saviors of Bosnian Muslims.

The Bosnian Serbs' avidity for lies and deception, along with the naivety of certain UN leaders, were potent weapons in their arsenal. Despite the International Association for the Prevention of Genocide, Crimes Against Humanity and War Crimes applying to indict General Janvier for "repeatedly and systematically impeding the necessary assistance to protect both the safe area of Srebrenica and the populations present there" in 1995, he was appointed a Grand Officer of the Légion d'Honneur the following year.

This was, sadly, in keeping with history: the United Nations – so notable in its successful work with children, refugees and world health – has repeatedly and tragically failed in its original central task of abolishing state-on-state conflict. The inconclusive stalemate of the Korean War had marked both the beginning of a willingness to provide peacekeeping forces in many twentieth-century international conflicts and the impotence that would all too often accompany it. Justifications or explanations for this abound. The unchanging state of the permanent five members of its Security Council despite eight decades of global change must bear some responsibility, but the ultimate conclusion is clear. By the mid-1990s, many suspected the UN of hindering peace efforts in the former Yugoslavia rather than accelerating them. NATO taking over the Kosovo campaign from the UN in September 1999 would become a key moment in the history of both organizations.

The Bosnian Serbs' respite from bombing was brief: NATO attacks resumed on 5 September when it became clear even to the UN that the Bosnian Serbs had not kept their commitments to move heavy artillery pieces out of range of Sarajevo. NATO's focus remained on military targets, a soft-touch approach designed as much to maintain coalition unity as to defeat the besieging troops on the ground. The Americans also fired more than a dozen Tomahawk land attack cruise missiles at Serb targets in Bosnia from the USS *Normandy*. This additional firepower, while concentrated on military targets, represented increasing aggressiveness and determination on the part of the Clinton administration.[36]

In total, Operation Deliberate Force consisted of 3,400 sorties, including 750 attack missions against fifty-six ground targets.[37] The campaign could not continue indefinitely, however; the strict rules of engagement and restricted target list meant that NATO soon ran out of facilities to bomb. Fortunately, the tide had begun to turn against the Bosnian Serbs. Battered by NATO from the skies, their forces were unable to withstand fresh assaults from Bosnian forces on the ground, especially after the Croatian Army – which was now receiving American support – joined the fight against them.

All the warplane-dropped bombs in the world cannot shift the possession of a territory, of course, only boots on the ground can do that, but an effective land campaign can build on what is achieved from the air. So it was with the Bosnian and Croatian forces, as they swept through towns in western Bosnia in the late summer and early autumn of 1995, defeating the Serbs as they went.[38] For two weeks, even as diplomatic negotiations continued, destruction rained down from the skies and was exploited on the ground.

It is here we can see a most striking example of the evolution in warfare. The operation in Bosnia demonstrated the growing capability of airpower, as for the first time precision-guided munitions became more important than conventional bombs and missiles.[39] "In Desert Storm, only two percent of all weapons expended during the air war were precision guided munitions," a senior Defense Department official later observed. "In Bosnia, they accounted for over 90 percent of all ordnance expended by US forces ... The bomb damage assessment photographs in Bosnia bear no resemblance to photos of the past, where the target, often undamaged, is surrounded by craters. The photos from Bosnia usually showed one crater where the target used to be, with virtually no collateral damage."[40]

When the Bosnian Serb forces finally withdrew their heavy artillery from the hills around Sarajevo and lifted their long and terrible siege of the city, Operation Deliberate Force was suspended on 15 September 1995, although the lack of trust in Serb promises meant that it was not formally declared over until the 21st, while Bosnian and Croatian forces continued their offensive in the west.

Unfortunately, Western leaders took the wrong lessons from Operation Deliberate Force. As the journalist Misha Glenny observed, "Milošević did not acquiesce in the face of bombing – he wanted it to go ahead so that he might be relieved of the responsibility of bringing the Bosnian Serbs into line. The illusion that he had crumpled after a short sharp bombing campaign would lead to unreasonable expectations on the part of statesmen who should have known better, four years later in Kosovo."[41]

That said, the NATO bombing campaign did bring the Serbs to the table at the peace conference. Ambassador Richard Holbrooke, Assistant US Secretary of State for European Affairs, convened a conference in the somewhat unlikely setting of the Wright-Patterson Air Force Base near Dayton, Ohio. A three-week diplomatic marathon involving the presidents of Serbia, Croatia and Bosnia ensued, and on 21 November 1995 the negotiators reached an agreement, signing the Dayton Peace Accords in Paris on 14 December.

The diverse entities of the former Yugoslavia – a "country" through which ran ethnic and sectarian fault lines from centuries of conflict –

had been splintered into new nations divided along ethnic and religious lines, and it would splinter further in the years to come. People who had lived in peace and close proximity together for half a century had suddenly resorted to genocidal viciousness and horrific cruelty, as was also being seen contemporaneously in Rwanda. Bosnia was partitioned, with separate territory for Muslims (that also included Croats) and Serb Orthodox Christians. But peace in the Balkans would fail to last even half a decade: General Wesley Clark, who had aided Holbrooke in working for the Peace Accords, would soon return as NATO's supreme allied commander–Europe to oversee a very different kind of war in the region.

After the war in former Yugoslavia it seemed as if warfare had evolved so far and so fast that future conflicts would be won or lost almost on airpower alone, so long as one side was totally dominant there. All that boots on the ground would do was occupy territory that had already been won by the aerial bombing of the superior air forces. As we shall see in the next two chapters, however, it was not to be that simple, as the experience of Vietnam had already shown.

Rwanda: the war of the machete

In October 1993, only two days after the US Army's disaster in Somalia, a UN peacekeeping mission was established just over a thousand miles away in Rwanda. The United Nations Assistance Mission for Rwanda (UNAMIR) had a mandate to oversee the peace in a country plagued by internal racial tensions. UNAMIR consisted of contingents from Belgium, Bangladesh, Tunisia and Ghana, despite the UN rule that former colonial powers do not act as peacekeepers in their ex-colonies.

During the First World War, Belgium had taken control of Rwanda, and the Belgian colonists had favored the Tutsi tribal population over their Hutu counterparts, believing the former to be closer to Europeans in race and character. The introduction of compulsory identification cards in the 1930s had created a formal caste system that ended social mobility for Hutus. Furthermore, the two tribes looked different: Tutsis tended to be taller and slimmer with Ethiopian features, while Hutus were shorter and stockier, with darker skin.

By the early 1990s, the Belgians had been out of power for three decades but the damage done by their governance remained immense. A civil war had been fought within Rwanda in 1990, with an uneasy peace agreed in 1993. UNAMIR, unlike its counterpart in Somalia, arrived in Rwanda on a traditional goodwill mission with Hutus and Tutsis alike abiding by agreements, with UN troops there to oversee proceedings. This would be a mission solely concerned with keeping the peace – no one was permitted to engage in fighting. But the gentle aims of dispassionate peacetime management were left in tatters by the reality of the situation that greeted UN forces: not only was violence pervasive and burgeoning throughout the country, but radio broadcasts associated with the Rwandan government were actively urging Hutus to decapitate their Tutsi neighbors.[42]

Belgian peacekeepers of UNAMIR were notable in two regards. The best equipped and trained of UNAMIR forces, they also aroused severe suspicion in the Hutu population, who accused them of harboring historical pro-Tutsi bias. In January 1994, Canadian Brigadier General Roméo Dallaire, the Commander of the UN peacekeeping forces, received information from a half-Hutu, half-Tutsi informant who alleged that Hutu militias were being trained and armed with the intention of murdering Tutsis, moderate Hutus and those Belgian peacekeepers they believed favored the former two groups.[43] Dallaire relayed this information to UN headquarters, but was ignored. The UN Department of Peacekeeping Operations felt that Dallaire was overstepping the boundaries of his mission and, rather than ordering UNAMIR to raid the training camps, it endeavored to put Dallaire "on a leash."[44]

On the evening of Wednesday, 6 April 1994, a jet carrying President Juvénal Habyarimana of Rwanda and his fellow ethnically Hutu President Cyprien Ntaryamira of Burundi was shot down as it came in to land in Kigali, the Rwandan capital. The assassination ignited a conflagration. Within hours, Hutu roadblocks had been erected to prevent Tutsis fleeing, and moderate Hutu politicians were hunted down and killed. Eight hundred thousand people would be killed within the next hundred days – many of them hacked apart by machetes.[45]

The efficiency and scope of the violence were astonishing. Fourteen hours after the President's plane had been shot down, Agathe Uwilingiyimana, the interim head of state, was sexually assaulted and murdered. Her ten Belgian peacekeeper bodyguards were found dead, their bodies showing signs of torture. Early reports had falsely stated that the soldiers' genitalia had been mutilated and placed within their own mouths. The reality was little better: they had been tortured by machetes and beaten with rifle butts before being killed.

Dallaire requested permission from the UN to use lethal force to try to restore order, but the organization – fearful of replicating Somalia on a mass scale – refused. Continually declined pleas to try to prevent the genocide was not Dallaire's only problem; he was equally conscious of the fact that the UN forces stationed in Rwanda had severely limited capabilities, with rations, ammunition and fuel all in short supply. The Tunisian and Ghanaian contingents did not even have body armor.[46] Furthermore, tensions were developing within the forces themselves. Following the murder of its ten peacekeepers, Belgium announced that it would withdraw from the mission.

By mid-April, barely a week after the President's plane had been shot down, the United States promulgated its new official policy towards Rwanda, insisting that the entirety of the UN mission withdraw from Rwanda.[47] The lesson the Clinton administration had drawn from the debacle in Mogadishu was for minimal involvement on the ground in an African civil war. During the last ten days of April, the estimated death toll rose to half a million. Just weeks before the campaign of violence concluded, a multilateral force of French and Senegalese troops commenced Operation Turquoise, clearing the way for refugees to escape to nearby Zaire (today the Democratic Republic of the Congo). Only in late July – after the genocide had occurred – did the United States send troops to aid refugees.

With the backdrop of 800,000 dead, mass sexual assault, political terror and genocidal intra-ethnic violence, the troops were tasked solely with providing medical care to refugees in Zaire. When weighed against the deaths of 10 percent of the country's population in the space of little over three months, global inaction appears unconscionable.

Despite the well-publicized horrors, no country on earth had offered to send troops to Rwanda to stop the fighting.[48] The lack of clear, achievable aims and security guarantees was the fault of the UN, but the horror was probably beyond the capability of any one country – even the United States in its hegemonic heyday – to prevent. There is a moral conundrum here: to what extent are nations responsible because they are powerful, even when their own national interests are not directly threatened?

Those who had taken part in the Tutsi genocide now found themselves able to seize control of the camps to which the Tutsis had fled, and infiltrated foreign camps in order to launch fresh campaigns against the Tutsis of Zaire. This was recognized in Washington. Although refugee camps had become, in the words of President Clinton, "havens for the killers," he did not see that it was America's responsibility to do anything about it, and in terms of sheer Realpolitik he was probably right.[49] Yet great, civilized, liberal democracies such as the United States have never been actuated by Realpolitik alone.

Operation Turquoise, whatever its intentions, protected genocide perpetrators. In the words of one Rwandan diplomat, "the genocide continued even within the Turquoise zone."[50] The UN's humanitarian mission would therefore sow the seeds of the 1996–7 First Congo War, in which a quarter of a million would die. It wasn't just death that haunted the survivors left in Rwanda. The mass rape of Tutsi and moderate Hutu women – by one estimate, a half-million within three months – had cast an even longer shadow upon the fractured nation.[51] Though some women would self-induce abortions, an estimated 15,000 *enfants de mauvais souvenir* (children of bad memories) were born in the aftermath of the genocide. HIV and AIDS deaths also flourished – in the midst of the conflict, patients suffering from AIDS had been released from hospitals and recruited into "rape squads" so that Tutsi women who survived the genocide might suffer "slow, inexorable death."[52]

The comparison between Somalia and Rwanda that so heavily influenced US thinking was based on false equivalences. "The lessons we thought we had just learned in Somalia simply did not apply in Rwanda," Madeleine Albright admitted almost a decade later. "Somalia was something close to an anarchy. Rwanda was planned mass murder."[53] Richard Clarke, who served on the National Security Council, had directed a task force that had suggested creating a protection zone in Rwanda, an idea that the United Nations had rejected. "If the UN had adopted the U.S. proposal, we might have saved some lives," he stated later. "The U.S. record, as compared to everyone else's record, is not something we should run away from ... I don't think we should be embarrassed. I think everyone else should be embarrassed by what they did or did not do."[54] In the story of the Rwanda massacre, there is more than enough embarrassment to go around.

In this instance, the evolution of warfare proved more of a horrifying devolution. That the weapons used overwhelmingly in the genocide were simple, crude devices wielded in close face-to-face massacres made the scale of the resulting death toll so much more galling. Rather than eradicating enemies through extermination camps, machine guns or gas chambers, Hutu extremists had butchered hundreds of thousands – frequently their acquaintances and neighbors – with machetes. The Rwandan genocide represented a strange and terrible synthesis – warfare that had regressed to an almost prehistoric form, yet on a very modern mass scale. Conflict is constantly evolving, but not necessarily progressing.

Kosovo: warfare without casualties (on one side)

The death throes of Yugoslavia led to yet another bloody slaughter in the Balkans. Just three years after the Dayton Accords in 1995, Kosovo would play host to the most intense and sustained military operation that Europe was to see between the Second World War and the Russian invasion of Ukraine.[55] Conflict seemed to evolve in that NATO would fight the Kosovo campaign without any casualties, and it was certainly a far cry from what had happened in Rwanda, but its positive lessons failed to be learned by Vladimir Putin.

Kosovo was an ethnic Albanian, majority-Muslim enclave within Serbia that had been a possession of successive empires from the Roman era up until the collapse of the Ottoman imperium after the

First World War. Granted to Serbia in 1918, it enjoyed a degree of political and social autonomy as part of Yugoslavia under Marshal Tito. Following Tito's death, Serbia's President Slobodan Milošević exploited the grievances of the Orthodox Serb minority within Kosovo, who chafed under the majority rule of Muslim Albanians. In 1989 he ended the province's relative autonomy, imposed direct rule from Belgrade and purged Kosovan cultural, educational and media institutions of Albanian influence.

Harassed by Serb authorities, Kosovar Muslims launched a nonviolent campaign of civil disobedience, courting world opinion in the vain hope that moral pressure might sway the Serbian autocrat. Milošević resolutely ignored it. The failure of the Dayton Accords to address their concerns persuaded many Kosovars that force alone would achieve their hopes for independence from Serbia, and displaced Kosovars around Europe banded together to found the Kosovo Liberation Army (KLA). The resulting campaign against Serbian military targets sparked a predictably vicious campaign of reprisals against civilians.[56] In towns and villages around Kosovo, Serbian rocket attacks and mass executions claimed the lives of hundreds of civilians.

The potential strategic consequences were grave enough to force Western leaders to pay attention, in a way they had not needed to in faraway Rwanda. "If Kosovo exploded," noted Michael Ignatieff, "other countries could well go up in flames with it. Kosovo stands in the center of a combustible region which stretches from Italy, through Greece, across the Eastern Mediterranean to the Turkish border with Iraq."[57] The United States, conscious of its inaction in Rwanda and earlier bloodshed in Bosnia, did not want idly to witness another genocide.

On 23 September 1998, the United Nations Security Council passed Resolution 1199 that warned of an "impending human catastrophe" in Kosovo and demanded that all parties "immediately cease hostilities and maintain a ceasefire," a resolution that was ignored by Belgrade. Lackluster UN action in recent years had done nothing to deter those seeking to commit crimes against their own people. Richard Holbrooke, the key architect of the Dayton Peace Accords, was dispatched by the Clinton administration to negotiate with Milošević, who batted away any interference and told him, "Kosovo is a part of Serbia. It is a *domestic* problem."[58]

The town of Račak felt the full force of Milošević's domestic problem-solving on 15 January 1999, when forty-five civilians, including women and children, were executed in reprisal for KLA attacks. Serbian denials of the shootings rang hollow, especially after an examination of the corpses revealed that, as Tim Judah records, their "clothes were bloody, with slashes and holes at the same spots as their bullet entry and exit wounds, which argues against government claims that the victims were KLA soldiers who were dressed in civilian clothes after they had been killed. All of them were wearing rubber boots typical of Kosovo farmers rather than military footwear."[59]

Despite this atrocity, on 31 January NATO offered to cancel a new campaign of airstrikes against Serbia if Milošević agreed to negotiations at the Château de Rambouillet near Paris. The conference began a week later, but soon ended in failure. The Serbian Foreign Minister sounded a refrain with ominous contemporary echoes, afterwards dismissing the talks as "all about geopolitics not about human or minority rights ... It was a pretext to expand NATO and the US presence to south-eastern Europe." Later he defiantly declared that his country "has never capitulated in all her history. Some think that history is a burden but we think it is a teacher."[60]

Holbrooke persevered one more time, warning Milošević on 22 March about the impending NATO assault. The Serbian leader admitted that he knew the consequences of his continuing campaign of ethnic cleansing in Kosovo, replying simply: "Yes, you will bomb us." As Holbrooke later recalled, "There was a long silence in the room ... I said, 'I want to be clear with you, it will be,' and I used three words I had worked out very carefully with the US military, 'it will be swift, it will be severe, it will be sustained.'"[61]

Sure enough, when NATO's Operation Allied Force started on the evening of 24 March, it boasted a force of more than 250 American military aircraft, as well as substantial contributions from the other member states. That night, B-52 bombers dropped their deadly payload of AGM-86C cruise missiles. Tomahawk missiles soon followed, launched from US Navy vessels in the Adriatic. The early strategy of an all-out forty-eight-hour assault followed by a pause to allow Milošević to capitulate was – considering Serbia's pugnacious stance – abandoned in favor of a "gradual, incremental, and phased approach."[62] The Americans in particular were confident that in the face of the NATO onslaught Milošević would quickly cave in. Madeleine Albright blithely stated in an interview, "I don't see this as a long-term operation."[63]

Yet the bombing campaign had to continue for over two months, partly because NATO failed to choose between focusing on Serbian command and control in the Belgrade area and, as the Supreme Allied Commander in Europe Wesley Clark wanted, focusing on the Serbian field army that had been responsible for the war crimes and the expulsion of the Kosovars.[64] More airstrikes followed, with planes launched from all over Western Europe to attack radar installations and other military targets.

The green, mist-shrouded valleys of Serbia and Kosovo seemed far from the Middle Eastern setting of the last great air campaign. As in the Gulf War, Serbia bristled with SAM batteries that could have wreaked havoc on an attacking air force. A month before the coming NATO air offensive, Serbian defense specialists travelled to Baghdad to discuss with their Iraqi counterparts how to try to defend against it.[65] To protect their planes from attack, NATO pilots were ordered not to fly below 15,000 feet (although not all of them obeyed). There would be almost zero tolerance for NATO casualties in this campaign, and the higher the altitude the greater the risk of collateral civilian casualties on the ground.[66]

The Serbian military threat was not restricted to ground-based systems. Milošević had a substantial, if somewhat antiquated, air force at his command, including, as the Iraqis had in the Gulf War, MiG fighters bought from the USSR. These proved no better at interdicting the Western air war than Saddam's air force.

Although Ronald Reagan had built up the US armed forces in the 1980s, the fall of the Berlin Wall had meant that politicians had been quick to capitalize on what was shortsightedly being dubbed the "peace dividend." In the decade after 1989, the US armed forces lost 36 percent of personnel, and the percentage of GDP spent on defense dropped from 6 percent to 3 percent.[67] Ignatieff believes that this at least in part explained why NATO engaged Serbia only from the relative safety of the air. "One reason why ground forces were not committed in Kosovo was that the US lacked the type of joint, mobile, rapidly deployable expeditionary force necessary for the task," he writes. "And even if such a force had been available, in the new political climate in which Western nations go to war, the military cost of a ground operation would have been prohibitively high. To some extent, America and its NATO allies fought a virtual war because they were neither ready nor willing to fight a real one."[68]

The campaign was led by General Wesley Clark, a cerebral soldier who had graduated top of his class from West Point and won a Rhodes Scholarship to Oxford. He was used to speaking directly. "If they tell me to bomb you," the Arkansas-born general told Milošević at one pre-war meeting, "I'm going to bomb you good."[69] As an American officer and Commander of US European Command, Clark was responsible to President Clinton as his commander-in-chief, but as the NATO supreme commander he was also accountable to the Secretary-General and the leaders of the member nations. The overcrowding within the operation's strategic decision-making would on occasion hamper him.

To Clark's great frustration, for example, political considerations entirely removed the possibility of any ground troops being used, despite his belief that the mere threat of it would change minds in Belgrade. Yet even if the Clinton administration had been willing to contemplate such a potentially unpopular and expensive escalation, the time taken in transporting and assembling the kind of three-to-one invasion force of Desert Storm would have allowed Serbia plenty of time to complete its campaign of ethnic cleansing against the Kosovars, thus undermining Clark's earlier clear threats of intensive, decisive and sharp attacks. Further complicating matters was the fact that the campaign would have to be fought with exquisite care to maintain the unity of a large and unwieldy coalition of allies, a situation reminiscent of the Korean and Gulf Wars.

There were two chief mechanisms for retaining coalition unity. Smaller munitions were used so as to minimize civilian casualties, and lawyers (known as "Judge Advocate General Officers") were employed to review potential targets to ensure that their destruction was consistent with the Geneva Convention.[70] Both factors, and the smaller number of aircraft available, made the job of the pilots and their commanders even harder than normal, and meant that it took NATO twelve days to complete the same number of strike sorties as had been conducted during the first twelve hours of Desert Storm.[71] From the beginning to the end of the decade, increasing paperwork, caution, legal issues known as "lawfare" and fragility in coalition relationships had dramatically changed the manner in which operations could be conducted.

Milošević's characteristically brutal response to the campaign was to expel nearly 2 million Kosovars from their country, creating a refugee crisis on a scale reminiscent of the closing scenes of the Second World War, a tactic that would be used with increasing frequency across Europe in the decades to come. His goal was both to force a humanitarian catastrophe on NATO countries that would affect their logistics and to remove the civilian population from which the KLA received shelter and support.[72] Simultaneously, he highlighted Serbian civilian casualties in Belgrade to try to evoke sympathy for Serbia. "Instead of fighting NATO in the air," as Michael Ignatieff put it, "he fought NATO on the air-waves."[73]

Despite this, NATO increased the intensity of its attacks on Serbian targets, and a race developed between the Serb forces driving the ethnic Albanians out of Kosovo and the NATO forces trying to stop him.[74] "I could actually see them burning houses," an American pilot later recalled. "It was extraordinary and horrifying."[75] Moreover there was no established front line, as in most wars, with front lines coming to be a matter of whichever villages were being attacked or defended by particular ethnic groups.[76]

The lack of a ground component to the campaign meant that it was harder to identify targets and call down air attacks on them, with Serbian forces enjoying flexibility in dispersing and concealing personnel and vehicles, using camouflage and other ruses. NATO pilots would strike what looked like a tank from the air, only to watch an inflatable rubber decoy collapse like a deflating bouncy castle.[77]

With relatively little to show for the initial stages of the air campaign and the situation materially worsening for the Albanian Kosovars, General Clark sharply increased the number of combat aircraft in theatre. In all, nearly a thousand aircraft, about 600 of which were US Air Force or Navy, were committed to the campaign.[78] He also requested the deployment of Apache attack helicopters to nearby Albania, which was granted with the greatest reluctance due to their vulnerability to Serbian SAMs and their association with ground warfare, the very idea of which was still anathema to his political masters.[79]

On 7 May a serious blunder led to a major diplomatic crisis when a NATO B-52 dropped three Joint Direct Attack Munitions at a building that had been identified as the headquarters of the Federal Procurement and Supply Directorate in Belgrade. In fact, the map being used was out of date, and the building destroyed was the Chinese Embassy, where four people died.[80] The Chinese said the precision munitions hit the Ambassador's office, the intelligence office and the attachés' office. As we have seen in this book, all wars are profoundly political, but this one was more than most, and bombing the Embassy of a powerful neutral country was humiliating for NATO and caused the US serious problems with China for years to come.

In the face of continuing Serb intransigence, the bombing campaign intensified, and Clark was finally permitted – or directed, depending on the source – to hit the kind of targets that he later asserted he had been eager to strike from the beginning. However, those targets had not been submitted at the outset and General Hugh Shelton, the Chairman of the Joint Chiefs of Staff, later described what Clark had initially developed as a "very weak battle plan, one without a strategic plan and corresponding targets" – a concern, Shelton wrote, that was "echoed by apprehensive calls ... from counterparts in England, France, Germany, and Italy," as well as from the US Air Force Chief of Staff.[81]

As Clark described it in his memoirs, *Waging Modern War*, he was struggling with the "divide between those in Washington who thought they understood war, and those in Europe who understood Milošević, the mainsprings of his power, and the way to fight on this continent."[82] Some in Washington recall it differently, however, with General Shelton noting that he felt compelled to develop an outline of the bombing campaign for Clark; he then flew to Clark's headquarters in Casteau, Belgium where, together with Shelton's four major NATO counterparts and the US Air Force Chief of Staff, they would review the revised concept. According to Shelton, Clark adapted his campaign superbly to what had been sent from Washington.[83]

Regardless of how it evolved, the bombing campaign now had coherence and purpose, and bridges, television stations and economic targets fell under NATO attack, in addition to military facilities and Serbian factories. More powerful munitions were employed, with the capacity to disable airfields and other Serb military facilities. The Serbian power grid also came under sustained assault, leaving much of the country in darkness. Even the Clinton administration's firm stance against using ground troops began to waver, partly thanks to the encouragement of Tony Blair, the British Prime Minister, and additional forces began to deploy to Albania. An earlier display of such resolve might possibly have brought the campaign to a speedier conclusion with fewer casualties and with more Kosovars remaining in their homes.

In early June, Shelton was able to get President Clinton to persuade the French to lift their objections to bombing strategic targets near Belgrade. "Less than forty-eight hours later," Shelton recalled, "Milošević hoisted the white flag and the war was over – the first time in history that a war was won by airpower alone."[84] Beforehand, as Clark ruefully noted, it had been "the only air campaign in history in which lovers strolled down riverbanks in the gathering twilight and ate at outdoor cafes and watched the fireworks."[85]

On 3 June, despite his army still being in the field, though much reduced, Milošević signalled his intent to discuss peace terms. The combination of increased air attacks closer to home and the threat of a possible ground campaign convinced him that he had little alternative. Even the Russians, unmoved by brutal attacks on civilians before and since, had finally withdrawn their tacit support of their long-term former ally. With NATO keeping up its bombing campaign throughout the negotiations, the Serbian government agreed to withdraw all its forces from Kosovo and to accommodate NATO peacekeeping forces to help with the return of refugees.

At 3:36 p.m. on 10 June 1999, after 34,000 sorties by 1,500 NATO airmen over seventy-eight days, and with no NATO casualties, the campaign came to a halt. The precise number of civilian casualties will never be known but is estimated at around 500. Tragically, a number of these were Albanian Kosovar refugees who were probably being used by the Serbians to screen their movements.

The total lack of NATO casualties, though a triumph, raised troubling questions in the minds of some observers. As Michael Ignatieff put it, the campaign resurrected "an ancient spectre in modern form: violence which moralizes itself as justice and which is unrestrained by consequences ... If one side of a future conflict is shielded from the reality of war and its consequences, why should it continue to be guided by restraint?"[86]

One incident in the immediate aftermath is worth mentioning. On 12 June 1999, without warning, Russian forces suddenly descended on and seized Pristina Airport in Kosovo, in a warning to NATO that also demonstrated their strategic interest in the region. A tense standoff developed, not only between General Clark and the Russians, but also between Clark and the British Commander of NATO's Kosovo Force, Lieutenant General Sir Mike Jackson, who saw the action as "a reminder that the Russians were still players on the world stage, that they still needed to be treated with respect." Clark, fearing something more profoundly sinister, wanted Jackson to block the runways to prevent any further Russian incursions, but Jackson refused, saying, "Sir, I'm not going to start World War Three for you." After a series of tense calls between Clark, General Sir Charles Guthrie, the UK Chief of Defence Staff, General Shelton in the Pentagon and the US National Security Advisor, Clark was instructed not to confront the Russians. Subsequent developments would validate Jackson's judgment, as the Russians later left the airfield and it was quickly put to use by NATO forces.[87]

In Kosovo, the apostles of airpower had another significant opportunity to demonstrate the ability of their fighters and bombers to achieve victory from the skies. President Clinton's caution (and acute political sense for what an American electorate would tolerate) made a ground campaign an impossibility. The result was a distilled version of combat that was almost clinical and sterile in its execution – at least to those outside the crosshairs. NATO involvement rendered the campaign more brittle than one that would have been carried out by the United States alone, although it was the Americans who flew the vast majority of the sorties and provided the most demanding capabilities, including combat search and rescue.

The Kosovo campaign was a success and halted the wholesale massacre of ethnic Albanian Kosovars by Serbian forces. Indeed, the Albanian refugees flooded back to their homes in what became the largest spontaneous return of refugees in Europe since the Second World War.[88] "We should be careful," Clark recorded, however, "about extolling the air operations in Kosovo as a pattern for future success." His caution was particularly apposite, as it had been the threat of Kosovar ground forces attacking from Albania into southern Kosovo that had compelled Serbian forces to mass in the final days of the air campaign, during which airpower proved most devastating.

In the years that followed, Milošević would be overthrown and captured at his villa in Belgrade after a thirty-six-hour standoff. Radovan Karadžić would be hunted for thirteen years and was eventually found masquerading as a spiritual healer under the name of Dragan Dabić, with long hair and a strange cover story. He offered acupuncture services and medallions that protected against harmful radiation in the air. "Dabić" claimed that all that was needed to cure male infertility was his own healing hands placed near the ailing appendage.[89] He had become overconfident – sitting in bars underneath official portraits of himself from a decade before – and was captured in 2008 and taken to The Hague, as Milošević had been before him, to face the International Criminal Court, where both men received life sentences.

The vital importance of seizing and retaining dominance in the air had of course been established during the Second World War, in such engagements as Operation Overlord, when the Luftwaffe managed to fly only 319 sorties over Normandy on D-Day, against the Allies' 13,368. It was to be further emphasized in almost all succeeding conflicts except those fought in jungles. Yet dominance in the air does not obviate the need for troops on the ground. Just as the Kosovars forced the Serbs to mass their strength, leaving them vulnerable to devastating air attack, in Afghanistan in 2001 the United Front forced the Taliban to mass with similar results. The centrality of air dominance in warfare is unquestionable, and in some engagements such as the Kosovo War it can even deliver victory.

Nor does air dominance go to the side with the larger air force; rather it goes to that with the superior (and faster) warplanes and the better-trained and motivated pilots and ground crews. This has been seen time and again in post-1945 conflicts, but especially in the Arab–Israeli wars – in one engagement in 1982 the Israeli Air Force shot down ninety Syrian planes for the loss of two of their own – as well as the Indo-Pakistan War of 1965 when the Pakistani Air Force won despite being outnumbered four to one (though that was not enough to prevail overall), and with the British Harrier fighters' performance in the Falklands.[90]

The lesson to be learned is that quality tends to trump mere quantity in the air, giving a vital competitive edge to those countries or alliances that can produce the state-of-the-art warplanes and their all-important spare parts. A pilot requires more than two years of training to fly an F-35 to combat-level efficiency, something very few air forces in the world can provide. The United States must remain capable of building the best fighter jet in the world, and at scale, and must be able to train the best pilots (or remotely and algorithmically piloted systems) to fly them. The future of Western civilization depends on many factors, but that is undoubtedly one of them.

Seven

The War in Afghanistan

2001–2021*

Never, never, never believe any war will be smooth and easy, or
that anyone who embarks on the strange voyage can measure
the tides and hurricanes he will encounter. The Statesman
who yields to war fever must realize that once the signal is
given, he is no longer the master of policy but the slave of
unforeseeable and uncontrollable events. Antiquated War
Offices, weak, incompetent or arrogant commanders,
untrustworthy allies, hostile neutrals, malignant Fortune, ugly
surprises, awful miscalculations – all take their seats at the
Council Board on the morrow of a declaration of war.

Winston Churchill, *My Early Life*, 1930[1]

O n a beautiful late-summer morning on 11 September 2001, nineteen men belonging to the terrorist group al-Qaeda hijacked four
civilian airliners flying from US airports. Two were flown into the World

---

* This chapter was written primarily by Gen. David Petraeus, with input from Andrew
Roberts. Since Gen. Petraeus was a principal figure in the history of the war in
Afghanistan, events in which he was involved will be narrated in the first person.

Trade Center towers in New York City and a third into the Pentagon in Arlington, Virginia. Passengers aboard the fourth aircraft, likely heading for the US Capitol building, fought back, causing the hijacker piloting the plane to crash it into the ground near Shanksville, Pennsylvania.

The incident marked the first attack against the US homeland or its territories since Japanese forces invaded Alaska in June 1942. The brazen assault killed nearly 3,000 people and shocked the American public, who burned with a fierce desire for revenge. In response, the Bush administration demanded that the Taliban regime governing Afghanistan surrender Osama bin Laden, the leader of al-Qaeda, who enjoyed sanctuary in that country. When the Taliban leaders declined to extradite him for trial in the United States, President George W. Bush ordered the Central Intelligence Agency (CIA) and the Department of Defense (DOD) to plan an invasion in order to bring the leaders of al-Qaeda to justice and to eliminate their sanctuary in Afghanistan. Little did anyone realize then that the terrorist attack on the United States would result in a twenty-year conflict – the longest war in American history.[2]

Planning the invasion[3]

Whether Americans were aware of it or not, the United States had been at war with al-Qaeda since at least 1996, when Osama bin Laden issued a *fatwa* declaring war "against the Americans Occupying the Land of the Two Holy Places [Mecca and Medina]."[4] The CIA certainly knew it, as the Agency had been fighting a "ghost war" against the terrorist organization since the early 1990s.[5]

Following the Soviet invasion of Afghanistan in late 1979, the Agency had worked closely with Pakistan to arm the Mujahideen (some of whose members later established al-Qaeda), but given the close association between the Taliban, al-Qaeda and the Pakistani Inter-Services Intelligence (ISI), the Agency needed another ally to help bring bin Laden to justice. They found one in Uzbekistan, whose government in 1999 had agreed to allow establishment of a CIA counter-terrorism training team and a communications monitoring station. They discovered another ally in Ahmed Shah Massoud, in Afghanistan's Panjshir Valley, north-east of Kabul. Massoud was a formidable warlord who was a prominent leader in the United Islamic National Front for the Salvation of Afghanistan (United Front, also known as the Northern Alliance), a group of Tajiks, Uzbeks, Hazaras and even some Pashtun tribes fighting the Taliban.[6] Although Clinton administration support for the United Front was tepid at best and Congress was uninterested in events in Afghanistan in the 1990s following the withdrawal of Soviet forces, the connections established with the United Front by the CIA would become critical after 9/11. In the meantime, the CIA Counterterrorism Center developed plans to supply the United Front with the wherewithal to battle the Taliban, if a day came when political support from Washington might be forthcoming.

That day came on 12 September 2001, when CIA Director George Tenet and Counterterrorism Center Director J. Cofer Black briefed the National Security Council (NSC) on the CIA concept of using Agency paramilitary teams and special forces to partner with Afghan irregular forces, should the Taliban decline to extradite Osama bin Laden. That weekend, President George W. Bush convened an extraordinary meeting of the National Security Council at Camp David, the presidential retreat. Tenet provided more details on the plan he and his team had put together. The groundwork had already been laid by covert CIA contacts with Massoud in the years leading up to 9/11 and by considerable preliminary planning during those years – though General Hugh Shelton, the Chairman of the Joint Chiefs of Staff (for whom I served as executive officer from 1997 to 1999), informed the group that any plan that involved US military boots on the ground would have to be developed from scratch, as none existed. Undeterred, the next day the President ordered the CIA and the DOD to begin preparations for an invasion.[7]

The immediate goals of the resulting operation – to be known as Operation Enduring Freedom (OEF) – were to topple the Taliban and destroy al-Qaeda. After those goals were accomplished, the Bush administration would focus on stabilizing Afghanistan and creating a free society ruled by a democratically elected government. Ultimately, only the first goal would prove achievable given the constraints imposed on the mission by successive US administrations. The second goal would prove elusive due to the limited cooperation that Pakistan would provide, while the third goal entailed the sort of nation-building that the President and his advisers had forsworn before coming into office and were reluctant to undertake. Moreover, OEF was to be "the opening campaign of a protracted, global conflict" against terrorist organizations with global reach and their state sponsors, and this would limit the resources the DOD was willing to commit to Afghanistan.[8]

Geography and the ethno-sectarian makeup of Afghan society would also seriously hinder the proposed US operations in Afghanistan. The Sunni Pashtuns, located in the south and east, were the dominant ethnicity and provided the majority of the support for the Taliban. Sunni Uzbeks and Tajiks dominated the mountainous north, while Shi'a Hazaras resided in central Afghanistan. Several other ethnic groups were scattered across the country. The United Front had almost no presence in the Pashtun-dominated areas of Afghanistan, making operations there problematic. The land-locked country could also initially be approached via only two main lines of communication running through Pakistan, one into eastern Afghanistan (through the Khyber Pass) and the other through southern Afghanistan. The only other option at the time was to go via airlift into Karshi-Khanabad airbase in Uzbekistan and from there on the ground or by helicopter into northern Afghanistan, which was separated by the Hindu Kush Mountains from Kabul and the south. Pakistani support was thus essential, and Bush pressured Pakistani President Pervez Musharraf to support US operations in Afghanistan, which Musharraf did, at least initially. Over time, however, US–Pakistan relations would prove complex, difficult and, at times, quite fractious and frustrating, as Pakistani leaders grew fearful of Indian penetration into the territory of their western neighbor (a vastly overblown fear). They valued the strategic depth they believed they gained from Taliban rule in Kabul, and thus allowed sanctuary on Pakistani territory for the Taliban and their associated insurgent groups.[9]

Given the time required to plan for the use of conventional ground forces and transport them into Central and South Asia, OEF would begin with a small deployment in late September of CIA paramilitary elements and special forces teams to link up with United Front fighters and support them with substantial airpower. The assumption was, however, that the coming winter would largely bring fighting to a halt, and by spring 2002 enough conventional forces would be in place to enable a more robust campaign in Afghanistan if that proved necessary. By then, the United States and the United Front would presumably have identified anti-Taliban partners among the Pashtun tribes in southern Afghanistan. After the defeat of the Taliban, US Central Command (CENTCOM) assumed that it would take three to five years to stabilize and rebuild the country. Nonetheless, only a small residual presence in Kabul was envisioned after the fall of the regime, for Secretary of Defense Donald Rumsfeld and General Tommy Franks, the CENTCOM Commander, were wary of placing large numbers of forces in Afghanistan, believing that doing so would trigger the kind of indigenous opposition that had stymied the Soviets in the 1980s.[10]

The United Front prevails

The first CIA team, led by Gary C. Schroen, a legendary Agency officer nearing retirement, arrived on the ground in Afghanistan on 26 September and linked up with Mohammed Qasim Fahim, a United Front commander operating in the mountains north-east of Kabul.[11] On 7 October, US and British aircraft and ships began airstrikes and cruise missile attacks on targets in Afghanistan, supported by four aircraft carrier battle groups operating from waters within range of Afghanistan. On the night of 19–20 October, the first special forces teams flew into northern Afghanistan, one linking up with Schroen and United Front forces in the Panjshir Valley, while the other linked up with a CIA team that had made contact with the United Front militia leader Abdul Rashid, aka Dostum.[12] The special forces and elements of the 160th Special Operations Aviation Regiment that supported their infiltration were grouped under Combined Joint Special Operations Task Force

(CJSOTF) Dagger, created around the headquarters of the 5th Special Forces Group, commanded by the very capable Colonel John P. Mulholland.[13]

US Army, Marine Corps and special operations units now descended on Central Asia and Afghanistan, their deployments slowed only by the immature base structure at Karshi-Khanabad and the other bases in the region. On 19 October, US Army Rangers deployed via helicopter off the USS *Kitty Hawk* in the Arabian Sea south of Pakistan and raided two objectives near Kandahar, a Taliban stronghold in southern Afghanistan. Three days later, special forces called in devastating airstrikes on Taliban positions on Bagram Air Base, a notable demonstration of US airpower in action. Between 21 and 25 October, other Taliban positions were hammered with airstrikes while Dostum's horse-mounted warriors charged Taliban positions.[14] These operations created the template for the following weeks, as United Front militiamen provided the ground elements that forced the Taliban to mass their forces and special forces-directed US airpower annihilated the Taliban positions, with the United Front then occupying the newly won positions.

On 9–10 November, after massive airstrikes had devastated Taliban positions in northern Afghanistan, forces led by Dostum and rival chieftain Mohammed Atta Nur seized Mazar-e Sharif, the first major city to fall to the United Front. It quickly became a hub for distribution of humanitarian assistance flowing into northern Afghanistan from the United States and other Western countries. The fall of Mazar-e Sharif also opened the floodgates of United Front cooperation, with nearly all of its various groups clamoring for US special forces teams and air support. Events now accelerated quickly, with all major cities in the northern part of the country quickly falling to United Front forces. The focus of operations then shifted to Bagram, site of the major airbase just north-east of Kabul and a gateway to the Afghan capital. After twenty-four hours of concentrated airstrikes that destroyed twenty-nine tanks and inflicted several thousand casualties on the Taliban, United Front militia advanced to within a few miles of the capital. The Bush administration wanted to delay the occupation of Kabul until an international peacekeeping force arrived, but that desire was overtaken by events on the ground. On the night of 12–13 November, the Taliban abandoned Kabul without a fight and began a withdrawal towards their heartland in Kandahar, in the southern part of the country. The remaining organized Taliban presence in the north was confined to Kunduz, north of Kabul, which fell to the United Front on 24 November after Pakistani aircraft ferried senior Taliban leaders and several hundred Pakistani citizens out of the area.[15]

Eliminating the Taliban presence in southern Afghanistan would prove more difficult, as the United Front lacked a robust presence in that part of the country. Nevertheless, the anti-Taliban militias in the area proved sufficient when coupled with US special forces, Marines and airpower. On 14 November, US forces and a small CIA contingent flew into the south and linked up with a group of anti-Taliban fighters north of Kandahar organized by Hamid Karzai, who was khan of the Popalzai Pashtun tribe and whose father had been assassinated by the Taliban. The group entered the village of Tarin Kowt and three days later drove off a counter-attack, killing 300 Taliban fighters with the help of concentrated airstrikes launched from the USS *Theodore Roosevelt* in the Arabian Sea. The group then slowly advanced towards Kandahar, calling in airstrikes on the Taliban they encountered along the way. Another group of US forces and CIA officers inserted into an area south-east of Kandahar near the Pakistani border on 19 November linked up with Gul Agha Sherzai, a former governor of Kandahar Province until ousted by the Taliban, and his 800 fighters, who then moved north-west towards Kandahar, their path cleared after a week of airstrikes eliminated significant Taliban opposition.[16]

Commanding the growing force in Afghanistan proved to be a significant challenge. A Combined Forces Land Component Command (CFLCC) headquarters, commanded by Lieutenant General Paul T. Mikolashek, became operational on 20 November in Camp Doha, Kuwait, but it was not sufficiently staffed to be a war-fighting headquarters; it was also too far from Afghanistan and had broader responsibilities throughout the CENTCOM area of responsibility. Therefore, on 23 November, Franks approved Mikolashek's plan to deploy a portion of the US Army's 10th Mountain Division headquarters to Karshi-Khanabad airfield to establish a forward headquarters. Though ably led by Major General Franklin "Buster" Hagenbeck, the small headquarters (much of the 10th Mountain Division's staff was deployed on an operation in Kosovo) would prove insufficiently robust and inadequately empowered to handle the multiple command-and-control challenges that would emerge over the next several months when it was moved to Afghanistan; in particular, it would not be given the authority needed to control the operations of the various special operations units.

On 25 November, the 15th Marine Expeditionary Unit (MEU), part of the 1st Marine Expeditionary Brigade commanded by Brigadier General James Mattis, flew into Objective Rhino, a dirt airstrip southwest of Kandahar that had been seized in an airborne operation by US Army Rangers a month earlier and had since been overwatched by US Navy SEALs. CH-53 helicopters ferried Marines and their equipment from the USS *Peleliu* in the Arabian Gulf, followed a couple of days later by Seabee Navy construction units that transformed the airstrip into a forward operating base (FOB). From there, mounted patrols fanned out to reconnoiter Taliban and al-Qaeda concentrations and contact local tribes. In the second week of December, the 26th MEU flew from the USS *Bataan* into newly established FOB Rhino and from there moved to occupy Kandahar Airport and to construct a second FOB in country.[17]

By early December, the Taliban were in full retreat. As the collapse of the regime approached more rapidly than had been anticipated, the US State Department quickly organized a conference in Bonn, Germany, with representatives from the United Front and three influential groups of Afghan expatriates (the Rome, Cyprus and Peshawar Groups). Foreign representation included delegates from India, Pakistan, Iran, Russia and several European countries. The United States was represented by special envoy Ambassador James Dobbins, Deputy Assistant Secretary of Defense for Near East and South Asian Affairs William J. Luti and senior National Security Council staff member Zalmay Khalilzad. All were willing to work together (at least for the moment) to stabilize Afghanistan. Not invited were representatives of the defeated Taliban, some of whom appeared willing to enter into negotiations.[18] Their absence was likely a mistake, in hindsight, as convincing the Taliban to lay down their arms and join in governing Afghanistan would be the key issue confronting the new Afghan government and the US-backed coalition supporting it for the next two decades.[19] The conference decided on an interim authority headed by Karzai, who would convene a *loya jirga* (a grand assembly of Afghan leaders) within six months to establish a transitional government that would create a constitution, which would subsequently be approved by another *loya jirga*. The final step would be elections for a president and a parliament. The entire process was expected to take a little more than two years.[20] The Bonn conference ended on 6 December with an agreement reached on an interim government not a moment too soon.

After receiving news of his selection as head of the interim authority, Karzai negotiated for the surrender of the Taliban in Kandahar. On 7 December, senior Taliban leaders and the remaining Arab followers of bin Laden fled and opposition in the city collapsed. Gul Agha Sherzai took the initiative and moved 500 militiamen into the city and seized the Governor's palace. US special forces personnel moved into the city, to the delight of the locals, who cheered their arrival, to forestall factional infighting. Karzai reluctantly appointed Sherzai governor of Kandahar Province to keep the peace. With Kandahar now secure, Karzai relocated to Kabul to prepare to assume his new duties as head of the interim authority. Once his administration had been established on 22 December, all armed groups in the country would presumably transfer their allegiance to it.

With the Taliban defeated, the focus shifted squarely to destroying the remnants of al-Qaeda and capturing Osama bin Laden and his top lieutenants, who were reported to have moved from Kabul to positions in the Spin Ghar Mountains south of Jalalabad, the major city of eastern Afghanistan (halfway between Kabul and the Khyber Pass border crossing with Pakistan). On 30 November, a CIA-sponsored group (Team Juliet) of Afghan militia attacked al-Qaeda positions in the Milawa Valley, 28 miles south of Jalalabad near the Pakistani border.

Airstrikes eliminated al-Qaeda's communications equipment and heavy weapons and forced the remaining fighters to withdraw to Tora Bora, a mountainous cave complex constructed during the Soviet–Afghan War that sat some 10,000 to 12,000 feet above sea level. Team Juliet was soon reinforced by a US special forces team and other special operators from Task Force Sword.[21]

Given that the destruction of al-Qaeda was the most important US national priority in Afghanistan, the plan for achieving it was, in hindsight at least, inadequately conceived and under-resourced. Mutually antagonistic Afghan factions observing the restrictions of Ramadan did not have the motivation to close with and destroy the enemy, while airstrikes alone could not accomplish the task. CIA officer Gary Berntsen, who had coordinated the three Afghan groups to fight al-Qaeda, now realized their limitations and requested the insertion of a Ranger battalion to help finish the fight. Brigadier General Mattis also proposed deploying Marines to seal off the Pakistani border. CENTCOM, however, decided to continue to use the formula that had worked to date: Afghan militia backed by special forces and airpower. That approach did not work at Tora Bora. After ten days of aerial bombardment, from 7 to 17 December, bin Laden and his remaining followers escaped south across the border into Pakistan.[22] This outcome would haunt the United States for the next decade and beyond, as bin Laden would continue to inspire (and, to varying degrees, guide) al-Qaeda operations around the world, including in Iraq and Yemen, as well as in Afghanistan and elsewhere, until his ultimate death in 2011.

On 20 December 2001, the UN Security Council passed Resolution 1386, establishing the International Security Assistance Force (ISAF) to secure Kabul in the wake of the Taliban collapse. However, CENTCOM lacked a substantial plan on what to do next. Rumsfeld did not want a large commitment of US forces in Afghanistan, but without such support Karzai's interim administration would struggle. Regardless, any US forces would require a robust base structure, and CENTCOM began by turning Bagram and Kandahar airfields into major logistics hubs, but with constraints, which meant that conventional units were only slowly deployed to augment the special forces elements in Afghanistan.

Meanwhile, Karzai worked to fill newly vacant provincial and district administrative positions; however, sorting out the allegiances of various candidates proved difficult. In the process, Karzai and US forces alienated Jalaluddin Haqqani and his Pashtun tribal network that had traditionally exercised sway over eastern Afghanistan, missing a major chance at reconciliation. Indeed, the Haqqani Network, as it would become known, would prove as problematic over time as would the Taliban, even joining with the Taliban in conducting operations against coalition and Afghan forces and establishing a joint overarching Shura that coordinated their activities at senior levels. The Shura was a national gathering of local, national and religious leaders for all ethnic, sectarian, tribal and other societal elements.

Early in March 2002, Operation Anaconda, the last major combat operation of the opening phases of OEF, took place in the Shahi Kot Valley west of Khost in eastern Afghanistan. The operation was initially planned as a repeat of earlier ones, with special forces and other special operations forces (SOF) working with local militia to target enemy concentrations with airstrikes. When intelligence of larger concentrations of Islamic Movement of Uzbekistan (IMU) and al-Qaeda fighters surfaced, a force was assembled under Major General Hagenbeck to clear them.[23] This was "a force that drew elements from eight countries, two U.S. Army divisions, two Special Forces groups, other special operations elements, a hodgepodge of aviation units, and a variety of clandestine organizations," not to mention a few hundred hastily trained Afghans who were designated as the main effort.[24] It was another rapidly improvised team, made worse by the inadequate communications and command-and-control facilities at Hagenbeck's base in Bagram, as well as by insufficient authority over the various SOF units and CIA elements, not to mention the absence at Bagram of representatives from the Combined Forces Air Component Command.[25] Few of the commanders or their planners were worried, however, as they assumed the enemy would capitulate when confronted with superior firepower.

That assumption was proved wrong. Islamic Movement of Uzbekistan and al-Qaeda fighters had positioned themselves on the high ground overlooking the valley, prepared to fight. When the operation kicked off on 2 March, the air assault forces (the 1st Battalion, 87th Infantry and the 2nd Battalion, 187th Infantry – from the 10th Mountain and 101st Airborne Divisions, respectively) and the Afghan militia force deployed into a killing zone. The first days of the resulting battle were marked by intelligence failures, friendly-fire issues, difficulty in maneuvering at high altitude, poorly coordinated close air support and, to be sure, exceptional bravery and courage under fire by those on the ground. Nonetheless, Afghan militia proved incapable of closing with and destroying the enemy, leaving the task to US conventional forces that were poorly positioned to do so. After several days of confused fighting, Hagenbeck was able to reorganize his forces, employ additional conventional units and establish an air coordination center that gained control of the numerous close air support aircraft circling over the battlefield. Yet, by this point, snow showers brought action to a halt, and the remaining enemy forces were able to withdraw from the area. Operation Anaconda concluded on 18 March with a thorough sweep of the battlespace, which located abandoned weapons and equipment but few enemy dead. The operation ended the major combat operations phase of Operation Enduring Freedom, but not as conclusively as had been hoped. Attention now turned to solving the manifest challenges facing the coalition in Afghanistan, which were increasingly political.[26]

Treading water, 2002–2005[27]

The military operations that removed the Taliban from power and degraded al-Qaeda had been breathtaking, but in the spring of 2002 the question of what to do with Afghanistan in the aftermath of major combat operations could no longer be ignored. Little thought had been put into what would happen once the invasion succeeded; the pace of the war had outstripped the pace of policy. In a meeting of the war cabinet in late September 2001, President Bush had asked the assemblage, " 'So who's going to run the country [Afghanistan]?' There was silence."[28] More thought had been put into the question after the Bonn conference, but there were still policy vacuums in many areas. Non-governmental and international relief organizations began to arrive to deliver humanitarian aid and reconstruction support, but their efforts were uncoordinated. The Office of Military Cooperation–Afghanistan began training a new Afghan National Army (ANA), despite lack of funding, insufficient volunteers and inadequate facilities. When Lieutenant General Dan McNeill and the XVIII Airborne Corps headquarters flew to Bagram to establish the overall headquarters in Afghanistan in May, its personnel were capped at 400 and the command was told not to construct permanent facilities in the country.[29]

In June, an emergency *loya jirga* selected Karzai as head of the transitional administration that would govern Afghanistan until presidential elections were held in 2004. The Bush administration now concluded that, given the defeat of the Taliban, Operation Enduring Freedom could become an economy-of-force effort as the administration's focus shifted to Iraq.[30] This decision overlooked the fact that the Taliban had not been destroyed; rather their leaders (and many fighters) had fled to sanctuaries in nearby Pakistan, there to rebuild their shattered forces. Nevertheless, Afghanistan receded so far to the margins of policy discussions that in October 2002 President Bush forgot who was in command in the country.[31]

The irony was that even as resources were diverted to preparing for, and then conducting, the war in Iraq, the Bush administration's goals in Afghanistan expanded. The initial agenda of defeating the Taliban and destroying al-Qaeda had morphed into nation-building and the establishment of democracy in Afghanistan. This was, at that point, unavoidable. And, by then, Bush had come to recognize the need for the performance of such tasks, later writing, "Afghanistan was the ultimate nation building mission. We had liberated the country from a primitive dictatorship, and we had a moral obligation to leave behind something better."[32] Afghanistan would ultimately become a nation-building project on steroids, though not until after the completion of the Surge

in Iraq in 2008. The \$143 billion spent by the United States on recon-
struction and military assistance in Afghanistan in two decades was
roughly equal to the amount of aid dispensed to all of Europe by the
Marshall Plan between 1948 and 1951, adjusted for inflation.[33]

The creation in 2003 of Provincial Reconstruction Teams, which
coordinated reconstruction and humanitarian support, was part of this
effort, but these largely military teams lacked the civilian expertise
required to sustain their operations over years of effort.[34] Moreover, the
new maximalist goals were not in synch with the minimalist approach
to resourcing the mission, and the approach often reflected inadequate
appreciation of the history, politics and culture of the Afghan people. As
the official US Army history of the conflict observes, "It would take years
before the Bush administration realized that a number of prominent
Afghans within the Kabul government placed personal, tribal, and
ethnic goals above the formation of a democratic, fully representative
central government."[35]

Indeed, one of the major problems in the immediate aftermath of the
fall of the Taliban was the empowerment of the same warlords who had
ushered in civil war in Afghanistan after the withdrawal of Soviet forces
in 1989 and the collapse of the Soviet-supported Afghan regime in
1991. While the Afghan people desired accountability for past war
crimes, US policy allowed the warlords an indefinite reprieve. Former
warlords assumed office in twenty provinces, nearly 60 percent of the
total. And rather than truth and reconciliation, the Afghan people were
subject to the rule of men who were in some cases more abusive than
the Taliban rulers they replaced.[36]

ISAF, meanwhile, established a rotating six-month national
command with Britain, Turkey and Germany/Netherlands providing
the first three headquarters. When NATO took control of ISAF on
8 August 2003, there were 5,000 troops from thirty countries in Kabul,
and some 13,000 US troops elsewhere in Afghanistan. The US headquarters in Afghanistan was downsized from three to two stars, and the
smaller staff did not have the capability to untangle the web of challenges it faced. To President Bush, these numbers seemed reasonable,
but, as Defense Secretary Rumsfeld later observed, "in retrospect, our

rapid success with low troop levels created false comfort, and our desire to maintain a light military footprint left us short of the resources we needed."[37] A three-star headquarters was reintroduced in October 2003 under the command of Lieutenant General David Barno, an accomplished infantry officer and former commander of the US Army's Ranger Regiment, and troop numbers grew slowly as the counter-insurgency approach he implemented in conjunction with US Ambassador Zalmay Khalilzad replaced the counter-terrorism focus of military operations, albeit still with seriously inadequate forces.

Another Afghan *loya jirga* approved a new constitution in 2004, one that had many good qualities (such as recognizing women's rights) but provided for an overly centralized Afghan government that ran contrary to the history of governance in the country.[38] On 9 October 2004, Karzai was elected president, demonstrating considerable charisma, impressive presence and substantial gravitas, but over time proving to be a decidedly mercurial leader.[39] The following September, Afghans voted for representatives for a national assembly and, when it met for the first time in December, the future of Afghanistan looked bright. The economy was growing, children (girls as well as boys, unlike under the Taliban) were in school, reconstruction was proceeding apace and Afghan refugees were returning home from Pakistan and neighboring countries.

But beneath the surface, problems lurked. The Taliban were already beginning to contest security in much of the south and east of the country and were expanding into other areas as well. Production of poppies that provided opium, once banned under the Taliban, increased ninefold in just one year after the Taliban defeat, undermining attempts to reform the rural economy. (As I would later observe while serving as the commander in Afghanistan, "It is very difficult to establish the rule of law in a country if its major export crop is illegal.") German and Italian commitments to fund police and judicial reform and send advisers to Afghanistan went largely unfulfilled. The program to disarm, demobilize and reintegrate militia forces that were no longer needed took several years – far too long – to accomplish. And the detainee population grew even as the coalition lacked adequate numbers of translators and interrogators to glean information from prisoners and prepare for release of those who posed no threat to security.

Meanwhile, the US Congress appropriated more than $38 billion in humanitarian and reconstruction assistance to Afghanistan from 2001 to 2009, which did considerable good but also fuelled corruption throughout the Afghan bureaucracy as the amount exceeded the government's absorptive capacity.[40] American commanders and their intelligence staffs struggled to understand the complex dynamics of Afghan society, and there were not enough civilian experts in the country to assist them. Karzai's government was weak and ineffective, partly because warlords who had gained stature with the fall of the Taliban actively competed with or co-opted the central government in the scramble for power and resources, furthering the corruption that was eating away at the fabric of Afghan society.

By October 2006 when NATO took responsibility for the entire country, ISAF numbered 31,000 troops from thirty-seven countries throughout Afghanistan, with 20,000 US troops conducting counter-terrorism and security-force training missions in separate commands. UK General David Richards, the highly respected incumbent ISAF commander, was elevated to four stars in recognition of the expansion of ISAF's mission and extended several additional months in command. Even with the increase in troop numbers and Richards' impressive work to pull together the disparate elements of the command however, ISAF found it difficult to maintain security throughout the country due to a significant rebound in Taliban strength and national caveats that limited the utility and performance of many NATO contingents, which had not signed up for intensive combat operations.[41] In February 2007, Richards was succeeded by US Army General Dan McNeill, another exceptional officer, who had been the first three-star commander in Afghanistan in 2002; nonetheless, the US troops in the country remained under a separate chain of command. Thus, six years into the war, despite successive impressive commanders, there were still inadequate resources and less than complete clarity on the strategy, with the organizational structure needed to pursue it still elusive as well.[42]

The Achilles' heel – training and equipping Afghan security forces

The greatest military deficiency was an inadequate program for building Afghan security forces. As I neared the end of my second tour of duty in Iraq as commander of Multinational Security Transition Command–Iraq in the autumn of 2005, Secretary Rumsfeld directed me to swing through Afghanistan on my way home to assess the progress of the program to establish the Afghan National Security Forces (ANSF). What I found was sobering. In comparison to Iraq, Afghanistan was vastly more challenging, despite a considerably lower level of violence (although that was increasing ominously). Most significantly, Afghanistan faced a situation in which its enemies enjoyed sanctuary in neighboring Pakistan. Moreover, Afghanistan also lacked the extraordinary natural resources and revenue of Iraq and had an anaemic economy with limited prospects, endemic corruption and an immature or broken infrastructure, along with a population three-quarters of whom were illiterate. Additionally, the vast distances, the Hindu Kush Mountains, challenging winter weather and the lack of a robust road network made operating in the country exceedingly difficult. Given all this, Afghanistan would, I reported to Secretary Rumsfeld, be the longest of the long wars, not Iraq.

Beyond that, the train-and-equip mission in Afghanistan, which had begun some two years before the serious effort began in Iraq, was failing, in part due to the focus of resources, attention and priority on Iraq. There was, in addition, a shortage of training teams to advise and assist Afghan brigades once in the field. And, beyond those shortfalls, there were other major deficiencies in the training program in Afghanistan. The police, for example, who were going to deploy into active counter-insurgency operations and were being trained by contractors, never fired weapons in their eight-week training program. They did, however, march for an hour in the morning and an hour in the evening (the latter was supposedly for physical fitness and discipline, though I noted that there were more relevant tasks in which those objectives could be achieved). Moreover, the metrics being used focused on the

number of security personnel *trained*, not the number actually in the
ranks, which, alarmingly, was declining due to casualties and desertions. In sum, the train-and-equip effort was actually losing ground, not
gaining it. Despite my report, the shortcomings were not fully addressed
for several years, and by the end of 2008, though the Afghan National
Army was on paper 70,000 strong, only 40 percent of its units were even
remotely combat ready.[43] And even those that were moderately capable
were highly dependent on US enablers such as air support and medical
evacuation, capabilities the Afghan military would never possess in
sufficient numbers.

The Taliban strike back

None of this would have mattered in the long run had the Taliban and
Haqqani Network elements remained a defeated force; however, given
their sanctuaries in Pakistan and support from various difficult-
to-identify outside sources (likely in the Gulf States and Pakistan), they
were able to reconstitute and strike back against the Afghan government,
its security forces and its coalition partners. Frustratingly, the Pakistani
Army and intelligence service tolerated the Taliban and the Haqqani
Network and allowed safe haven for the groups in the south-western and
western parts of the country. In addition to some funding from outside
sponsors, the Taliban also used the drug trade (Afghanistan was and still
is the leading cultivator of poppies for opium) to fund their operations.
Thus, by the summer of 2006, the Taliban were on the offensive, and by
the end of 2008 their attacks were crippling the Afghan government and
economy. After that point, defeating the insurgents was unlikely to be in
the cards, especially given the sanctuaries to which they could retreat in
Pakistan when put under pressure.

Signs of the return of the Taliban (and its Haqqani affiliate) were
increasingly alarming. In one attack on a US outpost at Wanat in north-
east Afghanistan on 13 July 2008, upwards of 300 Taliban troops –
aided by some of the local population – breached the perimeter, killed
nine American soldiers and wounded twenty-seven others before being
driven off by Apache attack helicopters and close air support.[44] British

forces in Helmand and Canadian forces in Kandahar were likewise hard-pressed by Taliban attacks and required reinforcement from US forces to maintain their positions. With US attention and priority given to the Iraq War, the Chairman of the Joint Chiefs of Staff Admiral Michael Mullen observed to members of the House Armed Services Committee, "In Afghanistan, we do what we can. In Iraq, we do what we must."[45]

By the autumn of 2008, then, the campaign in Afghanistan was under-resourced and at risk of strategic failure, and its top commanders knew it. The Commander in Afghanistan, General David McKiernan, requested additional forces to secure Pashtun areas in southern Afghanistan prior to the upcoming Afghan elections scheduled for August 2009, and began to build the base structure necessary to receive them. In response, the Bush administration in its final months approved the deployment of several additional ground brigades as well as a second aviation brigade – 21,000 troops in all – but left several other requests pending for the incoming administration of Barack Obama. Meanwhile, the shortcomings of NATO forces became apparent in Regional Command South, which rotated command among capable commanders from the UK, Canada and Netherlands, though those countries found it increasingly difficult to acknowledge that they were actually executing a counter-insurgency operation because the mission was no longer politically palatable to domestic audiences.[46]

The Afghanistan Surge, 2009–2011[47]

President Barack Obama entered office in January 2009 with the intention of winding down the war in Iraq and supporting what he and his backers perceived to be the "good war" in Afghanistan. The success of the Surge in Iraq had reduced the need for forces there, enabling a shift of focus and resources to Afghanistan. The President thus tasked Bruce O. Riedel, a respected former senior CIA officer, to lead a strategic review. Before the review was complete, however, the Pentagon asked for 30,000 more troops for Afghanistan to ensure security for the upcoming presidential election. After some debate, informed by the review, the number was pared down to 17,000, and Obama approved the request.

Delivered to the President in March, Riedel's review called for limiting the mission, which would be "to disrupt, dismantle, and defeat al-Qaeda in Pakistan and Afghanistan, and to prevent their return to either country in the future." The administration should treat Pakistan as an integral part of the problem. Recognizing the unlikelihood of shutting down the cross-border sanctuaries, the coalition should instead build up the ANSF so they could prevent a Taliban victory. Coalition forces could then draw down and depart Afghanistan. But first the Taliban had to be degraded to provide an opportunity to execute the strategy.

To achieve this goal, the review called for the execution of an integrated counter-insurgency strategy. The strength of the Afghan Army would be increased to 134,000 and of the Afghan police to 82,000 over a period of two years. The United States would send 4,000 more trainers to Afghanistan, on top of the other troop increases already approved. This would bring the number of US service personnel in the country to 68,000 after all the reinforcements had arrived in country, which would take some months. There would also be a surge of civilians to theatre to handle the multitude of nation-building chores.[48]

President Obama understood the magnitude of the effort required to turn the war around. "What was clear," he later wrote, "was that the U.S. commitment the Riedel report was calling for went well beyond a bare-bones counterterrorism strategy and toward a form of nation-building that probably would have made sense – had we started seven years earlier, the moment we drove the Taliban out of Kabul."[49] He was not wrong.

In the view of Secretary of Defense Bob Gates – who was so highly respected that President Obama had kept him in position even though he had been selected by his Republican predecessor – the new strategy required a different commander to execute it. General McKiernan, who had commanded the ground forces during the invasion of Iraq and later US Army Europe, had commanded ISAF since June 2008; however, in the view of Gates and other senior officials, he had not adapted sufficiently to the unconventional and complex situation in Afghanistan. In May, Gates removed him from command of ISAF, and a month later replaced him with General Stanley McChrystal, who had earned high regard for his impressive command of Special Operations Forces for five years in the war zones and then served as the director of the Joint Staff in the Pentagon.[50] Gates also instructed McChrystal to create a three-star headquarters in Afghanistan to handle the operational campaign and day-to-day operations, replicating the command set-up in Iraq and allowing the four-star ISAF headquarters to focus on the overarching strategic effort. The creation of NATO Training Mission–Afghanistan (NTM–A), also led by a three-star American general, which replicated the training and equipping organization in Iraq, followed five months later. All these organizational changes had been long overdue.

McChrystal was given sixty days to conduct a strategic assessment of the situation in Afghanistan and report back through Gates to the President. Tellingly, the US Embassy was not directly involved in the assessment process, a shortcoming that would affect the resulting discussions in Washington. After an extended tour of the area of operations, McChrystal concluded that the insurgency could be defeated only if governance improved, but this required improving security and protecting the population against intimidation and violence. This would be done by conducting population-centric counter-insurgency operations using the traditional "oil spot" approach, one area at a time, with the resulting security then expanded. The south would see the main effort. In practice, this meant starting in the Helmand River valley, where US Marines and British forces were already stationed (and embattled), and then expanding the operations east into the challenging Kandahar Province, birthplace of the Taliban leader and movement. The other commands would operate from the same counter-insurgency playbook, adapted to the local conditions in each area. And McChrystal would emphasize "courageous restraint" in the use of firepower in order to reduce civilian casualties, which were turning Karzai and the Afghan population against the coalition.[51]

To be successful, a surge of civilian and military resources to Afghanistan required a willing and able partner on the ground. However, the Obama administration viewed the Afghan government as hopelessly corrupt, and the US Special Representative for Pakistan and Afghanistan, Ambassador Richard Holbrooke (who had overseen the negotiation of the 1995 Dayton Accords that ended the civil war in Bosnia), even worked to undermine Karzai in the upcoming presidential election. To avoid a problematic run-off election, Karzai needed to gain more than 50 percent of the popular vote, and to achieve this threshold some of his supporters resorted to stuffing ballot boxes. The election in August 2009 went to a run-off anyway, which Karzai nevertheless won by default on 2 November when his opponent, Abdullah Abdullah, withdrew from the race with encouragement from the US government. According to Secretary of Defense Bob Gates, "It was all ugly: our partner, the president of Afghanistan, was tainted, and our hands were dirty as well." The election was, in Gates' estimation, a "clumsy and failed putsch."[52] It also undermined significantly Holbrooke's relationship with Karzai and, to a degree, that of the US Ambassador, Karl Eikenberry (a former commander of US forces in Afghanistan as a three-star general).

To focus combat power on population centers, commanders withdrew forces from remote combat outposts that now served little function. The process had not yet been completed when on 3 October 2009 a band of 300 Taliban fighters attacked Combat Outpost Keating in Nuristan Province, in north-eastern Afghanistan. The Taliban nearly overran the poorly sited position in a nine-hour battle, but timely air support, Apache attack helicopter runs and the bravery of the soldiers on the ground ensured they were driven off. The attackers killed eight Americans and three Afghan soldiers and wounded twenty-seven others, while sustaining dozens of casualties themselves.[53]

Meanwhile, McChrystal's stark assessment, delivered to Gates at the end of August, warned of strategic failure in Afghanistan if nothing changed. Unfortunately, but not surprisingly, McChrystal's assessment leaked to the Washington Post two weeks after its arrival in Washington. Soon thereafter, McChrystal recommended an increase of 40,000 troops to enable a partially resourced counter-insurgency strategy (a number that "truly alarmed" Secretary Gates), as well as a doubling of Afghan security forces to 240,000 soldiers and 160,000 police – figures that, as commander of CENTCOM, I strongly supported, as did Admiral Mullen, the Chairman of the Joint Chiefs.[54]

President Obama was now in a bind. He could either support the request for more troops by his commander on the ground or be accused of under-resourcing the war in Afghanistan, which would put the burden of potential defeat on his shoulders. Both the strategy and the reinforcements met with skepticism from Vice President Joe Biden, the President's political advisers, some National Security Council staffers and US Ambassador to Afghanistan Karl Eikenberry – who unhelpfully sent a long cable on 6 November, just before the final strategic review meeting chaired by President Obama, criticizing the proposed counter-insurgency strategy and asserting that Karzai was not an adequate strategic partner.[55] That leaked as well, and it soured the critical relationship between Eikenberry and President Karzai for the remainder of the Ambassador's time in Kabul. From 13 September to the end of November, in nine meetings of the National Security Council, the President and his national security team wrestled with the issues at stake in Afghanistan and the impact of McChrystal's request for more troops. The President, Vice President Biden and their staffs believed the military was boxing in the President to force upon him their chosen course of action.[56] The issues were hashed and rehashed until every detail had been dissected in a "process that had," in Gates' view, "gone on way too long."[57] Counter-terrorism-plus – a strategy focused on destroying al-Qaeda favored by the Vice President and some others – would not work without a willing and capable partner on the ground in Afghanistan, especially given that the major threat was not terrorist elements but the Taliban and Haqqani insurgent groups, who could be dealt with only by a largely counter-insurgency campaign that would have to be led by coalition forces for the next two to three years, at the least. On the other hand, no conceivable amount of nation-building would be likely to turn Afghanistan into a Western-style democracy at reasonable cost or in an acceptable period of time.

In essence, the administration tried to split the difference. The option chosen was a partially resourced counter-insurgency campaign (although the President and his staff would not refer to it as such) to defeat al-Qaeda, protect the population centers, degrade the Taliban, stabilize key ministries of the Afghan government and build Afghan security forces to the point that ISAF could transfer security functions to them. Obama approved an increase of 30,000 US troops, with another 10,000 to come from NATO allies. Afghan security forces would be increased, but not up to the 400,000 that McChrystal had recommended.[58]

On Sunday, 29 November, I was called from Tampa to a short-notice meeting in the Oval Office at 5 p.m. The session included the President, Vice President, White House Chief of Staff Rahm Emanuel, Secretary Gates, Admiral Mullen, Vice Chairman of the Joint Chiefs General James Cartwright and National Security Advisor Jim Jones. After explaining his decision on the strategy and the additional troop numbers, the President informed the group that when he announced the build-up to the public he would also announce that the additional forces would begin to withdraw in eighteen months. The decision was a complete surprise to me – and, I think, to the other uniformed leaders in the room – and we thus did not have time to consider the implications before signalling our assent.

Two days later, before an audience of cadets at the United States Military Academy at West Point, the President made his announcement, including the end date for the surge of reinforcements to Afghanistan, which would begin to redeploy by 1 July 2011. In retrospect, that decision, which had never been discussed during the actual policy review, was misguided and would dilute the impact of the new strategy. There was nothing wrong with establishing a tentative withdrawal timeline as a planning tool, but the public announcement of it sent a signal to the Taliban that they could hunker down and wait out the surge, which was going to be temporary and drawn down within eighteen months, apparently regardless of the conditions on the ground.

*   *   *

Surge operations began in January 2010 with the opening phase of Operation Moshtarak (Dari for "together") in the Helmand River valley. From there, it would spread east to Kandahar Province. Initial operations focused on route clearance; however, beginning with the second phase in February, British and US Marine Corps units would maneuver by air to clear select towns, beginning with Marjah, and revitalize local governments to link them to regional and national resources. The effort became associated with the regrettable catchphrase of providing cleared towns with a "government in a box," which was widely derided when such efforts fell short of expectations. Despite tactical successes in clearing various towns in Helmand, however, the Taliban did not completely disappear, as many operatives went to ground and others retreated to other districts in Helmand Province or to Pakistan, awaiting the day when the Americans and their NATO partners would leave.

As part of the new strategy, the CIA and the military also ramped up the number of drone strikes and counter-terrorism raids in Afghanistan and Pakistan. Strikes on Pakistani soil were limited to the western tribal areas, where Islamist extremist and Haqqani elements were located and thus had no effect on the Taliban sanctuaries in Baluchistan (with one exception after the surge was drawn down). In Afghanistan, however, there were fewer restrictions, and Task Force 714, the element of SOCOM operating in country, accelerated its campaign to kill or capture Taliban leaders and terrorist operatives.[59] Unfortunately, increased bombings and night raids sometimes inflamed tensions with the Afghan people and with President Karzai, who viewed the raids as excessive and "oppressive."[60]

The main effort in April switched to Kandahar Province, but unfortunately General McChrystal would not be around to direct the remaining surge operations. On 22 June, reporter Michael Hastings published an article in *Rolling Stone* titled "The Runaway General," a scathing, firsthand account of McChrystal and several of his immediate staff members on a trip in Europe making disparaging comments about the Vice President and other senior officials.[61] Coming after two previous incidents that the President had seen as McChrystal challenging his authority, this was more than unfortunate. That evening,

McChrystal flew to Washington, where President Obama accepted his resignation.

Given that I had participated in the Afghan policy reviews as commander of US Central Command, had served four years in three different positions in Iraq and had commanded there during the Surge in 2007–8, I emerged as the choice for Obama and Gates as McChrystal's replacement. I accepted (when the President asks you to take on a tough assignment, the only real possible answer is "yes"), and only a few hours after accepting McChrystal's resignation Obama announced that he would submit my nomination to the Senate for confirmation.[62] Following an extremely swift confirmation process, I was on a plane to NATO headquarters and then to Kabul within a week.

Deploying the surge forces into Afghanistan required a herculean logistical effort. Unfortunately, the supply routes running through Pakistan into Afghanistan were subject to protection and extortion rackets lining the pockets of warlords, criminal bosses and corrupt government officials.[63] One piece of good news was the establishment several months earlier of the Northern Distribution Network (NDN), a series of commercial supply routes running through Russia (which, for the moment, cooperated with the effort), the Caucasus and Central Asia to Afghanistan that we had coordinated when I was commanding CENTCOM. The NDN not only increased logistical throughput to northern Afghanistan, but reduced Pakistan's leverage over the coalition given that alternative supply routes now existed that did not transit its territory. The northern route had limitations, however, given the unwillingness of certain countries to allow combat vehicles, weapons and munitions to transit their territory and also the challenges of transiting through or around the Hindu Kush Mountains to central and southern Afghanistan.[64]

Between 2008 and 2011, US forces in Afghanistan increased from 28,000 to slightly more than 100,000 troops, and NATO forces increased from 32,000 to over 50,000.[65] With all the surge forces finally in country, the area around Kandahar began to stabilize, especially in the autumn of 2010 after the 2nd Brigade, 101st Airborne Division (TF Strike), under the command of Colonel Arthur Kandarian, a highly decorated, exceptional combat leader, cleared the problematic Arghandab River valley west of Kandahar in a series of aggressive operations, and other elements cleared and held the troubled districts south of the city. SOF night raids also took hundreds of Taliban fighters off the battlefield. By March 2011, the Taliban could no longer contest Kandahar city and its outlying districts, bringing large-scale maneuver operations in that pivotal province to an end. Afghan security forces now numbered 260,000 and, although quantity did not equal quality, they were now mostly partnered with coalition troops to enable them to be more effective. Despite this progress, Afghan Army and police numbers remained too low to secure the entire country. I sought to make up for this shortfall by backing innovative programs that would encourage local Afghans to secure their own communities.

One of these proved critical. Inspired by a paper written by special forces Major Jim Gant, regarded as the "ultimate counterinsurgent," published online in October 2009, Village Stability Operations (VSO) entailed embedding a twelve-man special forces team in an Afghan community, often supported by enablers such as interpreters, military working dog teams, medical teams, civil affairs personnel, psychological operations teams, intelligence analysts and cultural support teams. SF teams would assist in the recruitment of Afghan Local Police (ALP), a minimally trained and armed, community-based security force selected by the leaders of the local tribes.[66] The SF teams also coordinated security with Afghan Army and police in the area, as well as with coalition forces and Afghan commandos. Team leaders would work with local government and tribal authorities, and also synchronize development projects administered by international, non-governmental and national agencies and organizations. I worked hard to convince Karzai to back the program, a critical enabler of its success.[67] VSO and ALP enabled the Afghans to clear rural communities and hold them against a return of the Taliban, in essence dramatically increasing the areas in which security was established. I viewed the program as so important that I ultimately assigned two infantry battalions to augment the special forces and enable them to cover twice as many VSO sites, with the program at its height employing 30,000 local police.

Although criticism would be aimed at the "industrialization" of the program, especially as the surge began to wind down, VSO and ALP were so successful that a copy of Gant's paper was later found among the documents seized in Osama bin Laden's compound in Abbottabad, Pakistan, with notations in the margins, along with a directive targeting Major Gant for assassination.[68]

I also introduced a number of other initiatives to complement the military aspects of the surge, some of them borrowed from the Surge campaign in Iraq and modified to adapt to the conditions in Afghanistan. A reintegration concept and process ultimately supported by Karzai attempted to bring the reconcilable elements of the Taliban into harmony with the Afghan government in Kabul (similar to the successful reconciliation effort during the Surge in Iraq). The program ultimately reached roughly 25,000 members of the Taliban (compared with 103,000 reconciled in Iraq), but the conversions in many cases proved to be temporary given the lure of cross-border sanctuaries in Pakistan and the difficulty of sustaining security gains over time. We established a counter-narcotics task force, as well, though battling that scourge was a Sisyphean effort due to the strong global market for opium and the difficulties of establishing a successful crop-substitution program. And I established a rule-of-law task force under Brigadier General Mark Martins, a brilliant, exceedingly selfless Judge Advocate General officer who had headed a similar effort during the Surge in Iraq, focusing on much needed improvements to the Afghan judicial and prison systems and integrating Afghan traditional concepts of justice with the overarching legal structures of the country.

Battling corruption in the Afghan government was a particular challenge, which I tried to address by establishing an anti-corruption task force (Joint Task Force Shaffafiyat (Transparency) under Brigadier General H. R. McMaster, an exceptional officer who had proven himself in several previous combat tours in Iraq, including heading several critical strategic assessments during the Surge in Iraq and another when I took charge of Central Command. This achieved some notable successes such as forcing the removal of the Afghan Air Force Chief of

Staff (personally renting ramp space at Kabul Airport), the Surgeon-General (stealing/selling pharmaceutical drugs) and the Kabul Military Hospital commander (stealing fuel from generators powering expensive diagnostic equipment, among other crimes). But given the pervasive nature of corruption in Afghan society and the tsunami of aid that flooded Afghanistan during this period, this effort was bound to fall short.[69] International aid created too many opportunities for corruption among elites lacking accountability and sufficient self-discipline. Perhaps no better – or worse – example existed than the Kabul Bank, founded in 2004 as the newly elected Afghan government came to power. By 2010 the United States had deposited nearly $1 billion in the bank to pay the salaries of the Afghan Army and National Police. But the bank nearly collapsed that year from widespread embezzlement and fraud, with hundreds of millions of dollars disappearing into accounts and sketchy real estate investments in Dubai, in several cases linked to the families of senior Afghan officials.[70]

Still, by the end of 2010, nine years after the invasion, with the additional soldiers, diplomats, CIA operators, development workers, other civilian personnel and funding, the effort in Afghanistan finally had the "inputs" right for the first time. After years of inadequate attention and resourcing, the coalition effort had developed the right overarching strategic concepts (begun by General McChrystal and completed under my tenure, together with our NATO and US civilian counterparts), the right organizational architecture, nearly the right level of personnel, the right preparation of the forces and leaders and considerably improved military equipment and assets. All of this would enable significant gains over the subsequent two years or so, before the drawdowns of forces began to limit further progress. But the failure to have adequately taken advantage of the first nine years in Afghanistan would haunt the mission until the final withdrawal in 2021.

With the Taliban under pressure from the surge, Ambassador Holbrooke explored the possibility of bringing them to the negotiating table. Karzai received backing from a conference of tribal elders in the summer of 2010 to establish a High Peace Council to undertake negotiations. Appeasing the stakeholders in the process, however, proved to be difficult. Pakistan demanded a role in the negotiations to ensure its interests were considered. Tajiks, Uzbeks and Hazaras feared an agreement that might give too much power to the Pashtun majority. And many American policymakers and leaders in Afghanistan believed that achieving a meaningful agreement with the Taliban was impossible, especially given the Taliban understanding that the Americans wanted to leave and planned to begin doing so in the summer of 2011. Tragically, as the first feelers went out, Holbrooke died from an aortic rupture.[71]

In mid-May 2011, Afghan officials opposed to talks with the Taliban leaked the existence of preliminary negotiations to the press. Predictably, Pakistani officials were incensed that they had been left out of the process, and the Taliban suspended the talks. After tempers cooled, talks resumed in Doha, Qatar, in August. Negotiators discussed the opening of a Taliban diplomatic office there, but Karzai threw cold water on the idea, demanding a seat at any future negotiations. The effort once again fizzled out. In June 2013, the Taliban finally opened an office in Qatar, but when their representatives unveiled an "Islamic Emirate of Afghanistan" flag at the opening ceremony, reflecting their unwillingness to accept the flag of Afghanistan, tempers once again flared and the office was closed.[72] The skeptics had been proven right; if there was to be an agreement with the Taliban, it would likely be on their terms, unless the United States and its coalition partners reversed their stated plans to leave.

The only other way the war would come to an acceptable conclusion was to build Afghan security forces that could, over time, hold their own against the Taliban. This issue was pressing, for at the NATO summit in Lisbon in November 2010, in addition to extending the ISAF mandate, the alliance countries also voted to transition all areas to Afghan control by the end of 2014. Despite the importance of its mission in setting the conditions for that transition, however, the NATO Training Mission–Afghanistan (NTM–A) still operated with a critical shortage of advisers and trainers. And hands-on training was essential, for 86 percent of Afghan recruits were illiterate.[73] To deal with that reality, NTM–A Commander Lieutenant General William Caldwell instituted a literacy program to teach Afghan recruits to read and write, even as they were introduced to basic military and police tasks. He also created training programs to ensure Afghan soldiers and units operated in accordance with the counter-insurgency doctrine being implemented by the coalition.

As the surge approached the first drawdown of several thousand troops in the summer of 2011, Afghan security forces topped 300,000: a total of 170,000 in the Army and Air Force and more than 136,000 in the National Police. NTM–A continued to build Afghan security forces as the date for transition approached. With the ANSF approaching 350,000 soldiers and police in the summer of 2012, it was clear that Kabul would continue to require international financial backing to support that force level. But it was essential to create a force that could not only defeat – or at least hold its own against – the Taliban, and later downsize it to a level that Kabul could support. Some Afghan units, especially the commandos trained by US special operators, were well equipped and very capable, but the majority were plagued by inadequate discipline, low morale and desertions, in addition to sustaining significant casualties. Afghan forces also remained ominously dependent on US enablers such as close air support, tactical airlift and medical evacuation, even as the date for the first transitions of security tasks in select locations approached.

With the beginning of the drawdown two months away, on 2 May 2011, US special operations forces achieved a major objective of the war by launching a heliborne raid into Pakistan that located and killed Osama bin Laden at a compound near Abbottabad. Operation Neptune Spear was the culmination of more than a decade of intelligence work that finally pinpointed his location. Navy Seals flew into Pakistan in two specially outfitted Army helicopters, backed up by a quick-reaction force flying in a CH-47 helicopter and another CH-47 helicopter carrying fuel for the return journey. One of the Army aircraft crashed on landing, but its occupants emerged from the wreckage unhurt. The operators then entered and cleared the compound room by room, killing Osama bin Laden, his son Khalid and two accomplices, Ahmed al-Kuwaiti and Abrar al-Kuwaiti, along with the latter's wife, Bushra, who had attempted to shield her husband from gunfire. The assault team recovered vast quantities of documents, computer hard drives, CDs, DVDs, memory cards, thumb drives, videos and other material that would provide an intimate look at al-Qaeda's past operations and also those planned for the months and years ahead. Bin Laden's body was flown back to the SOF compound at Jalalabad, where his identity was confirmed by physical features and then facial recognition analysis.[74] The body was subsequently transported to Bagram and later flown in a Marine Osprey to the USS *Carl Vinson* in the Arabian Sea where, after Islamic funeral rites, the body was committed to the deep.[75]

Pakistani officials were upset by the violation of their sovereignty, but their protestations that they would have captured and brought bin Laden to justice, had the Americans given them notice of his location, strained the limits of credulity. That bin Laden could live little more than a mile from the Pakistani military academy without drawing the notice of the ISI also stretched the imagination, although there were never any indications that Pakistani authorities had been aware of bin Laden's location.[76]

On 22 June 2011, President Obama announced the beginning of the drawdown of surge forces in Afghanistan. Nearly 10,000 troops would be withdrawn from Afghanistan by the end of the year, and another 23,000 by the following summer. The surge had achieved its goals of degrading and pushing back the Taliban, buying time to increase the size and capacity of the ANSF, to further develop key Afghan government institutions and to commence transition of security tasks to Afghan forces in select locations. But what remained unclear was whether the Afghan government and its security forces could deal with the Taliban after full transition of security responsibilities, which was planned to take place at the end of 2014. No one at the time realized that US involvement in Afghanistan, then in its tenth year, had only reached its halfway point.

The Afghans take the lead, 2012–2014[77]

The surge of 2010 and 2011 was a tactical and operational success and bought time for the ANSF to grow and mature, but it did not change the underlying political and strategic dynamics of the war in Afghanistan. The Afghan people remained largely distrustful of a corrupt and inefficient government that often seemed, especially in rural areas, to do little to improve their everyday lives. Pakistan targeted some insurgent groups in the Federally Administered Tribal Areas, but predominantly those operating inside and against Pakistan rather than the Taliban and Haqqani Network that were the main antagonists in Afghanistan. Those groups were able to rebound from the setbacks in Afghanistan, using sanctuaries inside Pakistan or remote areas in Afghanistan to regroup after the beating they had sustained during the surge.

As provinces transitioned to Afghan control, provincial reconstruction teams also withdrew to make way for Afghan takeover of governance and reconstruction activities. While inevitable, the assumption of these responsibilities by the Afghan government increased opportunities for the pervasive corruption.

After the US command in Afghanistan had transitioned detention facilities and control over SOF raids to the Afghan government in spring 2012, Obama and Karzai began contentious negotiations for an Enduring Strategic Partnership Agreement, providing authority for US personnel to remain in Afghanistan in an advise-and-assist role.[78] Obama also designated Afghanistan a major non-NATO ally, giving it priority in obtaining security force assistance. Obama initially committed to keeping 9,800 troops in Afghanistan, but on 27 May 2014 he announced that he intended to withdraw most US troops (except for a small security force assistance office) by the end of his second term in office in early 2017. The transition process ended on 31 December 2014 when a non-combat NATO training mission replaced ISAF, and Operation Enduring Freedom morphed into Operation Freedom's Sentinel, a mission that focused on training, equipping and advising Afghan forces, as the Afghan forces assumed full control of the security of their country.

Removing vast quantities of coalition equipment from Afghanistan during the drawdown strained logistical routes through Pakistan, a situation exacerbated by an incident in late November 2011, when an Afghan special forces company supported by US special forces personnel raided the village of Maya in eastern Afghanistan at night and came under fire from nearby Pakistani forces positioned on the high ground just across the (less than distinct) border. When a low-level pass by an F-15E Strike Eagle failed to stop the fire, the US Commander on the scene called in airstrikes, killing more than two dozen Pakistani soldiers and wounding nearly a dozen more.[79] In response to the incident, the Pakistani government closed for some seven months the routes running through the two official border crossings with Afghanistan to the port of Karachi. To meet the withdrawal timeline, a great deal of equipment left Afghanistan via airlift, some was transferred to the ANSF and some deemed not cost effective to be shipped out. Four thousand vehicles, a thousand of which were mine-resistant ambush-protected vehicles, were destroyed in place.[80]

During the transition, US SOF continued to conduct robust counter-terrorist operations to keep the Taliban off balance, capture or kill their leaders and degrade their capabilities. The Obama administration, however, reduced by half the number of drone strikes in Pakistan (from roughly fifty in 2012 to twenty-five in 2014) as its policy came under fire from groups condemning civilian casualties connected with the strikes. The Taliban responded to the coalition drawdown and shift to security force assistance by targeting coalition trainers and advisers with "green on blue" insider attacks that ate away at the trust between Afghan and coalition forces. There were two insider attacks in 2008, ten in 2009 and 2010, fifteen in 2011 and forty-two in 2012.[81] The Haqqani Network and the Taliban also conducted spectacular raids on US bases such as FOB Salerno (1 June 2012), FOB Fenty (2 December 2012) and Camp Bastion (14 September 2012), the latter raid resulting in the destruction of six Marine AV-8B Harriers, a C-12 fixed-wing transport, three Marine MV-22 Ospreys, one C-130 Hercules and a British Sea King helicopter, before fourteen of the fifteen attackers were killed.

With the surge forces gone by the autumn of 2012 and the remainder of ISAF and US combat forces in the process of departing the country, the Taliban ramped up their attacks and began to expand their offensive operations from the south and east into other parts of Afghanistan. Recognizing the importance of local security efforts, the Taliban focused on attacking the ALP, which for the moment held its own against its assailants. By the end of 2014, however, US special forces were also being drawn down and had to shift supervision of VSO sites and the associated ALP to their Afghan counterparts. Neither program would survive the transition. With the withdrawal of US special forces teams that provided the supervision which held the programs together and established accountability for their participants, the ALP also became susceptible to abuse from local strongmen, and in many cases preyed upon the population it was meant to protect. Without these programs in place, the Afghan government had little chance of holding the rural areas of the country.[82]

Until the drawdown of ISAF in 2014, the vast majority of intelligence, surveillance, reconnaissance, fire support, engineer work and logistical support had come from coalition forces. Replacing these capabilities proved a formidable challenge. The ANSF needed to grow above 350,000 soldiers and police, a difficult goal given an annual attrition rate of some 30 percent. While the build-up of ANSF units was significant, the provision of Western technology created sustainability issues that were never resolved. For instance, members of the US Congress insisted that the Afghans "buy American," which saddled the Afghan forces with US helicopters, transport planes and close air support aircraft which were so sophisticated that they could not maintain them without the support of an army of civilian contractors; yet those aircraft became the lynchpin of the entire defense concept.

The overall concept for the defense of Afghanistan envisioned modestly trained and equipped Afghan forces securing major population centers and critical infrastructure, with approximately 35,000 well-trained and well-equipped commandos serving as reserve forces to respond to attacks around the country – using US-provided C-130 Hercules transport aircraft and UH-60 Black Hawk helicopters flown by

Afghans to transport them and provide emergency resupply and medical evacuation support. These forces would be supported by close air support aircraft and a constellation of US drones that would provide a rapid response to Taliban attacks. The system worked as long as Western support, including maintenance, remained robust, but such support was subject to the vagaries of domestic politics in the United States and its NATO allies.

The ANSF successfully protected the national elections on 5 April 2014, the first in which Karzai could not run due to term limits. As no candidate gained a clear majority, a run-off was held on 14 June, with Ashraf Ghani gaining 55 percent of the vote against his opponent, Abdullah Abdullah, although the election was again marked by irregularities and fraud. On 19 September, and after considerable drama in Kabul, Ghani assumed office as the new president of Afghanistan in a power-sharing deal that made Abdullah the nation's chief executive – a quasi-prime minister with unclear lines of authority. The deal kept the peace but led to even greater government dysfunction.

Withdrawal and endgame, 2015–2021

With a new government in Kabul and the withdrawal of the vast majority of Western forces from Afghanistan, the war entered a new phase, one highly dependent on the ability of the Afghans to secure their own territory and people. Ultimately, they could not do it. The Taliban made a steady comeback in areas vacated by US and NATO forces as the Afghan forces proved incapable of holding many areas without outside assistance. The troops were brave enough – the 66,000 dead Afghan soldiers killed during the war attest to that. But they fought for an often corrupt and incompetent government that never gained the trust and confidence of local communities, which had historically determined the balance of power within Afghanistan. For its part, the Obama administration refused to commit to substantial use of airstrikes to support the ANSF against the Taliban, the one action that could have forestalled, and perhaps even beaten back, the growing Taliban momentum.[83] US close air support was increasingly restricted to targeting clearly identified al-Qaeda elements and protecting American personnel whose security was in jeopardy, even as it was clear that the greatest threat to Afghanistan's security was not al-Qaeda but the Taliban and Haqqani insurgents.

In the absence of outright coalition military victory, both the Obama and later Trump administrations attempted to revitalize negotiations with the Taliban, to no avail. With the withdrawal of most Western troops from Afghanistan, neither administration possessed the leverage necessary to convince the Taliban to come to terms that would leave the government in Kabul in power. There was one attempt to change this dynamic. On 21 August 2017, President Donald Trump gave a major speech on Afghanistan policy, increasing US forces in the country to 15,000, assuring an extended commitment and pledging a conditions-based drawdown of the remaining US forces. The announcement failed to sway the Taliban negotiators, though, and the policy did not survive beyond the term of one of its chief proponents, US National Security Advisor H. R. McMaster, who departed the position in April 2018.

Events then moved quickly as the Trump administration attempted to achieve a deal with the Taliban prior to the end of the presidential term. Negotiations commenced in Abu Dhabi in December 2018; significantly, the Afghan government was not a party to the talks, an extraordinary concession to the Taliban. After prolonged wrangling, on 29 February 2020 US and Taliban negotiators signed a deal which provided for the withdrawal of US forces by 1 May 2021 and the immediate release of over 5,000 Taliban detainees being held by Afghanistan. In return, the Taliban promised not to attack US forces during their withdrawal and not to allow Afghanistan to become a terrorist sanctuary. The former promise was broadly honored (though US forces were no longer on the front lines); the latter promise proved hollow.

Intra-Afghan negotiations began in Doha on 12 September 2020 but went nowhere due to lack of shared negotiating space and no inclination by the Taliban to make any concessions, since they had already gained what they most sought – the withdrawal of American forces from Afghanistan. The Afghan government, under pressure from the

Americans, reluctantly released the 5,000 Taliban prisoners, knowing full well they would soon return to the battlefield – as they did. The government was willing to discuss a power-sharing arrangement, but the Taliban clearly aimed to achieve by force of arms what they could not achieve through negotiations: regime change. As far as the Taliban leaders were concerned, they had defeated the Americans, and the Afghan government would be next. As one Afghan negotiator later reported, "They were so arrogant. They thought they were there just to discuss the terms of [our] surrender."[84]

In 2021, the newly installed Biden administration reversed in short order a number of the Trump administration's international decisions, such as the withdrawals from the Paris Climate accords and the World Health Organization, but it chose to honor the terms of the agreement the Trump administration had inked with the Taliban, albeit with one minor tweak. Biden committed his administration to a complete withdrawal of US forces, but extended the deadline to 11 September 2021. The date had obvious symbolic significance, but it all but ensured the Taliban would exert maximum pressure on the ANSF at the height of the summer campaign season, with the US focused on securing its forces' withdrawal rather than on helping Afghan forces, with some exceptions.

Critical though the departure of the remaining 3,500 US troops was, even more significant was the exodus of the 17,000 Western contractors who maintained the sophisticated US aircraft for the Afghans and were unable to remain without the security, medical care and quick-reaction capabilities provided by the US Army and Air Force. Absent these maintenance personnel, the complex aviation and fire-support systems that enabled the ANSF to function quickly deteriorated and ultimately proved incapable of providing adequate support to beleaguered Afghan forces.

As US forces departed Afghanistan in the spring and summer, the Taliban launched an offensive that gathered momentum and began to overwhelm the ANSF, who came to realize that with degraded aviation readiness there would be no reinforcements, resupply or close air support for them when the Taliban attacked. With the military under pressure and surrendering or retreating around the country, the Afghan government collapsed with astonishing rapidity when President Ghani and many of his closest advisers flew north out of Afghanistan on helicopters.[85]

Kabul fell shortly afterwards on 15 August, before the United States could evacuate all of its citizens, not to mention the many tens of thousands of Afghan nationals who had worked for the US armed forces and Embassy in a variety of roles and who were now vulnerable to retaliation. As the Taliban advanced, Westerners and Afghan nationals rushed to Hamid Karzai International Airport in Kabul to catch whatever flights they could out of the country. With the airport overrun by desperate Afghans, a task force from the 82nd Airborne Division, commanded by Major General Christopher Donahue, an exceedingly capable former SOF commander, hastily deployed to Kabul to provide security around the airfield, a mission coordinated with the Taliban, who deployed their own troops to add an outer layer of security around the airfield's perimeter. One suicide bomber nevertheless penetrated the cordon on 26 August and detonated his explosive belt, killing 170 Afghan civilians and thirteen US military personnel. A US drone strike three days later targeted what analysts assessed to be a car bomb, but instead killed ten innocent civilians.

The withdrawal was concluded by midnight on 30 August, with Donahue the last soldier to board the final C-17 departing Kabul. In total, the airlift managed to evacuate 122,000 people, making it one of the largest airborne evacuations in history, albeit one that had been chaotic in execution and had left behind well over 150,000 Afghan former battlefield interpreters and their family members whose lives were in jeopardy because of their service with US and coalition forces or in Afghan units and institutions and who qualified for the Special Immigrant Visa established for them.[86]

As Winston Churchill famously said, wars are not won by evacuations. With Western forces out of the country, the Taliban returned to power in Afghanistan, and despite their protestations that they had changed since the days of their draconian rule in the 1990s, their actions quickly put such notions to rest. The regime immediately reinstituted the harsh version of sharia law that denied basic rights to women (including high school for girls and college for young women),[87] suppressed freedom of speech and assembly and privately invited leaders of al-Qaeda back into the country (in fact, on 31 July 2022, a US drone strike in Kabul killed al-Qaeda leader Ayman al-Zawahiri, who had taken over the group after the death of bin Laden in 2011).

The Afghan people, plunged into poverty and, in many cases, near starvation, today stumble onward into a bleak future under the control of their Taliban overlords – who may well find that it is easier to be insurgents than counter-insurgents, as various resistance forces and Islamic State extremists undermine security and as Western support is dramatically reduced.

Assessment

As the saying goes, success has a thousand fathers, but failure is an orphan. The failure of the United States in Afghanistan had multiple causes, and more than its share of fathers. The most critical failure, as is usually the case in lost wars, was one of policy and strategy. As was noted at the outset of Chapter 3, Carl von Clausewitz observed that "The first, the supreme, the most far-reaching act of judgment that the statesman and commander have to make is to establish ... the kind of war on which they are embarking; neither mistaking it for, nor trying to turn it into, something that is alien to its nature."[88] The Bush administration invaded Afghanistan to destroy al-Qaeda, but when the remnants of that group retreated across the border into Pakistan, President Bush changed the mission to one of nation-building and support for the nascent Afghan government. While unavoidable and necessary, this mission was never properly analyzed or resourced, especially after the United States had invaded Iraq in March 2003. The period between 2002 and 2006 was the best opportunity to create a resilient Afghan state with enough security forces to hold its own against a resurgent Taliban. Due to lack of focus, resources and strategy, however, that opportunity was squandered.[89] As a result, the Taliban were able to reconstitute their forces and return to the fight against Afghan security forces that ultimately lacked the numbers and capability to protect their state once US and coalition support was withdrawn.

The war in Afghanistan spanned four presidential administrations, but the strategic commitment made initially by President Bush did not survive even his own administration. Each subsequent administration lacked consistent and coherent commitment throughout its time in office. None of those administrations was willing to provide the resources required to achieve its strategic goals. John Paul Vann, a prominent American Army and, later, civilian adviser in Vietnam, said, "Security may be ten percent of the problem, or it may be ninety percent, but whatever it is, it's the first ten percent or the first ninety percent. Without security nothing else will last."[90] Nation-building was difficult enough without facing a virulent insurgency; attempting to build a modern Afghan state while facing a tough and resilient Taliban proved to be impossible. The Taliban understood this and were willing to wait out the United States and its allies in an extended war of exhaustion, just as the Mujahideen had the Soviets.

Once mired in conflict in Afghanistan, the US government and military found themselves woefully short of expertise. The failure to appreciate sufficiently the history, culture, politics, religion and tribal dynamics of Afghanistan and the region led to numerous mistakes in government formation, economic development and military strategy and operations. The result was the creation of an overly centralized Afghan government that ran contrary to Afghan history and culture and led to massive corruption fuelled by tens of billions of dollars in development projects in a state lacking the capacity to absorb such extravagant sums in a short period. And, of course, it also resulted in the development of military forces highly dependent on air support, firepower and centralized logistics that were unsustainable without an army of contractors to support them once the US insisted on providing American aircraft and weaponry and refused to continue to fund the refurbished Soviet/Russian aircraft the Afghans could maintain on their own.[91]

The United States and its NATO allies were also never able to convince Pakistan to eliminate Taliban and Haqqani sanctuaries on its soil. And the US never pressured Pakistan too hard, as American leaders knew that US and NATO forces were dependent on the lines of communication through Pakistan into Afghanistan and they were also concerned about the security of Pakistan's nuclear arsenal in the event that the Islamabad government collapsed. For its part, Pakistan viewed an Afghanistan that bent to its will as providing strategic depth against its existential enemy, India (though it is not clear that that concept is now being validated, especially as Taliban leaders appear to be allowing the Pakistani Taliban sanctuary in Afghanistan). The longstanding ties between the Pakistani intelligence service and the Haqqani Network and Taliban, and the limits of Pakistani Army capabilities, meant that eliminating their sanctuaries in Pakistan was always going to be difficult. But failure to do so ensured an extended conflict that enabled the Taliban's war of exhaustion against the West.

The Afghan government must also share a considerable part of the blame for its own collapse. Elections could not build the necessary foundations of the state or ensure the filling of positions based on merit. The officials in Kabul and their hand-picked subordinates in the rural areas and some provinces were never able to convince enough of the Afghan people that they would be better off under their rule than they would be with the Taliban in power. Moreover, the Taliban could claim they were fighting for Islam and against the foreign occupiers of their country, whereas the Afghan government had limited legitimacy outside of Kabul and the major cities. Too often, the predatory ANSF were as much a challenge as was the Taliban insurgency.

Near the end of his tour in Afghanistan, Major Jim Gant, the brilliant but ultimately tragic special forces officer who authored the paper that led to the creation of Village Stability Operations (and who was, in many respects, the John Paul Vann of Afghanistan), reconsidered the goals of the program in the area where he was located. "Sir, I need to tell you that there is no government of Afghanistan here," Gant emailed me from Mangwal in eastern Afghanistan. "The district center is seven kilometers away, but it might as well be seven thousand." We sought to connect the tribes to the government and the ANSF, but the latter were organizations that Gant viewed as incapable and hopelessly corrupt.

When I urged Gant to write a follow-on to his paper "to discuss how to connect the tribes to the government," a few days later Gant wrote back simply, "It cannot be done."[92]

It is no coincidence that the United States lost its two longest wars – those in Vietnam and in Afghanistan. In both countries, it propped up unpopular and corrupt regimes, eventually employed various counter-insurgency techniques (although not very skillfully in Vietnam) in an attempt to stymie popular support for insurgencies and was defeated by enemies enjoying cross-border sanctuaries and fighting to tire the American people and force the withdrawal of US armed forces from the conflict. But Vietnam was largely a war of choice, and the departure of US forces a necessity to heal the alarming fissures that were growing in the American body politic. Vietnam was, in the end, unsustainable. But Afghanistan was not. By contrast, in fact, the United States went into Afghanistan by necessity after a brutal and premeditated attack on the homeland. And departed from the conflict voluntarily, when it might have been possible even at the end to achieve a commitment that was doable in terms of blood and treasure and sustained for as long as it took – however frustrating and unsatisfactory it might have been. The reality was that the vast majority of the American people, although tired of hearing about "forever wars," had not sacrificed a great deal during the twenty years of the Afghanistan War, which was waged by volunteers representing less than 2 percent of the American population. That was in stark contrast to the Vietnam War, which had to resort to an increasingly unpopular draft to generate the enormous forces deployed in South-east Asia.

By 2021, the year of the American withdrawal from Afghanistan, US casualties were near zero, American military aid to Afghanistan totaled a little more than $3 billion annually, and the overall cost of the war was some $25–30 billion, not a heavy lift for a superpower with an over-all defense budget of over $750 billion that year.[93] With continued American support, the Afghan government could have remained in control of the major urban centers, while the ANSF matured, or at least survived, under the tutelage of US and NATO advisers.

Afghanistan was, to be sure, a difficult and frustrating situation;

however, the conflict could have been *managed* if not fully resolved, and Afghanistan did not have to fall. By withdrawing completely from the country, the United States eroded the foundation on which the ANSF rested and also gave up any leverage it might have had in negotiations with the Taliban. Thus, in the end, the US was able to withdraw from its forever war in Afghanistan, but only by losing the conflict, allowing the country to become an extremist safe haven once again and condemning some 40 million Afghans to a future of repression, deprivation, severely circumscribed opportunities and, very likely, continued violence.

Eight

The Iraq War

2003–2011*

I am neither an optimist nor a pessimist. I am a realist.
And the reality is that Iraq is all hard, all the time.
But hard is not hopeless.

Gen. David Petraeus during
the Surge in Iraq

T he terrorist attacks on the United States on 11 September 2001 led
not just to the invasion of Afghanistan and a global war against
Islamist extremists, but to renewed focus on the unresolved issues with
Saddam Hussein's Iraq as well. The invasion that would be launched
there in March 2003 and the subsequent conflict would present the
United States and its coalition partners with their greatest challenge to
date in the post-Cold War era.

The invasion of Iraq, initially seen by the Bush administration as
a showcase for high-tech forces and rapid, decisive operations, would

---

* This chapter was written primarily by Gen. David Petraeus, with input from Andrew
Roberts. Since Gen. Petraeus was a principal figure in the history of the war in Iraq,
events in which he was involved will be narrated in the first person.

quickly evolve into a grinding counter-insurgency campaign for which the US Army and Marine Corps initially were neither intellectually nor organizationally prepared. As with conflicts throughout history, the ability of the respective warring sides to adapt to the dynamics of the battlefield would do much to determine which side prevailed.

As I write, twenty years after the initial invasion, we are still grappling with the effects of the war, which necessitated an overhaul of military doctrine and renewed study of and training in counter-insurgency, saw the advent of new battlefield capabilities (especially drones, counter-terrorism force capabilities and intelligence fusion), imparted major cautionary lessons about regime-change operations (reinforced by the outcomes of the various Arab Spring uprisings) and brought about significant changes in regional and global dynamics. Over time, in particular, the strategic effect of toppling the Ba'athist Sunni regime in Iraq would live on in the form of an emboldened and more powerful Shi'a Iran, with its hybrid forces unleashed throughout the Middle East in an attempt to dominate the so-called Shi'a Crescent – stretching from Iran through Iraq and Syria into southern Lebanon – and to create ever greater problems for Israel and many of the Sunni Arab countries of the region as well.

Background

The Bush administration decided for a number of reasons to topple Saddam Hussein and the Ba'athist regime that had ruled Iraq brutally for nearly twenty-four years. But primarily it focused its public case for the war on a flawed intelligence assessment that the regime possessed a hidden arsenal of weapons of mass destruction and the means to deliver them, as well as the capability to produce more.[1] President George W. Bush also highlighted the importance of ending the rule of a dictator who had invaded Iran in 1980 and Kuwait in 1990, and who continued to threaten Kuwait as well as Iraq's Kurds and Shi'a Arabs, all this despite Saddam's forces in Kuwait having been destroyed by a US-led coalition during Operation Desert Storm in 1991.[2] For Bush, being able to introduce democracy into the Arab–Islamic world as part of a broader response to Islamic terrorism in the aftermath of 9/11 was significant too.

In planning the military campaign to destroy Saddam's regime, however, the administration made a number of assumptions that would prove unfounded and would undermine the prospects for success. Foremost among these was the belief that the Iraqi people would welcome US forces as liberators and would continue to regard Americans as such for a while thereafter. This was not an unreasonable assumption given the enmity with which the Saddam regime was regarded by the vast majority of Iraqis, and it did prove true for a signif-icant percentage of the Iraqi people in the early months of the operation. However, a sizeable number – particularly in Sunni Arab areas – came to resent US interference in Iraqi affairs, the turmoil that followed and the opening it created for Iraq's Shi'a majority to take control of the coun-try. Moreover, even among Iraqis supportive of the US invasion, many remained skeptical of American intentions, and few wanted to see foreign forces occupy their country.

Early on, US policymakers also relied on initiatives to establish liberal democracy and a market economy to cure Iraq's ills. But, as in Afghanistan, they ignored the culture and history of the Iraqi people and their traditions of governance, justice and economics – not to mention underestimating how momentous such changes would be.[3] The problems created by these unfounded assumptions would be compounded enormously by decisions made by the US-established Coalition Provisional Authority two months into the war.

The setting

It is instructive here to reflect briefly on the history of Iraq, which became a country upon the break-up of the Ottoman Empire after the First World War. Formed from the Ottoman provinces of Mosul, Baghdad and Basra, Iraq traced its ancient history to Mesopotamia, the Land of the Two Rivers – the Tigris and the Euphrates – that run the length and breadth of Iraq, from the north and the west, respectively, to the south, and give rise to the extensive agriculture of the region along and between those great rivers.

The northern third of the country is centered on Mosul, Irbil and oil-rich Kirkuk, with the mountainous northern part inhabited by Kurds, one of the largest ethnic groups in the world without a state of their own. The rest of the region is predominantly Sunni Arab, but contains significant numbers of Shi'a Arabs, Turkmen, Christians (of various denominations), Yazidis and other ethnic and sectarian group- ings. The Iraqi Kurds had largely governed themselves following the Gulf War in 1991, when the United States supported a Kurdish uprising and kept Iraqi troops out of the northern region (not least by the estab- lishment in 1991 of a no-fly zone policed by US-led air forces, just as a similar no-fly zone prevented Iraqi aircraft from attacking the Shi'a in the southern half of the country).[4]

Sunni Arabs dominated in the western and north-central parts of the country, mostly desert outside the populated Tigris and Euphrates River valleys, while Shi'a Arabs largely populated southern and eastern Iraq, including the oil-rich area around Basra at the northern tip of the Persian/Arabian Gulf. In the center of the country sits the capital city of Baghdad, which in 2003 contained nearly 6 million people – mostly Shi'a and Sunni Arabs but also small numbers of Assyrian Christians, Faili Kurds and others.

Since his ascension to power in 1979, Saddam Hussein had ruled Iraq's polyglot population with an iron fist, including throughout the bloody eight-year war with Iran that began with an Iraqi attack on Iran, as well as during the Iraqi invasion and subsequent occupation of Kuwait in August 1990 that ended in defeat at the hands of the US-led coalition some eight months later. The American-led invasion in late March 2003 that toppled Saddam's regime in less than three weeks would radically change the dynamics in Iraq and unleash sectarian, ethnic and tribal tensions long kept in check by Saddam's brutal rule. Neither the dynamics of Iraqi society nor the country's traditions were sufficiently understood by leaders in the Bush administration at the outset, and this left them inadequately prepared for the aftermath of regime change.

The plan

The operational plan for the invasion of Iraq, dubbed Operation Iraqi Freedom, relied heavily on the concept of a swift build-up of forces and rapid execution of an offensive campaign to overthrow the Iraqi government. This would be in considerable contrast to the six-month build-up of forces in Saudi Arabia in 1990 and the subsequent five-week air campaign before ground forces were launched in February 1991 to liberate Kuwait. Gone was the force structure of three corps and 380,000 troops envisioned in previous war plans. Instead, General Tommy Franks, influenced by the swift success of the invasion of Afghanistan in late 2001 and under pressure from Secretary of Defense Donald Rumsfeld to pare down the forces employed, would rely on a light-footprint approach to the war in Iraq.

Rumsfeld had issued a number of his trademark "snowflakes" (directives that showered down on his staff like snow and had to be answered expeditiously) and forced six iterations of the war plan. At the end of the process, Franks had pared down the ground-troop list to about 130,000 troops in all: most of a US Army corps and a Marine Expeditionary Force (MEF) that together comprised two Army mechanized infantry divisions (one of which would ultimately be delayed arriving in theatre and miss the fight to Baghdad), a Marine division and an accompanying air wing, an air assault infantry division, one-third of an airborne infantry division, an independent airborne infantry brigade and an attack helicopter regiment, plus various additional aviation, artillery, air defense, engineer, communications, intelligence and logistical elements. To these forces were added roughly 30,000 coalition troops, primarily a British armored division and its associated support units, and small contingents from Australia and Poland.[5]

Adding to the inherent friction associated with any such complex undertaking, Rumsfeld delayed approval of some of the forces (among them the 101st Airborne Division [Air Assault], which I commanded. In fact, I had to declare an emergency deployment readiness exercise in order to get the division's 5,500 vehicles, thousands of containers and

254 helicopters to the port in time to meet the vessels that were inbound and would take the assets to Kuwait). Rumsfeld also did not approve certain logistical forces for the 101st that were integral to the performance of critical functions until so late in the deployment process that their vehicles and equipment were not in Kuwait by the time of the invasion.[6]

The shortcomings of the plan and its associated force structure would become apparent early in the operation as order collapsed in Iraq in the wake of the toppling of the regime, though there had been warnings of inadequate "boots on the ground." In his testimony to the Senate Armed Services Committee on 25 February 2003, as forces were arriving in Kuwait to prepare for the invasion, for example, US Army Chief of Staff General Eric Shinseki offered his view – based on his experiences as commander of the US peacekeeping forces in Bosnia – that several hundred thousand troops would be required to stabilize Iraq in the wake of regime change. Two days later, however, Deputy Secretary of Defense Paul Wolfowitz dismissed such warnings, telling the House Budget Committee, "Some of the higher-end predictions that we have been hearing recently, such as the notion that it will take several hundred thousand U.S. troops to provide stability in post-Saddam Iraq, are wildly off the mark. It is hard to conceive that it would take more forces to provide stability in a post-Saddam Iraq than it would take to conduct the war itself."[7] Wolfowitz's statement might have been true if assumptions had turned out to be valid that the Iraqis would support a war of liberation and that its institutions (especially local police and municipal governments) would remain intact after the toppling of Saddam and his senior officials. That would not prove to be the case.

In contrast to the thirty-eight-day air campaign that preceded ground operations during the Gulf War in 1991, air and ground operations in Iraq would begin nearly simultaneously, preceded only by special operations forces that infiltrated into Iraq just ahead of the invasion.[8] Shock and awe would substitute for mass. The plan called for the US V Corps and I MEF to attack north from Kuwait on the west side of the Euphrates and east side of the Tigris, respectively, to destroy the regime's center of gravity in Baghdad, supported by an attack by the 4th Infantry Division (Mechanized) southwards from Turkey and by the British 1st Armored Division, which would attack north from Kuwait to seize Basra, the major city in southern Iraq.

At the last minute, however, the Turkish parliament denied permission for deployment of a US division through its territory, so the ships carrying the 4th Infantry Division's armored systems, helicopters, vehicles and containers were redirected by sea to Kuwait. The two-pronged attack from north and south was thus reduced to a single operational thrust from the south, led on the west by the US Army 3rd Infantry Division (Mechanized), commanded by Major General Buford "Buff" Blount, followed by the 101st Airborne Division (Air Assault), which I commanded, and on the east, by the 1st Marine Division, commanded by Major General James Mattis. The combat elements of the 82nd Airborne Division, commanded by Major General Chuck Swannack, were to be retained at an airfield in Kuwait as the theatre reserve, with the possibility of an airborne operation to seize the major airport on the west side of Baghdad. The entire operation, from deployment through invasion to regime change, was projected to take less than six months.

While considerable effort had been devoted to planning the combat operations, much less had been put into planning post-regime-change operations to stabilize the country, leaving a huge void that would haunt coalition leaders as the months wore on. Administration officials and military planners both assumed that Iraqi people and local officials would largely cooperate with efforts to remake their government and economy and that significant assistance would be offered by the international community and non-governmental organizations, the latter an assumption that overlooked the degree of ill-will generated by the Bush administration's rush to war.

The little planning for these tasks that had taken place was conducted primarily by the Office of Reconstruction and Humanitarian Assistance (ORHA), led by retired US Army Lieutenant General Jay Garner and established only two months prior to the onset of the invasion. However, Garner, who had led an effort in the Kurdish area of northern Iraq after the Gulf War, had little time to assemble his team and prepare plans to mitigate humanitarian disaster, stabilize the Iraqi government and oversee reconstruction of essential infrastructure. In the end, his organization lasted only a month after the onset of hostilities, as Secretary Rumsfeld grew frustrated with it and decided to replace it with another element, the Coalition Provisional Authority, leading to further confusion after the fall of the regime in Baghdad.[9] Hindsight of course is 20/20, but there were voices raising concerns at the time, mine among them. When, during the final gathering of commanders before the invasion I asked if the ORHA leaders could provide additional information about what would happen after we got to Baghdad and toppled the regime, I was told, "You just get us to Baghdad, Dave, we'll take it from there."

Prior to the invasion, Saddam Hussein had apparently been more concerned with preventing an uprising among an unruly population than with preparing the Iraqi armed forces for an invasion that he did not believe would take place – or, if it did, would stop short of Baghdad as the international community stepped in to force a ceasefire.[10] The Iraqi armed forces had, moreover, been severely degraded by years of under-resourcing, inadequate training and insufficient staffing. Two actions Saddam had taken would, however, bedevil US forces long after the invasion – stockpiling arms and ammunition in and around urban areas throughout the country, and strengthening paramilitary groups such as the Saddam Fedayeen, a force of more than 30,000 lightly armed volunteers who were loyal to Saddam and reported directly to one of his sons. The Fedayeen would present a unique set of challenges to US commanders as their lines of communication lengthened from the Kuwaiti border all the way to Baghdad (some 350 miles), and the dispersed weapons caches served to arm a nascent insurgency once the Ba'athist regime collapsed.

The invasion

Air attacks on Iraq commenced on 19 March 2003, and the ground invasion was launched less than twenty-four hours later on the 20th. Although not every aspect went according to plan – a massive dust storm brought the advance to a halt for several days well over halfway to Baghdad, and logistical challenges caused additional delays – the operation succeeded well beyond the most optimistic expectations of coalition commanders. The 3rd Infantry Division (Mechanized), advancing towards Baghdad west of the Euphrates River, and the 1st Marine Division, advancing north-east from Kuwait, ultimately crossing to the east of the Tigris, made rapid progress against Iraqi forces that were surprised by the absence of a lengthy air campaign and were largely outmatched. Ground convoys of the 101st Airborne Division (Air Assault) followed the 3rd Infantry Division, and the 101st also conducted a large air assault operation to position infantry forces and the bulk of our division's fleet of more than 250 helicopters just south of the 3rd Infantry Division elements. Meanwhile, the British 1st Armored Division advanced through southern Iraq towards the Al-Faw Peninsula, the port of Umm Qasr, Basra and the Rumaila oil fields, meeting only sporadic opposition.

The first significant resistance came in Nasiriyah, in south-central Iraq, where regular Iraqi forces and Fedayeen militia defended the city against Marine attacks to secure the vital bridges spanning the Euphrates River.[11] The 2nd Marine Expeditionary Brigade (Task Force Tarawa) seized the bridges after a sharp fight, while the 3rd Infantry Division, west of the Marines, headed north for the Karbala Gap, which was defended by the Medina Armored Division, a formation of the Iraqi elite Republican Guard. To open that battle, the V Corps commander, Lieutenant General William S. Wallace, ordered a deep attack against the Medina Division by thirty-two Apache attack helicopters of the 11th Attack Helicopter Regiment. On the night of 23–24 March, the Apaches maneuvered to engage their targets east of Karbala but were hit by a barrage of fire from anti-aircraft weapons, machine guns and small arms as they passed over inhabited areas. The strike achieved little other than the destruction of a few air defense systems and gun trucks, an unacceptable trade-off considering the damage suffered by the attacking helicopters. One helicopter crashed on take-off, another was shot down and all but one of the remainder sustained damage from multiple hits, effectively taking the unit out of action for the remainder of the offensive.[12]

As the American units advanced north, they continued to run into unexpected opposition from members of the Fedayeen paramilitary forces. Dressed in either black outfits or civilian clothes, these militia elements established numerous ambushes with RPGs, mortars and small arms. Undetected by high-tech intelligence, surveillance and reconnaissance systems, the low-tech Fedayeen frustrated American combat and logistics forces, leaving Lieutenant General Wallace to declare to a reporter while visiting me at the 101st Airborne Division headquarters, "The enemy we're fighting is a bit different than the one we war-gamed against, because of these paramilitary forces. We knew they were here, but we did not know how they would fight."[13] Likely because his comment truthfully acknowledged that not everything was going according to plan, Secretary Rumsfeld was displeased with Lieutenant General Wallace's admission, which reflected an accurate assessment of the situation. This may have been a factor that precluded Wallace's extension of command in Iraq when his tour as commander of V Corps was complete in June. This resulted in the rotation back to the United States of an exceedingly competent and experienced corps commander and his replacement by a two-star general who was promoted to three stars in Iraq to take the position. America's most important operation would thus be commanded by the junior three-star general in its ranks.

Days before the commencement of operations, special operations forces had been inserted into western and northern Iraq to hunt for Iraqi long-range SCUD missiles. Other special operations forces had been positioned well in advance in the north to coordinate military operations with the 65,000-strong Kurdish Peshmerga forces that would operate from the largely autonomous Kurdish region of northern Iraq, and to target the terrorist group Ansar al-Islam near the border with Iran in the eastern part of the Kurdish region. Meanwhile, the American presence on the ground in northern Iraq convinced Turkey at least to open its airspace to US aircraft. That allowed nearly a thousand paratroopers of the 173rd Airborne Brigade, based in Italy, to parachute on to Bashur Airfield, north-east of Irbil, on 26 March. Within three days, the rest of the brigade and its vehicles were on the ground, followed by the tanks and infantry fighting vehicles of the 1st Battalion, 63rd Armored regiment that were airlanded after the paratroopers had secured the airfield. The paratroopers and their armored augmentation then advanced on Kirkuk, destroying elements of four Iraqi infantry divisions along the way, and took control of the city on 10 April.

A huge sandstorm that blew into southern Iraq on 24 March forced an operational pause for several days by the forces attacking to the north in execution of Cobra II, the name given to the overall ground campaign. US forces used the pause to secure their lines of communication and replenish fuel, ammunition and other supplies for the next phase of the operation, while also resting their troops after the nearly 300-mile advance in the early days of the invasion. Lieutenant General Wallace also determined that major urban areas initially bypassed would need to be secured in order to extend vulnerable lines of communication back to Kuwait, and he tasked the 82nd Airborne Division with securing Samawa, and a reinforced Marine Task Force Tarawa to clear Nasariyah. The 101st Airborne Division, under my command, was tasked with clearing Najaf, a major urban center of over 500,000 people (and the site of the holiest shrine in Shi'a Islam), as the Iraqi elements in it threatened the V Corps' lines of communication. Together, these operations did a great deal to secure extended lines of communication back to Kuwait that would otherwise have been very vulnerable.

As the sandstorm dissipated on 27 March, the main body of the 1st Marine Division headed north-east for Diwaniyah and a crossing of the Tigris River near Kut, while the 3rd Infantry Division advanced west of Najaf towards the Karbala Gap and a critical crossing over the Euphrates River (Objective Peach), which it captured after a stiff fight and a failure by Iraqi forces to completely destroy the bridges spanning the river.[14]

Meanwhile, my team and I were employing the 101st Airborne Division's attack helicopters to engage enemy elements ahead of the 3rd Infantry Division in deep night attacks. We also positioned the 101st's three infantry brigades in the desert west and then north of Najaf prior to ordering them to seize that city. When the several-day battle for Najaf was complete on 3 April, I radioed my boss, Lieutenant General Wallace, that I had good news and bad news. "The good news," I reported, "is that we own Najaf." "What's the bad news?" he asked. "The bad news," I responded, "is that we own Najaf. What do you want us to do with it?"

Wallace suggested that I contact the ORHA leaders (who were still finalizing preparations of their elements in Kuwait) for help. My subsequent call resulted in no assistance. At that point, it became clear that ORHA would be of no immediate use in administering areas of the country that were being liberated. As had often been the case historically, the US Army and Marine Corps would have to pick up the slack, in this case by leaving forces in the major urban areas they seized to ensure security and order.[15] I subsequently had to leave a brigade to secure Najaf and areas we seized to its north. We also had to commit forces to stabilize Karbala after the fight to secure that city, home to well over 400,000 people and also the location of two extremely important Shi'a shrines, and again after seizing Hillah, another substantial city to the east of Karbala (and the site of ancient Babylon).

The endgame began on 3 April as US forces destroyed the last resistance south of Baghdad and advanced on the Iraqi capital. As V Corps considered whether to send the infantry-heavy 101st Airborne and its more than 250 helicopters to seize Baghdad International Airport, the Commander of the 3rd Infantry Division, Major General Blount, radioed Lieutenant General Wallace, affirming, "Sir, we trained for this . . . We prepared for this ... We're ready for this. We need to go now." After a pregnant pause, Wallace replied, "Have a good fight. Victory 6, OUT."[16] Colonel Will Grimsley's 1st Brigade, 3rd Infantry Division hit Baghdad International Airport (Objective Lions) with an artillery barrage and then maneuvered its armored and mechanized forces through sporadic enemy resistance to reach the facility on the west edge of Baghdad by 11 p.m. After two days of often intense fighting, Grimsley's troops cleared the airport of its Special Republican Guard defenders.[17] In southern Iraq, British forces seized Basra by 7 April with little resistance.

Leery of becoming caught up in urban combat, Wallace had intended for the 3rd Infantry Division to conduct a series of armored raids, or "thunder runs," into the center of Baghdad to weaken Saddam's hold on the city, with the goal of collapsing the regime within thirty days. In the end, it didn't take nearly that long.

On 5 April, the 1st Battalion, 64th Armor, under the command of Lieutenant Colonel Rick Schwartz, launched a thunder run into the heart of Baghdad from its position south of the city, encountering heavy fire and tough fighting but meeting little organized resistance. Despite some losses, the Task Force attacked north and then turned west onto the airport road, joining the other forces gathered at the airport by nightfall. Meanwhile, the 3rd Brigade, under the command of Colonel Daniel B. Allyn, attacked along the western outskirts of Baghdad, ultimately isolating the city from the north by the evening of 6 April. For the next sixty hours Iraqi forces attacked the 3rd Brigade positions, leading to "some of the most intense fighting of the invasion."[18]

Given the relative ease of the first thunder run into the heart of Baghdad, Blount ordered the 2nd Brigade commander, Colonel David G. Perkins, to attack again, but this time using the full might of the brigade. Perkins maneuvered his brigade into Baghdad on 7 April and, after fighting through tough resistance, reached the government center (what would become the "Green Zone"). This time, instead of withdrawing, he and Major General Blount made a crucial decision and stayed. Wallace was concerned that the corps was not ready to seize Baghdad completely, but he did not order a withdrawal, understandably loath to give up ground that Perkins' brigade had fought hard to take.

Meanwhile, the 1st Marine Division, east of Baghdad, crossed the Tigris River and attacked west into the eastern part of the city to link up with Colonel Perkins' brigade in the center. On 10 April, Allyn's brigade north of Baghdad attacked south into the city, joining Perkins' brigade as well. With the better part of two divisions' worth of combat power occupying the heart of the Iraqi capital, the regime collapsed. The remaining Iraqi soldiers discarded their uniforms and went home (most taking their weapons with them), and Saddam and his senior Ba'ath Party leaders fled.

Consolidation

Although the regime had crumbled and coalition forces now occupied two of Iraq's three largest cities, the northern part of the country, including the most populous Sunni areas and the city of Mosul, remained to be secured. The 1st Marine Division continued its movement north into Diyala and Salahadin Provinces to secure the cities of Baqubah, Samarra, Tikrit and Bayji, ahead of the arrival of the 4th Infantry Division (Mechanized), which had landed in Kuwait and was making its way north to relieve the Marines. Meanwhile, US special operations forces, supported by considerable airpower, largely destroyed the elements of the Iraqi V Corps still defending in the north and entered Mosul on 11 April. More than 1,000 largely dismounted members of the 26th Marine Expeditionary Unit were then flown in from the Mediterranean to assume control of the city; however, they proved insufficient in number, especially after a large demonstration against a Ba'ath Party loyalist who had declared himself mayor threatened the Marines guarding the governance center in the heart of the city. The resulting death of seventeen civilians touched off further upheaval in Mosul and forced the Marines' withdrawal to the outskirts of the city.

Well to the east of Mosul, Peshmerga forces had entered Kirkuk ahead of US special operations forces, though that situation was precarious until the arrival of the 173rd Airborne Brigade. Simultaneously, other special operations forces began to establish checkpoints and roadblocks on Iraq's western border, although much of Al-Anbar Province west of Baghdad remained ungoverned territory until the arrival of a brigade of the 82nd Airborne Division, followed by the US Army's 3rd Armored Cavalry Regiment in late April, though neither unit contained enough combat power to control the huge area assigned to it.

It became increasingly clear, at least to those on the ground, in mid-April that much more force was needed in northern Iraq to quell growing demonstrations, looting and lawlessness – and also to rein in the Kurdish Peshmerga that had encroached on Nineveh Province – as well as to provide the capacity to establish security, restore essential services and institute governance in that vast area of Iraq. Accordingly, Lieutenant General Wallace decided to deploy the 101st Airborne Division – which had the helicopters required to rapidly air-assault substantial forces to the province, starting with Mosul, the capital. That decision would leave in Baghdad only the 3rd Infantry Division (which was scheduled to be replaced by the 1st Armored Division in June) and would impact operations in the Iraqi capital in the months ahead. Nevertheless, the "Screaming Eagles" of the 101st made the jump north beginning on 22 April and we quickly established ourselves in Mosul and throughout Nineveh Province, as well as positioning small elements in each of the three major cities of the Kurdish region.

As I led the 101st Airborne Division into Iraq and watched many of the assumptions that had been made before the war invalidated, I rhetorically queried the reporter embedded with me, military historian Rick Atkinson, "Tell me how this ends?"[19] Now my division and I were responsible in part for providing an answer to that all-important question – one I was destined to hear again and again over my subsequent seven years in Iraq and the greater Middle East.

With the fighting now largely ended, US and UK commanders wrestled with challenges unforeseen by the pre-invasion wargames – which had broadly assumed local Iraqi cooperation with and assistance to the occupying forces and that additional reconstruction and governance activities required would be turned over to agencies outside of the Department of Defense and to international organizations. It was increasingly apparent that insufficient effort had been put into planning for so-called "Phase IV" operations that would stabilize Iraq following regime change. What planning there had been was overwhelmingly focused on humanitarian operations rather than on the establishment of wide area security, repairing critical infrastructure, re-establishing basic services and instituting governance – all issues that had been given inadequate attention by the planners and leaders at the most senior levels.

Of these, security was the most important issue. With the collapse of the regime, all order broke down and destructive looting ensued. Iraqis destroyed Ba'athist regime and government facilities and residences and ransacked public buildings, often stripping them down to nothing more than concrete shells, even taking roof tiles and plumbing fixtures and tearing copper wire from the walls. Meanwhile, Ba'ath Party operatives and others plundered the numerous arms caches that Saddam had dispersed throughout Iraq in the years after the Iran–Iraq and Gulf Wars, providing the wherewithal for a future insurgency. Given these developments, it quickly became clear that the coalition forces lacked not just the number of soldiers needed, but the rules of engagement, equipment and training for the kind of operations that would enable them to stop the mayhem, as nowhere near enough Iraqi police remained on the job to deter destructive vandalism and theft. Secretary of Defense Rumsfeld brushed the issue aside with the dismissive comment, "Freedom's untidy."[20] This seemed more than a bit flippant to those of us on the ground, especially given the challenges Rumsfeld had injected into the planning process by delaying the deployment of critical logistical elements prior to the invasion, some of which were just joining us with their vehicles and equipment.

Lacking a comprehensive plan for the situation in which they found themselves and receiving minimal guidance from their higher headquarters in Baghdad, individual division commanders employed a variety of techniques to chart the way ahead. In Nineveh Province, the 101st Airborne Division quickly established security, held provincial and district caucuses and elections, formed a multi-ethnic, multi-sectarian governing council, reopened the border with Syria, began training police and other local security forces and used confiscated Iraqi funds to jump-start the economy, promote business deals and repair local infrastructure – the latter being efforts that would later be funded by the US Congress as the Commander's Emergency Response Program (CERP). Other divisions did likewise, though some focused more heavily on security, which unquestionably was job number one and the foundation on which all else depended, but which was nonetheless typically solidified by other, non-military tasks. As the magnitude of the stability operations undertaking became clearer, the need for civil affairs, military police, engineer and medical units quickly dwarfed the US military's capacity and available force structure. Substantial numbers of private contractors were used to fill many of the gaps, but these substitutes came with additional challenges such as differing rules of engagement, capability shortcomings and issues of indiscipline and lack of accountability.

In retrospect, it is obvious that the planning for the post-conflict phase, the organizations established to perform the necessary tasks and the assumptions about Iraqi assistance were inadequate and flawed. And these shortcomings would be compounded enormously by subsequent decisions made in May 2003.

Losing the peace

In early May, frustrated by the lack of progress, the Bush administration decided to replace the ineffective ORHA with what would become known as the Coalition Provisional Authority (CPA), led by former US Ambassador L. Paul "Jerry" Bremer III.[21] Bremer was unquestionably very intelligent; however, he lacked any significant knowledge of Iraq or the wider Middle East and proved to be an unfortunate choice for the job. His first three major decisions after arriving in Iraq, taken without consultation with those of us who had been on the ground for two months already, all but ensured that a virulent insurgency would erupt in the months ahead.

Bremer's first decision was to order an extensive de-Ba'athification of Iraq society, which cut too deeply into the Iraqi social fabric. Instead of forcing from public life just the top *two* layers of the ruling Ba'ath Party – around 6,000 members of Saddam's regime and his top lieutenants – the order axed the top *four* layers, removing from their positions 85,000–100,000 Iraqis, most of whom had joined the Ba'ath Party as a prerequisite to employment and were advanced largely automatically. These included civil servants, doctors, engineers, university professors, schoolteachers and other professionals – the very people that US war plans had assumed would remain in their positions to provide continuity in government and to enable basic service restoration and reconstruction of Iraq. The order denied them government employment, pensions and, in many cases, their government-provided housing and car, as well as meaningful participation in civic life. With exceptions to the order controlled by a de-Ba'athification commission chaired by the highly sectarian Shi'a leader Ahmed Chalabi, an expatriate Iraqi who was closely connected to influential members of the Bush administration and who had returned to Iraq in the wake of the invasion, the process was draconian.[22] The CIA station chief told Bremer that if he followed through with the order, "By nightfall, you'll have driven 30,000 to 50,000 Baathists underground. And in six months, you'll really regret this."[23] His estimate was an undercount by an order of magnitude or more.

The second fateful decision was to disband all Iraqi security organizations, including the Iraqi Army. Bremer claimed that Iraqis despised these instruments of regime control and that he was just acknowledging the obvious, as most soldiers had discarded their uniforms and gone home.[24] However, the hatred levelled at the Republican Guard and regime loyalist organizations was not directed at the sizeable Iraqi Army, which was seen by many Iraqis as their only true national institution. Indeed, coalition leaders had planned to use that Army to help secure the country after the fall of the regime. Though Iraqi soldiers had eventually melted away in combat, many leaders had re-emerged after the collapse of the regime and a very considerable number of soldiers could have been easily recalled with the right initiatives. This would have alleviated the enormous deficit of boots on the ground when order broke down. Instead, Iraqi officers and soldiers alike were dismissed, deprived of their positions, their pensions and their place in society. It took five increasingly violent weeks of enormous demonstrations before Bremer agreed just to provide stipends and back pay, using money captured during the invasion, to the former members of the military. After an exceedingly tense and dangerous demonstration in Mosul the fifth week after the military was fired, in which many thousands of former soldiers with weapons gathered in the square around the governance center, I engaged with the demonstrators using a loudspeaker system and promised to fly immediately to Baghdad and convey their concerns if they dispersed peacefully, which they did. I subsequently flew to Baghdad and located the individual under Ambassador Bremer overseeing security sector reforms. "Your policies are killing our soldiers," I told him bluntly. "We must pay the former Iraqi soldiers stipends to enable them to take care of their families." Ambassador Bremer approved that recommendation within a week and it was implemented shortly afterwards, staving off a major crisis. Unfortunately, by the time stipends were announced, enormous damage had been done and many of the former members of the military had begun to take matters (and their weapons) into their own hands and entered the ranks of the nascent insurgency.

With these two decisions, Bremer had gone beyond regime change to destroy the Iraqi state and to create the political and military foundations for the insurgency. Unfortunately, President Bush and the National Security Council had not discussed nor considered the potential effects of these decisions.[25]

Bremer's third decision was to govern as a viceroy through an appointed Iraqi Governing Council (IGC), rather than holding local elections to build interim governance from the ground up, as some of us had already been doing in the areas for which we were responsible. This was a much more debatable decision, as it would have taken considerable time to set the conditions for free and fair elections countrywide and to conduct them properly. Nonetheless, this decision inevitably created enormous challenges as well. The Governing Council included representatives from most of the major ethnic and sectarian elements, as well as existing Iraqi political parties (several of which had been clandestine or in exile under Saddam), but left out a number of important players, such as Shi'a leader Muqtada al-Sadr and his adherents, as well as influential representatives of the Sunni community. Essentially, the Governing Council was automatically making winners and losers in the new Iraq by creating fiefdoms within the Iraqi government to provide jobs to their supporters. It was not a coincidence that people left out of the system viewed it as foreign-imposed and corrupt, and Shi'a militias beholden to the fiery cleric Sadr would soon pose as big a challenge to stability in Iraq as the growing Sunni insurgency. Moreover, Bremer began to put the brakes on democracy initiatives at local levels that I and others had been pursuing and instead focused on building a new Iraqi government from the top down through the creation of a Transitional Administrative Law. This inevitably lacked the necessary bottom-up legitimacy and sense of representation needed to unify the Iraqi people behind the new government. In addition, Bremer alienated the leaders of Iraq's tribes, a critical element of Iraqi society, by deciding that they were another relic of the past that had no place in the new Iraq, an action that would not be reversed until the beginning of the Anbar Awakening in mid-2006 and the conduct of the Surge in 2007. Bremer's decisions reminded us of the importance of ensuring that policies (and operations) did not create more enemies than they took off the battlefield by their implementation.

The first roadside bombs, or improvised explosive devices (IEDs), targeted American soldiers in April, followed shortly thereafter by the first car and truck bombs that could bring down the façades of entire buildings. Foreign fighters and terrorist operatives, sensing an opportunity to kill Americans, began to flock to Iraq in increasing numbers. US and other coalition forces battled to stem the rising tide of violence, but there simply were not enough security forces available to protect the population and also root out the operatives of a growing insurgency. The robust, seasoned land component headquarters that coordinated the invasion forces departed Iraq in June, and was replaced by a combined joint task force (CJTF-7) under the command of Lieutenant General Ricardo Sanchez – who was promoted in Baghdad and assumed command of V Corps (which formed the nucleus of CJTF-7) from Lieutenant General Wallace on 14 June.[26] Both he and his headquarters lacked the bureaucratic heft required to manage the many tasks associated with stabilizing the country, a situation made worse by unclear lines of authority and the development of a strained relationship between his headquarters and Bremer's. Throughout its existence, CJTF-7 struggled to develop a strategy and associated operational concepts to address the challenges it faced increasingly from the developing Sunni insurgency and extremist groups and the nefarious activities of Shi'a militia elements. Given the inadequate number of US and coalition forces, it was also increasingly clear that Iraqi military and police elements needed to be reconstituted. To address this task, Major General Paul D. Eaton was brought to Iraq from the US Army's Infantry Training Center to establish a Coalition Military Assistance Training Team, but it and its affiliated organization, the Civilian Police Assistance Training Team (headed by a British civilian official), were vastly under-resourced. Moreover, the goal of creating just nine brigades (a total of twenty-seven infantry battalions) over a period of two years was a far cry from what was needed to help defeat the insurgency and eventually take over the security tasks in Iraq once US forces withdrew.

Despite these headwinds, there were some successes. On 22 July, soldiers of the 101st Airborne Division's 2nd Brigade and special operations forces killed Saddam's sons Uday and Qusay in Mosul. The 1st Armored Division defeated the insurgency's first Ramadan offensive in Baghdad in October and November. And most importantly, special operations forces working with the 4th Infantry Division captured Saddam Hussein in a spider hole south of Tikrit on 13 December. These accomplishments temporarily took some wind out of the sails of the Sunni insurgency, which might have shrivelled altogether had Bremer and the IGC made an effort to placate the Sunni community by establishing a national process for reconciling with the lower-level former Ba'athists and to more aggressively rebuild the Iraqi Army. Sadly, they did not.[27]

The April uprising

Rumsfeld, meanwhile, took advantage of the relative calm to begin to draw down US forces in Iraq in the 2004 spring rotation, from the 125,000 soldiers still on the ground to just 95,000, while non-US coalition forces would drop from 35,000 to 25,000.[28] Since the Iraqi Army had been disbanded and there were still major shortages of police, however, there were simply not enough troops in country to protect the Iraqi people and deal with the growing insurgency, Shi'a militia challenges and rising criminality. We were, in essence, reducing numbers just when we should have been ramping up. In northern Iraq, the situation was particularly acute when the three infantry brigades of the 101st Airborne Division and its 250 helicopters were replaced by a much less capable reinforced Stryker infantry brigade and a small aviation element overseen by a modest headquarters under a one-star general in early February 2004, the overall element less than one-third our size. The situation was compounded when, shortly after I left Mosul in early February 2004, Ahmed Chalabi, the Iraqi head of the de-Ba'athification Commission, disapproved my long-pending recommendations for reconciliation with the many thousands of former Ba'ath Party members we had needed (and had been using) to run the Ministry elements, hospital, university and other government entities in Nineveh Province. With Chalabi's actions, Baghdad was doing the opposite of what we should have been doing, and further inflaming a deteriorating situation.

Meanwhile, General John Abizaid, the Commander of US Central Command, the headquarters that oversaw operations in Iraq and the greater Middle East, began to assess that US forces were becoming part of the problem as opposed to part of the solution. In his view, the longer US forces remained embedded among the Iraqi people, the more the insurgency would gain strength. This was not an unreasonable concern; I had often noted that "Every army of liberation has a half-life before it becomes an army of occupation." But the question was one of timing in various areas and whether adequate Iraqi forces existed to take over security tasks from the Americans. Regardless, by early spring 2004, US forces began a slow progress that, over the next two years, would result in a reduction of numerous positions inside Iraqi cities to consolidate on large forward operating bases mostly on the outskirts.[29] Again, this view was not without some logic as long as violence remained relatively low; however, over time (especially when sectarian violence escalated in 2006), the repositioning and handover of security tasks to Iraqi forces of still limited capacity would lead to a vacuum of security inside Iraqi cities. And that vacuum would, increasingly, be filled by Sunni insurgents, Shi'a militias (some supported by Iran) and the increasingly lethal terrorists of al-Qaeda in Iraq (AQI).

The deceptive calm in Iraq came to an end on 31 March 2004, when four American civilian contractors took a wrong turn and drove their SUV into Fallujah, a Sunni stronghold just west of Baghdad, where they were killed by an armed mob who hung their charred bodies from a bridge in the center of the city. Incensed by this brutal action, Bush and Rumsfeld demanded an operation to apprehend the perpetrators. The Marine commander, Lieutenant General James Conway, argued for patience, believing a full-scale assault would be counter-productive. Ordered nonetheless to attack, the Marines began an assault on the city. The operation turned into a public relations nightmare, however, with Arabic news media accusing US forces of war crimes and several members of the IGC threatening to quit if the offensive was not halted. Bremer got cold feet and ordered a ceasefire just when the Marines were on the verge of securing the city. Sanchez thus had to order the Marines to withdraw, leaving the city under the de facto control of Sunni insurgents, who claimed victory. In the next few months, al-Qaeda fighters led by Abu Musab al-Zarqawi, a particularly effective (and brutal) Jordanian jihadist and leader, would turn Fallujah into a terrorist haven, complete with bomb factories and torture chambers.[30]

To make matters worse, for the first (and only) time in the war, Sunni and Shi'a militants coordinated their attacks against coalition forces. The Jaish al-Mahdi (JAM), a substantial Shi'a militia beholden to Sadr, rose up in early April to attack US troops in Sadr City, a large, densely populated Shi'a area of north-eastern Baghdad, and to sever the coalition supply line that ran from Baghdad to Kuwait by destroying bridges and ambushing convoys along the route. The few battalions of new Iraqi security forces available, upon which the coalition strategy increasingly rested, proved wholly ineffective. As coalition forces defended against hundreds of insurgent and militia attacks in Samarra, Baqubah, Baghdad, Ramadi and Fallujah, as well as across south-central Iraq, their stocks of food, fuel and ammunition dwindled to dangerously low levels. The 1st Armored Division, which was in the process of redeploying at the end of its year-long deployment, was ordered to assist in restoring the situation. With its tour in country extended from twelve to fifteen months, the division sent battalion task forces to destroy militia concentrations in Najaf, Karbala and Kut, and to reopen the main supply route to Kuwait.[31] US forces counter-attacked to defeat Sunni insurgents and Shi'a militia forces and regain control of other Iraqi cities that had been besieged. But, just before his militia was destroyed, Sadr negotiated a ceasefire to temporarily end the fighting – a decision that enabled him to claim victory and that would come back to haunt the coalition in the years ahead.

The Sunni insurgency and Shi'a jihadists also targeted other members of the coalition in areas that had previously been calm. An attack on the Italian headquarters in Nasiriyah in southern Iraq on 12 November 2003, followed four months later by an attack by al-Qaeda suicide bombers on commuter trains in Madrid on 11 March 2004, ultimately led both Italy and Spain to withdraw their forces from the war. Honduras and the Dominican Republic also removed their forces from the Multi-National Division in south-central Iraq. As violence spread throughout Iraq in April, other national contingents responded by withdrawing into their bases, leaving their areas without effective security forces.[32] These attacks ended whatever chance there was that US forces could be replaced in Iraq by an international peacekeeping force.

As these battles raged, plans were finalized for the dissolution of CJTF-7 and its replacement by Multi-National Force–Iraq (MNF–I), a strategic-level, four-star headquarters that would be commanded by General George W. Casey Jr. and would be focused on politico-military affairs and overall strategy for the war, and the Multi-National Corps–Iraq (MNC–I), a three-star headquarters commanded by Lieutenant General Thomas F. Metz that would focus on operations on the ground to subdue the insurgency and stabilize the country. The two new headquarters were established in mid-May and order restored in time for CPA to transfer sovereignty to a newly established interim Iraqi government under Prime Minister Ayad Allawi on 28 June, after which CPA was replaced by a US embassy under Ambassador John Negroponte, a seasoned and highly respected diplomat.

CPA had been hobbled by inconsistent, constantly rotating staffing and lack of capacity throughout its tenure. On reflection, it appears to have been established because of Defense Secretary Rumsfeld's desire to control the effort in post-war Iraq through an ad hoc organization that reported to the Pentagon rather than establishing a US embassy, which would have meant having to share control of the war with Secretary of State Colin Powell, to whom the new US Ambassador naturally would report.[33] It had been an unwise decision. The prompt establishment of a US embassy in Baghdad, with experienced diplomats and development workers, would have been vastly better than the ad hoc team that was the CPA.

US credibility was further eroded in the spring of 2004 when news outlets published photos of prisoner abuse at Abu Ghraib, a former Iraqi prison just west of Baghdad that was being used for detainee operations by American forces. The exposure of the criminal acts by members of the 800th Military Police Brigade shocked the world and incentivized large numbers of Iraqis and others to join the insurgency. A subsequent US Army investigation substantiated widespread and systematic prisoner torture and abuse at the facility and resulted in the courts-martial and conviction of eleven soldiers.[34] In the aftermath of the scandal, public support for the war in the United States and among its coalition partners plummeted, and photos of the abuse released online would become indelible images of the US effort with repercussions throughout America's time in Iraq and beyond.

Urban battles

The United States, limited to that point by a short-war mentality, was now stuck in Iraq without a truly viable strategy to defeat the insurgency and disengage from the conflict. The focus on rapid, decisive operations had resulted in regime change at a modest cost in blood and treasure, but American policymakers and forces had been unprepared for the messy aftermath. Military leaders at that point, in essence, adjusted their tactics, unsure of what was required to stabilize Iraq and secure its people. Rapidly formed units of the Iraqi Civil Defense Corps, intended to be a stopgap until the establishment of a new Iraqi army, had lacked cohesion and melted away when they came under fire during the April uprising, as had the initial battalions of the Iraqi Army that had been hurriedly trained and equipped. US commanders had been focused on rooting out and apprehending Saddam's "former regime elements," but it was becoming increasingly clear that the infusion of Sunni foreign fighters and the growth of homegrown insurgent and militia forces, as well as the emergence of ethnic and intra-sectarian tensions, were changing the character of the conflict.

One of the most powerful of the new groups was Tawhid wal-Jihad, a terrorist organization led by Abu Musab al-Zarqawi that would evolve into al-Qaeda in Iraq (AQI). Zarqawi's strategy was to attack the Shi'a and their religious sites in an attempt to ignite a sectarian civil war, which would force Sunni Arabs to back AQI elements as their defenders of last resort. In large part, it worked. In August 2003, AQI truck and car bombs targeted the Jordanian Embassy, the United Nations headquarters in Baghdad and the Imam Ali Mosque in Najaf. These attacks killed hundreds of people as well as UN Special Representative in Iraq Sérgio Vieira de Mello and the prominent Shi'a cleric and politician Mohammed Baqr al-Hakim. From that point onward until the group's demise during the so-called Surge in Iraq from early 2007 to the summer of 2008, Zarqawi and his adherents wreaked havoc, tearing apart the very fabric of Iraqi society and plunging the country into a full-blown Sunni–Shi'a civil war. The blows they struck also meant that the expected involvement of international organizations in Iraq would not materialize, as the security threats were seen as too great to allow the establishment of such elements in the country. Even the United Nations headquarters in Baghdad closed for a period after the terrible attack in August 2003.[35]

Coalition hopes for a way out of the quagmire in Iraq in 2004 increasingly came to rest on nationwide elections for a new government planned for early 2005, with the hope that the outcome would be seen as more representative of all Iraqis and thus more legitimate in their eyes. This strategy rested on the assumption that all groups in the country would view the elections as legitimate, participate in them and respect the results. Sadly, that would not be the case with most Sunni Arabs. Rather than participate in a process that would inevitably give power to the Shi'a majority, significant numbers of Sunnis decided to opt out of voting and instead back the insurgency, which they viewed as legitimate resistance to a foreign-imposed order that had displaced the Sunni-led regime of Saddam Hussein. Additionally, the distribution of seats in the Council of Representatives by a *party*'s share of the vote rather than by election of *individual* politicians vying for individual seats would exacerbate civil tensions as Iraqis largely voted along ethno-sectarian lines.

The creation of effective military and police forces was also increasingly seen as a critical prerequisite for coalition withdrawal from Iraq. To oversee this effort, coalition leaders established the Multi-National Security Transition Command–Iraq (MNSTC–I) in the summer of 2004. I was promoted to lieutenant general and returned to Iraq to command it – and to implement recommendations I had provided to General Abizaid and Secretary Rumsfeld after an assessment in Iraq I conducted following the violence that April and the general failure of Iraqi forces. The development of all Iraqi security forces was now centralized under one command, but significant challenges remained, not the least of which was a lack of funding and advisers to embed with Iraqi Army and police forces, and the necessity of building our headquarters and multiple subordinate headquarters to oversee the tasks of helping the Iraqis rebuild their Ministries of Interior and Defense and hundreds of thousands of military, police and other security forces. The US Army ultimately created an advisory center at Fort Riley, Kansas, in 2005 to prepare adviser teams for deployment to Iraq, but institutional support for the advisory effort at the outset was late and inadequate, with personnel from US Army Reserve training units assigned to advise the new Iraqi units on operations in combat that few of the adviser team leaders, however energetic and determined, had ever conducted themselves in peacetime, much less in combat.

MNSTC–I focused on quickly adding combat capability to deal with the growing insurgency, which meant that Iraqi logistical capacity and institutional capability to oversee and administer the armed forces lagged well behind their growth. In the interim, Iraqi units would have to rely on contractors, fixed base logistics, local Iraqi civilian capabilities and coalition units for much of their supply, maintenance, medical care and non-infantry combat support capabilities. Additionally, after the transfer of sovereignty to the elected Iraqi government in early 2005, MNSTC–I lost its power of promotion and dismissal. Unfortunately, that led to an insidious rise of sectarianism within the ranks as Iraqi politicians meddled with the makeup of the officer corps, favoring those from their ethnic, sectarian, tribal and political groups.

In the meantime, the Sadrist presence in the holy Shi'a city of Najaf, the lack of coalition control in predominantly Sunni Arab Samarra (north of Baghdad) and a Sunni insurgent/extremist-controlled Fallujah remained threats that needed to be addressed. An early-August 2004 gunfight in Najaf between a Marine patrol and JAM fighters led to a larger operation involving a Marine expeditionary unit and two battalions of the 1st Cavalry Division to clear the city. The fighting in Najaf, particularly in the massive cemetery near the Imam Ali shrine, severely weakened the JAM and forced Sadr to retreat once again via a ceasefire brokered by Grand Ayatollah Ali al-Sistani, the supreme Shi'a religious figure in Iraq. In a welcome development, the impressive interim Prime Minister Dr. Ayad Allawi was supportive of the operation, as he was of subsequent confrontations in Samarra and Fallujah, north and west of Baghdad respectively. With his backing, the 1st Infantry Division also cleared Samarra in a multi-battalion operation in October.

Fallujah, which had festered under insurgent control since April, would require a much larger operation and prove the most challenging. By the autumn of 2004, more than 4,000 Sunni insurgents, extremists and foreign fighters had populated Fallujah after the collapse of the coalition presence in the city, declaring the imposition of sharia law and forcing Shi'a Iraqis to flee. US leaders, conscious of their failure to win the battle of the narrative in the April fighting, focused on ensuring accurate information in the offensive operation they would conduct by embedding reporters in combat units and ordering the early seizure of the city's hospital to preclude disinformation emanating from there about civilian casualties. Military shaping operations then isolated the city, killed key insurgent leaders and convinced the majority of the civilian population to evacuate until the completion of the operation, thereby significantly reducing civilian casualties.[36]

On the evening of 8 November, two Marine regiments reinforced by two US Army mechanized infantry task forces and six newly trained Iraqi Army and police battalions assaulted the city. The Marines and soldiers moved block by block, house by house in urban combat that tested their strength and endurance.[37] They eliminated organized resistance in five days, although Marines continued to clear small pockets of resistance in the weeks that followed. The fighting devastated the city's buildings and much of its infrastructure, however, dissuading many of the local inhabitants from returning. And disappointingly, with the exception of the 36th Commando Battalion, the Iraqi units proved relatively ineffective in close combat. It was clear that more rigorous training was required for the new units. Fifty-seven Marines and six soldiers were killed in the battle of Fallujah and 600 wounded. The insurgents lost nearly 2,200 killed and more than 2,000 captured in what was the largest battle of the Iraq War.

While the battle for Fallujah raged, Zarqawi's fighters and local insurgents took advantage of the relative dearth of coalition troops in Nineveh Province to seize Mosul and Tal Afar on 10 November. Iraqi forces established there largely collapsed under the pressure, while the one US brigade left in the area was seriously stretched. After regaining situational awareness, the elements of Task Force Olympia, the headquarters in charge of Nineveh Province, reinforced by a Stryker battalion and two newly established Iraqi Special Police commando battalions from Baghdad, counter-attacked and regained control of Mosul after tough fighting. The city's police force of 4,000, however, had been devastated by the battles and would largely have to be rebuilt.

Nevertheless, the operations across Iraq from August through November 2004 had established the preconditions for holding the national elections of early 2005, upon which the fate of the coalition's strategic plan rested. And a critical decision by the exceedingly capable US Ambassador John Negroponte, who acted decisively to shift $1.8 billion from other priorities to fund MNSTC–I's still under-resourced effort to recruit, train, equip and organize elements of the Iraqi

Ministries of Defense and Interior, would enable provision of increased numbers of Iraqi boots on the ground and supporting infrastructure that would prove crucial in providing security for the elections.

Clear, hold, build

The elections held on 30 January 2005 proceeded smoothly but they were not the magic bullet for which some had hoped. That idea rested on the assumption that all groups in the country would view the elections as legitimate, participate in them and respect the results. Sadly, that would not be the case with most Sunni Arabs.

An assemblage of Shi'a Islamist and Kurdish parties garnered the most votes but could not agree on a candidate for the most important position of prime minister. Ayad Allawi, the interim Prime Minister who had performed impressively under difficult circumstances, continued in office as a lame duck for four months of squabbling until the selection of his successor, Ibrahim al-Jaafari. Predictably, the political winners (mainly Shi'a and Kurds) would use the levers of government to exclude their enemies (mainly Sunni Arabs) from sharing in the power and resources of the new Iraqi state. And significant Shi'a control over the Defense and Interior Ministries (even though a Sunni would be the Minister of Defense) would lead to the creation of state-sponsored forces that would deliberately target the Sunni community. The conflict would thus continue largely unabated.

As the war carried on, General Casey began to adjust MNF–I's strategy.[38] Although he included the term "counter-insurgency" in the MNF–I mission statement and brought counter-insurgency advisers Kalev Sepp and Bill Hix to Iraq to establish a counter-insurgency academy in the spring of 2005, Casey had begun to conclude that the struggle had fundamentally changed into a competition for power and resources among a variety of Iraqi factions. Since, in his view, the war had morphed beyond an insurgency against the Iraqi government, Casey decided that counter-insurgency principles were somewhat less central to the campaign. He thus emphasized MNF–I's focus on building Iraqi Army and police and beginning to transition responsibility of security operations to them – a strategy neatly summed up by President Bush on 28 June in a speech to the troops: "As the Iraqis stand up, we will stand down."[39] In hindsight, however, this shift would be premature – especially over the following year, as the violence escalated precipitously in the wake of a significant attack on a Shi'a shrine and as the still young Iraqi forces were progressively degraded by that escalation.

Increasing numbers of Iraqi security forces were certainly helpful, but in the near term they were not fully capable and could not contend with the more numerous and capable insurgent and militia forces populating Iraq. Leaders of the Shi'a JAM, humbled by numerous defeats in 2004, reached out to the nefarious but capable Iranian Revolutionary Guards Corps (IRGC) Quds Force Commander Qasem Soleimani for training and arms. Meanwhile, the Sunni insurgency leaned more heavily on foreign fighters from across the Arab–Islamic world, jihadists who brought their extreme views on Islam with them to Iraq – and whose ranks included substantial numbers of would-be suicide bombers. The Iranians supplied their Iraqi protégés not just with additional small arms, ammunition, explosives and rocket-propelled grenades, but also, most worryingly, with explosively formed projectiles (EFPs), an extremely lethal form of improvised explosive device that could slice through the armor of a main battle tank. Ultimately, EFPs would kill more than 600 American soldiers in Iraq. The Sunni jihadists, meanwhile, relied heavily on suicide attacks, whether delivered by individual suicide bombers wearing explosive vests or by those driving cars and trucks full of explosives. The intent of both was to kill as many coalition soldiers and Shi'a Iraqis as possible. In retaliation, Shi'a militia groups such as the JAM targeted Sunni communities, carrying out operations that forced sectarian displacement and further stoked the flames of civil war.

During these darkest days of the war, there were initiatives that pointed a way forward. In May 2005, the US Army's 3rd Armored Cavalry Regiment was deployed north to Tal Afar as part of a campaign to control western Nineveh Province and Iraq's western border with Syria in order to stem the flow into Iraq of foreign fighters and jihadists. The talented regimental Commander, Colonel H. R. McMaster, partnered with a very capable police chief (a former senior Iraqi Army officer), Najim Abed Jabouri, whom he then appointed as the city mayor. For the next nine months, McMaster and his regiment conducted a comprehensive counter-insurgency operation to isolate Tal Afar, clear it of its extremists and protect the population from violence by establishing numerous combat outposts within the city, and then going on to repair damaged infrastructure, restore basic services and revive the local economy.

Marine Lieutenant Colonel Dale Alford, similarly versed in classical counter-insurgency practices, used a comparable approach with the 3rd Battalion, 6th Marines in Al-Qa'im along the Iraqi–Syrian border in Anbar Province. The central concept of these operations, known as clear-hold-build, would subsequently be endorsed in a major speech by Secretary of State Condoleezza Rice. This was much to the consternation of Defense Secretary Rumsfeld, wedded to the strategy of transitioning tasks to Iraqis that he still believed was working. Adopting a fully resourced counter-insurgency strategy such as that previewed by McMaster in Tal Afar would require far more resources, and Rumsfeld was not about to advocate sending reinforcements to a conflict he increasingly viewed as fundamentally unwinnable by American forces.

The promulgation of a new constitution for Iraq and new parliamentary elections in the autumn of 2005 led to further alienation of the Sunni community, which had voted overwhelmingly against the country's new foundational law. Adding to the challenges was the selection in May 2006, after an extended period of political turmoil, of Nuri al-Maliki as the prime minister. Maliki was the candidate of the Dawa Party (known for its Shi'a conspiratorial inclinations) and was supported, initially, by Muqtada al-Sadr. At the time the parliament settled on him as a compromise candidate, he was a figure, in the eyes of many, whose "strength was his weakness." Over time, however, repeating a pattern we have seen in Vietnam, Afghanistan and other countries, Maliki would aggressively gather the reins of power in his hands and transform into an increasingly dictatorial and mercurial figure rather than the unifying leader so desperately needed in a country through which ran so many of the ethnic, sectarian and tribal fault lines of the Middle East. Following the elections, General Casey continued the MNF–I strategy of transition, which now included the very helpful addition of civilian-led Provincial Reconstruction Teams to help with improving infrastructure, governance, basic services and the economy across Iraq. The plan was for the coalition force to gradually shrink as Iraqi forces progressively took control of security, with reductions of two brigade combat teams and other elements planned for early 2006.

The 2006 elections had given some on the outside reason for hope, based on the increased voter turnout. But the lack of full visibility into, and understanding of, social and security dynamics hid a darker truth: Iraqi society was a powder keg ready to explode.

Civil war

Coalition assumptions of progress in Iraq came crashing down on 22 February 2006 when AQI destroyed the Al-Askari Mosque in Samarra, a significant Shi'a shrine in a largely Sunni city north of Baghdad. The destruction of such an important Shi'a shrine by Sunni terrorists finally ignited the reaction Zarqawi had sought since the earliest days of the war. Across Iraq, Shi'a Arabs took to the streets demanding retribution and taking matters into their own hands by torching dozens of Sunni mosques and kidnapping, torturing and killing Sunnis. The killings were followed by deliberate efforts to displace residents of various neighborhoods along sectarian lines. This became, in certain areas, sectarian cleansing.

In Sunni and Shi'a communities alike, power devolved to militia, insurgent and terrorist groups with the most guns. Millions of Iraqis fled their homes, becoming refugees inside Iraq or in neighboring countries. As sectarian civil war escalated, MNF–I, which had lost a considerable amount of situational awareness of what was happening on the streets by consolidating forces on large bases and handing over security tasks to Iraqi security forces, did not immediately appreciate the magnitude of the chaos that was enveloping Iraq. Significantly, MNF–I did not change the strategic approach to the conflict or immediately reverse the decision to reduce the number of brigades, which was projected to decline from fifteen to just ten by the end of the year.

The killing of AQI leader Zarqawi in a US airstrike controlled by special operations forces on 7 June 2005, had given coalition leaders a temporary lift; however, the demise of Zarqawi did not prove to be a fatal blow to the organization, which was too robust to be defeated by the death of one man, however capable he may have been. The episode was a reminder that counter-insurgents must avoid the belief that killing or capturing the leader of an insurgent group will cause the group's collapse. That is true only in small, immature groups. In larger organizations, such as AQI, there are always subordinate leaders ready to fill the void should the head of the group meet his end.

As sectarian conflict racked the country, MNF–I responded in the summer and autumn of 2006 with Operations Together Forward I and II, attempts to stem the rising tide of violence in Baghdad. Involving tens of thousands of troops, these operations did temporarily reduce violence in the areas of focus, but ultimately failed due to the inability of Iraqi Army and police forces to hold the areas cleared by US troops after the Americans handed over security responsibilities to them and returned to the large US bases on the edge of Baghdad. Moreover, some Iraqi forces, such as the Iraqi National Police, became part of the problem as they targeted Sunni residents and sought to intimidate them and to retaliate for Sunni extremist and insurgent attacks. Iraqi Army and police units, increasingly composed of Shi'a Arabs, had little incentive to protect Sunni areas. And Shi'a areas remained largely untouched by security force operations, giving free rein to Shi'a militias to continue to intimidate the local populations and also to prey on nearby Sunni neighborhoods. As the autumn of 2006 arrived, Shi'a death squads in and out of uniform stalked many streets in Baghdad, and the bodies of the slain floated down the Tigris River and piled up in the Baghdad morgue. Meanwhile, suicide bombers targeted Shi'a mosques and markets, adding to the carnage. More than a hundred civilians were killed every day during the month of June, a rate that would only increase as the year progressed.[40]

In response to the violence, General Casey, backed by Rumsfeld and US Central Command Commander General John Abizaid, remained committed to a strategy of transition and the withdrawal of coalition forces from Iraq. Even as Operation Together Forward I commenced in the second week of July, the MNF–I staff planned the reduction of three brigades by the end of the year. Casey was convinced the problem in Iraq was political, which was absolutely correct, but he also believed that coalition forces could contribute little to the outcome of the Iraqi struggle over power and resources, which was arguable.[41] In the meantime, the lack of forces also prevented the creation of an operational reserve, even as the overall strategy pulled US forces out of neighborhoods, and several areas of Iraq became insurgent and militia safe havens due to the lack of troops to clear and hold them.

As the situation in Iraq deteriorated, President Bush, to this point wholly supportive of his Defense Secretary and Commander on the ground, came to realize that the strategy of transition was not working. The descent of Iraq into communal violence continued to invalidate the assumptions upon which the MNF–I campaign plan rested. Killing Zarqawi and other insurgent leaders, the mass participation of Iraqis in the election of a government, removing coalition troops to reduce the number of "antibodies" they might produce among the local population and the transition of security operations to Iraqi control had not resolved the issues underpinning the conflict or improved the state of security. Seeking alternative approaches, Bush became much more involved in strategic discourse, and tasked his National Security Advisor, Stephen Hadley, with gathering information and developing options. As the end of 2006 approached, more than 3,000 Iraqis were dying every month as a result of ethno-sectarian-motivated violence. Until security improved and the levels of violence were driven down, politics would not be operative as the primary driver for the distribution of power and resources in Iraq.

Ramadi: the Awakening

In mid-August 2006, the intelligence officer of Multi-National Force–West (MNF–W), Colonel Peter H. Devlin, concluded in a formal assessment that coalition forces were no longer capable of controlling Al-Anbar Province, the Sunni Arab province that stretched from just west of Baghdad to Iraq's borders with Syria, Jordan and Saudi Arabia. This was a highly disturbing conclusion, and it made its way to the press in less than a month. Around the same time, however, a small bright spot began to appear in Anbar: the sheer violence and depredations of AQI prompted a reaction among some of the tribes in the area, particularly in Ramadi, where the 1st Brigade, 1st Armored Division under the command of Colonel Sean MacFarland had arrived in June after a short stint in Tal Afar.[42] MacFarland was a gifted leader who had been employing many of the same techniques as McMaster had in Tal Afar, moving his troops off their large forward operating base and into local communities in smaller, mutually reinforcing combat outposts. There they would *partner* with local Iraqi security forces rather than transitioning control and turning neighborhoods over to them. More importantly, when approached by a local sheikh, Abdul Sattar Abu Risha al-Rishawi, who wanted to join forces with the Americans to battle the terrorists who had killed his father and two brothers, MacFarland and his key subordinates for tribal outreach, Lieutenant Colonel Anthony Deane and Captain Travis Patriquin, jumped at the opportunity. The brigade quickly enrolled hundreds of tribesmen in the local police, and then accepted the offer of other sheikhs to ally with the Americans as they saw the benefits of doing so.

Four days after the intelligence report about Al-Anbar Province was published in the press, Sheikh Sattar gathered together a coalition of tribes to begin what would become known as the Anbar Awakening, a movement that would gradually change the balance of power in Al-Anbar Province and eventually help turn the tide of conflict against AQI.[43] Continued outrages by AQI led forty-one sheikhs from seventeen tribes to ally with the coalition and go to war with AQI around Ramadi in September, a development firmly supported by MacFarland. The alliance was just what the coalition needed: local tribes provided superb intelligence information, and local police were incentivized to defend their home territory, with the coalition providing firepower, economic assistance and other support. In late November, a brief, sharp battle eliminated the AQI presence on the eastern outskirts of Ramadi. And in February and March 2007, shortly after the beginning of the Surge in Iraq, forces under Colonel MacFarland, and then the 1st Brigade, 3rd Infantry Division, which replaced MacFarland's brigade, painstakingly cleared every neighborhood and building in Ramadi – and then held them, increasingly aided by local security forces.

Initially, MacFarland had difficulty convincing his superiors in Multi-National Force–West to support the Awakening, as the leaders in his higher headquarters were intent on engaging more senior, seemingly more significant sheikhs who had fled to Amman, Jordan, rather than those who had remained in the province and were now doing the fighting and dying there. The tension between the two approaches would not be resolved until the following February when, newly installed as commander of MNF-I, I directed full support of the Awakening movement, seeking to expand it throughout Anbar and then throughout the other Sunni areas of Iraq, including those in Baghdad. In time, with coalition support, the movement would expand west and east throughout Anbar and then east and north into neighboring provinces, and, ultimately, all the way to Mosul in the north, radically altering the dynamics of the conflict against AQI and the insurgency. Additionally, some of the mid-level leaders and very substantial numbers of rank-and-file members of certain insurgent groups were "reconciled" through a process established by MNF–I and became supportive of the Awakening, a dramatic and very welcome political development.[44]

In response to the Awakening, al-Qaeda leaders formed the Islamic State of Iraq (ISI), led by Abu Omar al-Baghdadi (a pseudonym for Hamid Dawud Muhammad Khalil al-Zawi) and declared Ramadi as the capital of its new caliphate. The evolution of parts of AQI into ISI had no immediate consequences at the time; however, it would prove to be a significant development in the longer term. Seven years after the defeat of AQI during the Surge of 2007 and 2008, ISI and its successor

ISIS, the Islamic State of Iraq and al-Sham (the Arabic word for the Levant), would morph into the dominant jihadist force in the Middle East and eclipse al-Qaeda as the most powerful global terrorist brand.

The Surge

In the United States, meanwhile, the war in Iraq was very much on the ballot in the mid-term elections of November 2006. Reflecting a sense that the war was increasingly futile, the American people handed President Bush's Republican Party a resounding defeat at the polls, with the Democratic Party taking control of both houses of Congress for the first time in twelve years.

President Bush, still intent on winning the war despite the overwhelming desire of the American people to disengage from the conflict, took matters into his own hands by swiftly replacing Secretary of Defense Rumsfeld with former CIA Director Bob Gates. Two months

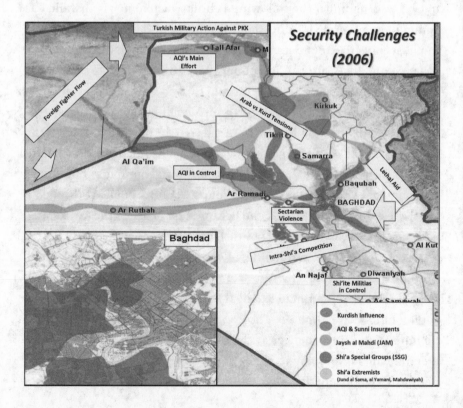

later, he announced a new commander for MNF–I, as well as a new ambassador, a new commander of US Central Command and a new strategy.

The Surge, as the new initiative for Iraq came to be known, featured the reinforcement of coalition forces by five US Army brigades and two Marine battalions, along with thousands of enablers such as aviation, engineer and military police units, as well as an additional division headquarters. Over time, the number would grow from the 20,000 announced by the President to nearly 30,000 troops in all, which would bring the coalition troop presence in Iraq to nearly 180,000 (of which 165,000 would be American), slightly higher than the number of troops in the initial ground invasion force. More importantly, these forces and those already on the ground in Iraq would be used differently, in accordance with a new counter-insurgency doctrine that made clear that control and protection of the population were the key to winning a counter-insurgency struggle. As part of that, defeating Shi'a militias would become an objective as important as defeating AQI and the Sunni-led insurgency – all critical tasks if the cycle of intercommunal violence in Iraq was to be broken.

The events over the next year and a half would change the strategic momentum in Iraq, leading to the defeat of AQI, the Sunni insurgents and the Shi'a militias, as well as a massive reduction in violence.[45] The results would provide Iraq and its leaders with an entirely new opportunity to resolve the challenges they faced.

The Surge was the result of President Bush's growing frustration with the strategy of transition, which had not halted the rising levels of violence in Iraq. Fuelled by this frustration, Bush had, by late 2006, come a long way in exercising the powers of his office since his initial decision to topple Saddam Hussein and subsequent "outsourcing" of the prosecution of the war to his Secretary of Defense. In response to his frustration with the situation in Iraq, three strategic reviews – by the National Security Council staff, a council of colonels convened by the Joint Chiefs of Staff and a small group under the auspices of the State Department – were conducted in the autumn of 2006 to examine alternatives to the existing strategy. Of these, it was the review overseen by

National Security Advisor Stephen Hadley that would prove critical to this change in strategy. Leveraging the input of experts from outside, as well as inside, government (including outreach on several occasions to me by Meghan O'Sullivan, Hadley's very impressive deputy for Iraq) and bolstered by a timely, influential think-tank study on how reinforcements employed in a fully resourced counter-insurgency could change the tide of the war in Iraq, Hadley's team presented a plan that offered the best opportunity for fulfilling the President's objective of decisively reversing the course of the war.[46] The President agreed, and after much back and forth with Casey, the Joint Chiefs of Staff (who were concerned that providing substantial reinforcements to Iraq would break the services – to which Bush responded, "I share your concern about breaking the military. The surest way to break the military would be to lose in Iraq") and the new Secretary of Defense, Bush went before the American people on 10 January 2007 to announce a change of strategy in Iraq and the provision of 20,000 troops to reinforce MNF–I.[47] The first Surge brigade would arrive in Iraq in late January, followed by one more each month until all were on the ground by late May.

I was nominated to command MNF–I, having already served more than two and a third years in Iraq, initially as a division commander during the invasion and the first year in Iraq and then as the first commander of the Multi-National Security Transition Command–Iraq (MNSTC–I). And since returning to the United States in late September 2005, I had, together with my Marine counterpart, Lieutenant General Jim Mattis, overseen the crafting of a new counter-insurgency doctrine for the Army and Marine Corps, which we published in December 2006.[48]

As commander of the US Army Combined Arms Center at Fort Leavenworth, Kansas, from which I oversaw the bulk of the Army's schools and training centers, I had sought to "shake up the Army," as the US Army Chief of Staff, General Peter J. Schoomaker, had instructed me to do when I asked for guidance before assuming my new position. Over the subsequent fifteen months, with full support from General Schoomaker and General Scott Wallace (now the Commander of US Army Training and Doctrine Command), I worked with the leaders of the staff college, Center for Army Leadership, Battle Command Training Program, combat arms schools and various training centers to overhaul every element of the year-long "road to deployment," the various actions taken to prepare leaders, staffs, units and soldiers for deployment to Iraq and Afghanistan. In essence, these initiatives led to considerably increased focus on counter-insurgency doctrine and operations during the schooling of commissioned, warrant and non-commissioned officers in professional development courses and also at the major combat training centers where units conducted the critical pre-deployment training exercises.

As I prepared to take command in Iraq, I was convinced that the "surge of ideas," the major concepts that would constitute the new approach in Iraq, would prove even more important than the surge of forces. The most significant of these were very different from the previous strategy. First and foremost, US forces would return to the Iraqi neighborhoods, generally in partnership with Iraqi security forces, and would "live with the people," as that was the only way to secure them. US forces also would take back control of security from the Iraqi forces.

Additionally, the Anbar Awakening would be fully supported and replicated throughout the entire country, augmented by a formal reconciliation process with insurgents and, over time, militia members. This was crucial, for as we often noted, "You cannot kill or capture your way out of an industrial-strength insurgency. You have to reconcile with as many of the rank and file as you can." Getting a Shi'a-led Iraqi government's support for reconciliation with Sunni Arabs who had tacitly or actively supported the insurgents was not easy, but it was ultimately achieved. (Of course, a substantial number of those reconciled were former Ba'athists with whom the Iraqi government could have reconciled years earlier had such a process been established.) Simultaneously, the exceptional special operators led by Lieutenant General Stan McChrystal and, later, Vice Admiral Bill McRaven, two extraordinary leaders, would capture or kill, if necessary, the "irreconcilables," the hard-core members of AQI and ISI, and the leaders of the major Sunni insurgent groups and Shi'a militias who were not assessed as reconcilable.

Detainee operations, not a trivial task (as there were 27,000 Iraqi detainees at the height of the Surge), would also have to be completely overhauled and transformed from serving as a "terrorist-training university" in which detainees were further radicalized by hard-core extremists in their midst, to an endeavor that focused on rehabilitating and training young Iraqis for return to their tribes. All that would be enabled by identifying and removing the serious extremists from the general detainee population and transferring them to maximum-security facilities that we constructed and establishing the education and training programs to reduce the recidivism rate of the rank-and-file detainees.

Finally, there would be even greater emphasis than previously on integration between the Multi-National Force–Iraq and the US Embassy team, including a significantly overhauled and fully integrated civil–military campaign plan developed by the staffs of Multi-National Force–Iraq and the Embassy and signed by both Ambassador Crocker and me.

My staff and I worked hard to ensure that our subordinate commanders understood and implemented the new concepts through the conduct of a rigorous schedule of activities on our "battle rhythm": daily morning operations updates, twice-weekly visits and patrols with tactical units conducting security operations, a weekly videoconference with President Bush and the entire national security team and another with Secretary of Defense Gates and his key staff members, a weekly meeting (together with the Ambassador) with Prime Minister Maliki, weekly attendance at Maliki's National Security Council gathering, regular meetings with the leaders of the various efforts and headquarters subordinate to MNF–I, weekly and monthly gatherings of key military commanders and exhaustive quarterly campaign-plan reviews that I hosted together with Ambassador Crocker and our respective military and diplomatic staffs, development personnel and intelligence leaders and key staff members. I also periodically published letters and counter-insurgency guidance to all the members of the Multi-National Force.[49]

Ryan Crocker, the highly experienced and exceedingly talented new US Ambassador, proved the greatest diplomatic partner any commander could have had. It was an enormous privilege to serve with him. Early on, we focused intently on overseeing the development and then implementation of the new comprehensive civil–military counter-insurgency campaign plan. The plan sought to ensure synchronization of all coalition activities across a wide range of areas, the primary focus of which would be improving security for the Iraqi people and then solidifying the security gains through various civil–military measures to improve daily lives and demonstrate to the local civilians that life would be better for them if they supported the new Iraq instead of fighting against it or supporting those who were fighting against it.

Ambassador Crocker and I also sought to work closely with Prime Minister Maliki, who initially had a very different vision of what US and coalition forces should do during the Surge. In fact, a few days after I assumed command of MNF–I in early February 2007 and before the arrival of Ambassador Crocker, Maliki's National Security Advisor, Dr. Mowaffak al-Rubaie, presented then-Ambassador Zalmay Khalilzad and me with a list of demands on behalf of the Prime Minister that included a *more* rapid transition of security authority to Iraqi security forces, an Iraqi veto on US military operations, further removal of coalition forces from Iraqi neighborhoods and other actions that were completely contrary to the approach we intended to follow. Taken aback, I replied that those actions would accelerate the implementation of a failing strategy; in essence, they would ensure that the overall effort would actually "fail faster." If Prime Minister Maliki was truly intent on pursuing his demands, I suggested that he present the list to President Bush via the weekly video teleconference scheduled for the next day with the PM, which the Ambassador and I would attend. But I warned al-Rubaie that if Maliki did so, I would be on the next plane back to Washington – and that I intended to "take the policy with me" – informing President Bush that the administration's goals in Iraq could not be achieved. Needless to say, I did not sleep well that night given, in essence, my threat to quit in my first week as the new commander.

To my enormous relief, Maliki backed down and I never heard anything more of his demands; nonetheless, the episode showed the tenuous nature of the relationship between senior coalition leaders and

Iraq's Prime Minister. I also recognized that once Ambassador Crocker arrived, a united US diplomatic and military front in Baghdad would be imperative, and we agreed that we would see Maliki only if the other were also present, a practice we religiously observed, even as we also continued the practice of maintaining offices so close together in the US Embassy that we shared the same reception area.[50]

When the Surge commenced, Bush personally took over the Washington front of the Iraq War, showing how highly he prioritized the effort in Iraq by chairing an hour-long meeting of the National Security Council in the White House Situation Room every Monday morning at 7:30 a.m., with Ambassador Crocker and me participating by secure video teleconference, as did Admiral William "Fox" Fallon, the new Commander of US Central Command. This was a dramatic initiative by the President (at an unheard-of time for Monday morning in Washington). Over the eighteen-month course of the Surge, the amount of direct contact between the President and his Commander and Ambassador in Iraq would be totally unprecedented.

After being confirmed by the Senate and before heading to Iraq, I met one on one with Bush in the Oval Office. Among the topics we discussed, I noted that the military was about to go "all in" on Iraq and I hoped the rest of the US government would do likewise. Bush assured me that he would see to it, and he did. The early Monday meetings were a powerful demonstration of Bush's commitment, sessions during which Crocker and I could regularly speak directly to the President on matters that concerned us. These would be in addition to the issues that I raised in a weekly memo I sent each Sunday evening, Washington time, to the Secretary of Defense, Chairman of the Joint Chiefs and Commander of Central Command, in advance of the Monday-morning meeting. The Monday meeting also sent a powerful signal to those in the military and the administration who might seek to undermine or impede the President's strategy, as Ambassador Crocker and I made it clear early on that we would not hesitate to raise matters directly with President Bush if we felt that necessary.[51]

My key subordinate commander at the operational level in Iraq was Lieutenant General Ray Odierno, a great soldier and leader who had earlier served in Iraq as a division commander and had taken command of MNC–I in mid-December. He was of like mind with me on what was needed and had already begun to plan the actions to be pursued with the arrival of the Surge forces. Ray and I saw eye to eye on the priority of securing the neighborhoods in Baghdad and the surrounding belts of territory used by insurgents as safe havens, while also targeting both Shi'a militias and Sunni insurgent and terrorist groups to tamp down the flames of civil war. As I would note later, if I was the strategic architect of the Surge, "General O," as he was called, was the operational architect, and I treasured my relationship with him and the leadership he provided.

Important though the additional forces were to the Surge, how they were used and positioned would prove even more critical. In accordance with the "surge of ideas," beginning with Operation Fardh al-Qanoon, or Enforcing the Law, launched in Baghdad on 13 February 2007, the objective of military operations would be to protect the population against ethno-sectarian violence. That would, in turn, make possible political progress that would more fairly distribute power and resources among the various sects and factions in Iraq. Hence the need to return to the neighborhoods, especially in Baghdad, the epicenter of the Sunni–Shi'a civil war, from the large bases on which US forces had been consolidating over the previous years.

As part of the Baghdad security plan, Colonel Stephen J. Townsend's infantry-heavy and highly mobile 3rd Stryker Brigade (in Stryker wheeled armored vehicles), the theatre reserve force, along with the other US units in the capital, would clear key Baghdad neighborhoods of insurgents, terrorists and militia members, one after the other. Townsend was another exceptional leader, and his forces did superb work in clearing extremely tough areas. However, instead of *handing over* cleared areas to Iraqi Army and police units, which had proven incapable of holding them in the past, US Army units now would *hold* the areas, *partnering* with Iraqi forces in doing so, and walling neighborhoods off with tall concrete barriers to inhibit the movement of insurgent and militia forces. (We sometimes jokingly noted to Iraqi citizens that "gated communities" in the US were very expensive, but we were establishing them free of charge for the Iraqis.) Checkpoints would be equipped with biometric identity devices used to screen residents and check vehicle occupants against a database of known malign actors. Instead of patrolling their zones of operation from large bases, US troops would establish combat outposts and joint security stations in those zones throughout Baghdad, ultimately seventy-seven in total, and operate from them. They would live and patrol with Iraqi Army and police elements, allowing the professionalism of the US soldiers to rub off on their Iraqi counterparts and enabling the gathering of human intelligence from local residents. Only when the security situation and the readiness of the Iraqi forces met specific conditions would the American elements begin to thin out and transition tasks to the Iraqis. Even then, US forces would retain a presence to ensure situational awareness.

My subordinate commanders and I also worked with Prime Minister Maliki and the Defense and Interior Ministers to rid the Iraqi Army and National Police of the worst of their sectarian commanders. Over time, in fact, every general officer in the National Police would be replaced and each police brigade would undergo intensive refresher training to reinstill a sense of professionalism in the force and remove sectarian actors, a process that included pulling the unit off-line to a training center for thirty days of receiving replacement personnel, retraining and re-equipping. The process was far from perfect, and Sunnis still viewed the National Police to a degree as an extension of Shi'a militias, but the combination of new commanders, better training and close partnership with US forces and military and police advisers began to change the dynamics on the ground.

Lieutenant General Odierno also recognized that securing Baghdad required gaining and maintaining control of the belts of territory surrounding the city, a necessity made even clearer by the capture of documents outlining AQI's plan for Baghdad.[52] He ordered divisions outside of Baghdad to focus their efforts on interdicting car bombs, weapons and other accelerants of violence before they could enter the capital. A new division headquarters (Multi-National Division–Center) was placed south of Baghdad to oversee forces that would dominate the routes leading into the city from that direction and the east. MNC–I would also direct operations to clear insurgent and terrorist safe havens outside the capital, such as the area between Baghdad, Taji, Lake Tharthar and Karmah.

Accordingly, only two of the Surge brigades went into Baghdad; the remaining three, as well as the two Marine battalions and the 13th Marine Expeditionary Unit that went into Anbar Province, were deployed to locations that enabled them to focus on denying AQI safe havens outside the city. Nevertheless, Baghdad remained the priority, with additional battalions from brigades outside the city attached to Multi-National Division–Baghdad (MND–B) to increase its operating strength as needed. At the height of the Surge, MND–B alone controlled twenty-five maneuver battalions, double the number patrolling the streets of the capital the year before.

The Counter-Terrorist Task Force under the command of Lieutenant General Stan McChrystal (a brilliant leader who commanded the Task Force for an unprecedented five years, building it into an extraordinary machine for relentlessly targeting high-value targets (HVTs), night after night) also substantially ramped up its efforts to locate and capture or, if necessary, kill key Sunni and Shi'a extremist leaders. The Task Force under McChrystal included special elements of the Army SOF and Navy SEALs, the Ranger Regiment, the UK's Special Air Service, the 160th Special Operations Aviation Regiment and other US special mission units. An airborne infantry battalion, Task Force Falcon, was also attached to it to conduct operations in select areas that contained insurgent sanctuaries, forcing the insurgent and militia leaders to move and communicate and, in so doing, become targetable by Task Force elements. A robust intelligence system supported the Task Force and integrated all forms of intelligence to provide specific, actionable intelligence on the patterns of life and locations of the HVTs. Task Force operators then conducted operations that targeted them, which often led to subsequent operations – often on the same night – as intelligence gathered on site was quickly processed and turned into targetable information. The Task Force scored innumerable successes against Sunni extremists and insurgents and Shi'a militia leaders (for whom an additional task force would be established halfway through the Surge), and also detained operators of the Iranian Revolutionary Guards Corps Qods Force, which were attempting to turn special groups of the JAM and other militia into a Hezbollah-like militia beholden to Teheran.

Operations by MNC–I and McChrystal's special operators were all part of the broader holistic strategy that I labelled the Anaconda Plan, which envisioned using every possible military and civil capability to deny the assets needed by the insurgents, extremists and militia members – in essence, to squeeze the life out of them.[53] It included six major parts: kinetic operations (conventional and special operations by coalition forces, as well as use of affiliated Iraqi forces for local security); political engagement (support of the Awakening, outreach to insurgent groups and militias, reconciliation initiatives and Iraqi legislation); intelligence (greater use of unmanned aerial vehicles (UAVs), signals and other forms of intelligence, and intelligence fusion centers); inter-agency actions (strategic communications, information operations, winning the battle of the internet, efforts to stem the flow of suicide bombers and lethal aid to Iraq from other countries); non-kinetic aspects (religious engagement, provision of basic services, education and economic initiatives); and reform of detainee operations. Derided as armed nation-building by some critics, the plan's overarching objective was rather to secure the population, that foundation without which nothing was possible; with it, other initiatives could be pursued that would reinforce and solidify the security gains. Over time, careful transition of security to Iraqi control would be conditions-based, a function of both the increasing capabilities of Iraqi security forces and improvements in the security situation, rather than a lever to force them to take on more challenges than they could handle. The goal was the provision of sustainable security and stability in the near term, leading to a secure, stable, free and democratic Iraq over the long term.

Early on, I led a process with General Odierno, Lieutenant General McChrystal and my exceptional deputy, the British Lieutenant General Sir Graeme Lamb (a long-time friend with whom I had previously worked in Bosnia, Iraq and our respective home countries) to help persuade our subordinate commanders to fully support the Awakening and, over time, the reconciliation initiative that built on it. I promoted these actions despite the understandable reluctance of some of our battalion and brigade commanders to sit down with Iraqis who had American blood on their hands; but these defectors were after all willing to ally with the coalition against a common enemy rather than continue to fight us, and we needed to reconcile with them. The progress in Al-Anbar Province was stunning, especially given the fact that less than a year earlier the intelligence officer of MNF-W had assessed the situation there as out of control. The Awakening was proving increasingly important. From its start in Ramadi, the movement had spread over time throughout Anbar and then to the Baghdad belts, reducing the ability of AQI to inject personnel, car bombs and other weapons into Baghdad. Defections from the ranks of the insurgency were encouraged further by a strategic engagement effort led by Lieutenant General Lamb. (In fact, I knew that Lamb would be a hugely important part of the effort to promote reconciliation, given his experiences in Northern Ireland; hence, one of the key requests I made of Prime Minister Tony Blair, when I stopped in London on the way to Iraq, was to extend Lamb's tour as the deputy commander of the Multi-National Force-Iraq. "Lambo's" contribution more than validated the wisdom of that request.) This involved promoting reconciliation with former enemies, and its success allowed MNF-I increasingly to focus its resources on the truly irreconcilable elements in the insurgency and AQI, as well as among the Shi'a militias.

In June, a new opportunity arose when a group of insurgents in Ameriyah, a district of western Baghdad that had been devastated by Sunni extremist and insurgent violence, indicated their willingness to fight AQI in coordination with the American forces in the area. Lieutenant General Odierno and I took advantage of this development to recruit local security forces – who came to be known as the Sons of Iraq – to help defend neighborhoods from depredations by Sunni terrorists and Shi'a militias alike. Over time, I gained Washington's approval to use US funds to pay Iraqis as local security contractors, with the Sons of Iraq working under the local US chain of command, which provided oversight and enforced accountability for their actions. Those who had joined the coalition effort in Anbar and elsewhere were quickly added to the program as well. In the autumn of 2007, MNF-I established a similar initiative for Shi'a militia members. Ultimately, at the height of the program more than 103,000 Iraqi men – the equivalent of 180 battalions of light infantry – were paid to help guard their communities, significantly adding to the combat power of coalition forces. Since most of the Sons of Iraq were Sunni, however, Prime Minister Maliki was less than pleased with having to support the initiative. Nonetheless, he did not curtail the program until well after the completion of the Surge, when security gains and the expansion and professionalization of the Iraqi security forces made it less critical – at least in the near term.[54]

The first five months of 2007 were very difficult for the coalition, with military casualties increasing and the insurgents, extremists and militias fighting hard to retain their control of neighborhoods to which US forces were returning. I had warned during my Senate confirmation hearing in Washington in January that the situation would get harder before it got easier, and now that warning was being validated – even as key metrics such as civilian casualties, suicide bombings and sectarian violence began to decline. By the summer of 2007, however, MNC–I forces had gained the strategic initiative and never relinquished it for the duration of the Surge.

By June, with all the Surge forces on the ground in Iraq, Odierno launched Operation Phantom Thunder, a full-scale offensive targeting extremists across the Baghdad belts in Anbar, Diyala, Salahadin, Babil and Wasit Provinces. The battle to clear the AQI sanctuary city of Baqubah, where the group had relocated its capital after the fall of Ramadi, was particularly intense. This was the first of a surge of operations that took the fight to AQI, eliminating its sanctuaries and safe havens. And rather than pause once the operations were completed, MNC–I continued to put pressure on AQI by advancing into areas further afield, into which the organization would likely retreat, always coordinating its actions closely with McChrystal's Counter-Terrorist Task Force, whose elements were simultaneously targeting the key leaders. The remaining elements of AQI gradually withdrew north to the greater Mosul area, where a future confrontation with coalition forces awaited.

While AQI had taken the brunt of coalition efforts during the Surge to date, that was about to change. Increasing pressure on the JAM had caused Sadr to flee to Iran, leading to a splintering of the organization into multiple "special groups," supplied, armed, trained and, to various degrees, directed by IRGC Qods Force operatives, while mainstream JAM still responded to orders from Sadr and his representatives. On 27 August, a gun battle between JAM operatives and shrine guards in the holy Shi'a city of Karbala, south of Baghdad, led to the deaths of fifty-two people and the wounding of 279 others, causing Grand Ayatollah Sistani to cancel the enormous pilgrimage to Karbala normally conducted during the Islamic holiday of Mid-Sha'ban. Prime Minister Maliki was incensed and responded aggressively, strapping on a pistol and personally leading a column of fifty-two vehicles to Karbala, where he and his security detail arrested scores of JAM fighters to the cheers of onlookers. Having overplayed his hand, Sadr declared a six-month ceasefire, which further reduced violence across Iraq as mainstream JAM elements took a knee and ceased operations.

The Surge was also helped enormously by dramatically increased numbers of US technological enablers – drones, blimps with optics, special signals intelligence capabilities, wheeled armored vehicles designed to counter IEDs, counter-rocket systems, etc. – and also a significant increase in Iraqi security forces in both quantity and quality, helped by the steady deployment of increasingly well-prepared coalition military and police transition teams. When Iraqi security force numbers were combined with coalition forces, private security contractors and the Sons of Iraq, the ratio of security personnel to population in Iraq by the end of the Surge reached roughly one to fifty, the minimum ratio deemed necessary to defeat an insurgency according to the counter-insurgency doctrine we had published in late 2006. For his part, Secretary Gates drove the US military and defense industries to produce the additional armed drones and new mine-resistant, ambush-protected (MRAP) vehicles, which could better withstand blasts from roadside bombs, the signature weapon of the insurgency in Iraq. MRAPs, in particular, saved countless lives in Iraq and, later, in Afghanistan.[55] Gates and Admiral Mike Mullen, who took over as chairman of the Joint Chiefs in the early autumn of 2007, were an exceptional team and absolutely relentless in ensuring that the military services provided everything humanly possible that our men and women in combat needed. They were aided enormously on Capitol Hill by equally committed and relentless members of Congress, especially John McCain, Joe Lieberman and Lindsey Graham, whom I dubbed "the Three Amigos." No one on Capitol Hill had our soldiers' backs more than they did.

Charge of the Knights

As Lieutenant General Odierno and I saw it in early 2008, shortly before he rotated back to the US, areas in northern Iraq were the last AQI strongholds that needed to be cleared, and Odierno crafted a major offensive to do just that. However, the operation would not take place as scheduled. Instead, an intra-Shi'a civil war would erupt in Basra and Baghdad in the spring of 2008, leading Prime Minister Maliki to use the Iraqi Army to go to war with the Sadrists – and coalition forces would be there with their Iraqi partners when the situations were at their most dire.

The prelude to this development was the tacit ceding of Basra to JAM control by Multi-National Division–Southeast (MND–SE), led by British commanders imbued with deep experience of "the Troubles" in Northern Ireland and increasingly pressured by policymakers in London to reduce casualties and set conditions to withdraw, despite the Surge being conducted by US forces. Basra in late 2007, however, was not Northern Ireland, and the deals cut by British commanders with JAM leaders in order to rush the transition of Basra to Iraqi control ultimately led to increasing violence and criminality by JAM in the most important city in southern Iraq. By early 2008, Basra had descended into near chaos as militias took over the streets and engaged in largely unchecked criminal activity.

In response, the Iraqi Commander in Basra, Lieutenant General Mohan al-Freiji, and his British adviser, Colonel Richard Iron, developed a plan for regaining control of Basra, which they briefed to me and my Iraqi counterparts in Baghdad in early March. I believed the assumptions underpinning the plan were somewhat overoptimistic and preferred to set conditions for several months before beginning the execution of the plan – focusing coalition resources on clearing AQI in northern Iraq before tackling the intra-Shi'a conflict in the south. But Prime Minister Maliki had different priorities, and the improvement in security in Baghdad had allowed him to focus on his political competitor, Sadr, and Sadr's militia, especially in Basra, the most important city in the oil-rich south.

On 22 March, Maliki informed me that he was deploying a division headquarters and several Iraqi Army brigades to Basra, and that he would personally lead the operation as he had done the previous August in Karbala. Though surprised by this decision, I pledged to do all that we could to support the operation. Dubbed Saulat al-Fursan, or Charge of the Knights, the operation came close to disaster. As the Iraqi 14th Division moved into Basra, it was unable to "set conditions" on the outskirts for several days, as Maliki had committed to doing so that we could reposition coalition assets to support the Iraqis. Rather, it got sucked immediately into a major battle. JAM elements hit it with multiple ambushes, and volleys of indirect fire also rained down on the headquarters Maliki had rashly established in Basra Palace, located in the center of the city. Without coalition advisers on site, the F-16 aircraft and drones that had been quickly shifted to Basra could not bring their ordnance to bear on JAM positions, which were difficult to distinguish from Iraqi security force elements in the dense urban setting. The situation only got worse as supply shortages began to cripple the Iraqi formations. In fact, the Iraqi Army lacked the logistical capability to sustain its forces away from their home bases, and they were unable to garner sufficient supplies in the local markets of Basra as they had often done in other areas of Iraq.

The impending failure of the operation portended the end of Maliki's government. Smelling blood in the water, political sharks circled in Baghdad until the Deputy Prime Minister, Barham Salih, a highly educated and exceedingly competent Iraqi Kurd I had come to know and respect highly during my previous tours in Iraq, called the Iraqi ethnic, sectarian, political and religious leaders together at his residence in Baghdad and rallied them to support their Prime Minister, who was literally under heavy fire in Basra, with the situation on the battlefield hanging in the balance. "Dr. Barham," as we referred to him, would eventually serve as the prime minister of the Kurdish Regional Government and then as president of Iraq. And during the battle of Basra, his leadership inspired leaders of various allegiances to put Iraq and their elected Prime Minister first at a pivotal moment for the Land of the Two Rivers. Despite understandable differences at times with a frequently sectarian Prime Minister, Dr. Barham had Maliki's back at the most critical of moments. His performance was exceptional.

Lieutenant General Lloyd Austin, the highly competent commander who had replaced Lieutenant General Odierno two months earlier as the commander of MNC–I, and I were determined to prevent defeat of the Iraqi forces. Austin quickly flew to Basra and immediately deployed his corps' mobile tactical command post, equipped to display the video "feeds" sent down from the drones and to provide the command-and-control systems needed to orchestrate the fight. Set up next to the British command post, the US and British headquarters quickly committed additional assets to reinforce Iraqi forces in Basra, even as JAM elements also rose up in Sadr City and across southern Iraq. A courageous British brigadier, Julian Free, who was the acting MND–SE Commander (while his boss was on his mid-tour leave skiing in the Alps), decided on his own, without asking London's permission, to send British advisers to rejoin and help their Iraqi counterpart units fighting in the city. I remain grateful to him for that gutsy decision.

Additional US advisers rapidly deployed and embedded in other Iraqi Army formations in Basra, enabling the provision of much needed close air support. Attacks by F-16 fighter jets, Apache attack helicopters and armed drones pummelled JAM positions, turning the tide of the battle. Coalition staff sorted out resupply and logistics requirements and surged emergency support, as well as Iraqi reinforcements, to the units on the front lines, even as additional Iraqi units, along with their embedded US Army and Marine Corps transition teams, arrived, including brigades from the well-trained 1st and 7th Divisions from the Sunni province of Al-Anbar, which had no issues with attacking and destroying Shi'a militias. The assembled Iraqi forces, with US enablers overhead, now cleared Basra in a deliberate process, seizing weapons caches, killing JAM fighters and dealing another significant blow to Sadr's reputation.

The scene of the heaviest fighting then shifted to Sadr City, the densely populated Shi'a neighborhood in eastern Baghdad of well over 1.5 million people, where JAM militiamen had risen in concert with their brethren in the south. JAM forces attacked Iraqi Army and police checkpoints encircling Sadr City and launched seemingly endless volleys of rockets and mortars on Baghdad's Green Zone, imperilling Iraqi government facilities and the US Embassy. As the fighting continued, the battle centered on the construction of a line of concrete barriers along Quds Street (Route Gold on coalition maps), which would enable US and Iraqi forces to control the south-western third of Sadr City and push JAM rocket teams out of range of the Green Zone. The wall also threatened JAM's access to the key Jamila market (a big source of extorted revenue for JAM), forcing JAM fighters out into the open in an effort to prevent its establishment. Colonel John Hort's 3rd Brigade, 4th Infantry Division became the coalition (indeed, the national) main effort, and his brigade was massively reinforced by fixed and rotary-wing air support, multiple launch rocket systems with precision munitions, drones, engineers and even special forces sniper teams.[56] The tanks and infantry fighting vehicles of Hort's mechanized infantry brigade, normally stored at his unit's bases and not employed on the streets of Baghdad, were brought into the action and used very effectively to attack the militia in areas in and around Sadr City.[57]

The forty-five-day battle for the south-western third of Sadr City and its surrounding neighborhoods was the most intense fighting seen in Baghdad since the early tough months of the Surge, and Hort's soldiers on the ground and precision assets in the air fought tenaciously, finally breaking the back of JAM, with more than seventy rocket teams destroyed and key militia leaders either killed or in exile in Iran.[58] In May, Maliki ordered the Iraqi Army to seize the remainder of Sadr City, and a month later he also ordered forces to clear the city of Amarah, near Basra in southern Iraq, tasks easily accomplished against negligible resistance, as was the clearance of other smaller cities in the Shi'a south. The JAM militia had now been defeated, along with the Sunni insurgents and the vast majority of AQI. US and coalition units would focus the final months of the Surge on the remnants of insurgent and militia elements around the country, as well as on further development of the Iraqi security forces and the various civil efforts that were solidifying the gains achieved since the commencement of the Surge in early 2007.

Aftermath

The drawdown of US forces in Iraq began in December 2007, and the final Surge brigade departed in July 2008. Ethno-sectarian violence had been driven down by over 85 percent over the previous eighteen months, and the number of attacks and other security metrics would continue to decline modestly over the subsequent three years. The ethnic, sectarian, tribal and political centrifugal forces threatening to tear Iraq apart in early 2007 had been brought under control, and the major centripetal force – the central government's distribution of the country's enormous oil revenues – was once again helping to keep Iraq together. The dramatic improvements in security in Baghdad and the rest of the country made politics, rather than violence, the focus of the competition for the distribution of power and resources.

The final showdown with the remnants of AQI in northern Iraq would not materialize until the spring of 2010 (with General Odierno promoted, back in Iraq, and at the helm of Multi-National Force–Iraq), when coalition operations defeated the group in its remaining northern strongholds. Driven across the border into Syria, the surviving elements of AQI combined forces with Syrian and foreign jihadists and rebranded themselves as ISIS, the Islamic State of Iraq and al-Sham. ISIS would remain there and exploit the explosion of Sunni grievances with the onset of the Syrian Civil War. Ultimately, ISIS would build its power there so substantially that it was able to return to Iraq beginning in late 2013, defeat the Iraqi security forces in Al-Anbar and Nineveh Provinces, and establish a caliphate that encompassed north-eastern Syria and northern Iraq (where the capital would be Mosul). Until its ultimate defeat in 2019, ISIS was the most dangerous Islamist extremist group in the world and attracted tens of thousands of would-be jihadists to Syria and Iraq, as it also sought to inspire, direct and support sensational terrorist attacks in Western Europe and the United States.

Meanwhile, over a year after the conclusion of the Surge in Iraq, a massive electoral turnout in provincial elections in January 2009 brought to power more representative governance at the local level. The trends seemed positive. However, the results of parliamentary elections in March 2010 proved much more contentious, as former Prime Minister Ayad Allawi's coalition won two more parliamentary seats than Prime Minister Maliki's coalition – but Maliki was able to retain his position nonetheless when Allawi was blocked by judicial decree from forming a government and Malaki was then able to generate the votes (and, curiously, both the US and Iran supported his candidacy). Maliki's machinations thus enabled his return as prime minister, but under clouded circumstances.[59] Regardless, through it all, the achievements of the Surge were generally sustained, although the progress in the quality of governance and provision of basic services never reached the levels Iraqi citizens (or we) had hoped to see.

Unfortunately, by late 2011, as the time approached for US combat forces to leave Iraq in accordance with an agreement reached in 2008 between Maliki and President Bush, complications emerged. Maliki frustrated US political and military leaders who were considering whether to request Iraqi support for a modest continued US military presence by being unable to provide a parliamentary-approved Status of Forces Agreement (SOFA) – which US President Barack Obama had unnecessarily established as a requirement for US combat forces remaining beyond the end of 2011. The result was the departure in mid-December 2011 of the residual US combat forces and also of General Lloyd Austin, the final four-star commander in the country (from 2021 the US Secretary of Defense).

Less than forty-eight hours after the departure of US forces, Maliki's government undertook the first of a number of destructive and highly sectarian actions, seeking to shore up his support in the Shi'a community, the base of his political strength. In the first of these, the Iraqi Judicial Council brought charges against Iraqi Vice President Tariq al-Hashemi, the senior Sunni Arab in the government, and his security force for alleged illegal killings. Maliki followed this several months later with charges against the next most senior Sunni in government and then against a prominent Sunni member of parliament. Tragically, these actions began to undo much of what the Surge and subsequent years of initiatives had accomplished, particularly in knitting back together the fabric of Iraqi society. The actions inflamed Sunni sentiment, led to large demonstrations – which were put down violently by predominantly Shi'a security force units – and reignited the sense of Sunni disenfranchisement and alienation that had imperilled Iraq from shortly after the invasion in 2003 until the spread of the Anbar Awakening, the creation of the reconciliation initiative and the establishment of the Sons of Iraq program during the Surge. Over and over again, Maliki used the levers of government to attack his political enemies, brutally suppress dissent and dissolve important Surge initiatives, once again driving Iraq's Sunni Arabs into resistance and essentially pouring fuel on the remaining embers of the sectarian civil war that had nearly destroyed Iraq in 2006 and the first half of 2007.[60]

US combat forces would not be present to save Maliki from the fire once Iraq again descended into near civil war. Nor would the once impressive Iraqi security forces, since, to ensure his administration would be safe from a coup, Maliki had hollowed them out by appointing politically reliable but militarily incompetent, and often abusive, officers (many of whom had been fired during the Surge) to command its formations. He thought he could do this without jeopardizing the security situation, given the reduction in violence since 2007, but the calm proved illusory. As Sunni outrage grew, many came to see ISIS as preferable, in some areas, to Shi'a-dominated Iraqi security forces.

Fallujah was the first city to fall to ISIS in January 2014. When confronted by an invasion in northern Iraq of largely truck-borne ISIS fighters five months later, the four Iraqi Army divisions in northern Iraq ignominiously collapsed, obliging the Iraqi government to turn to hastily raised Shi'a militias to defend the capital, thus giving the militias defeated in 2008 the authority to return to the streets in uniform, this time supported by the state. The creation of these Popular Mobilization Forces, many of which were beholden to Iran, would create a long-term problem for the Iraqi government – a problem that continues to create enormous challenges for Iraq to this day.

Meanwhile, the threat of ISIS-inspired genocide against the Yazidi people of northern Iraq in 2014 compelled the Obama administration to return US forces to Iraq, and also to deploy them to Syria, which had collapsed into civil war after Syria's Arab Spring uprising of disenfranchised Sunni Arabs threatened the government of strongman Bashar al-Assad.[61]

The ultimate war against the Islamic State, however, would be fought mostly via the Iraqi Army and local Syrian forces with US airpower, special operations forces, and advisory teams, the latter coordinating the air reconnaissance and precision-strike assets that enabled the Iraqis and later Syrians to defeat ISIS and eliminate its caliphate. In the end, that achievement would require five more years of combat operations, an extension of the war that, tragically, had not quite ended after the withdrawal of US combat forces from Iraq in 2011.

Lessons from the Iraq War

The Iraq War was America's most costly conflict since the end of the Vietnam War. The Bush administration embarked on regime change in Iraq for a number of reasons, but, on reflection – especially given the intelligence failure regarding weapons of mass destruction – one can ask whether any of those reasons represented an existential threat to the United States and its vital national interests. Although military operations quickly dispatched the Iraqi armed forces and ousted Saddam Hussein from power, the conflict continued as an unanticipated insurgency erupted and various sects and factions began a violent competition for power and resources.

The lessons from this experience are not new, but they are clear. As Clausewitz observed in *On War*, the political underpinnings of war are far more important than skillfully conducted military operations. The Bush administration put far more intellectual effort into fashioning a war plan than into determining how to stabilize and rebuild the Iraqi state once Saddam Hussein and the Ba'ath Party were forced from power.

As Winston Churchill observed in his history of the Great War, policymakers must never assume that the conduct and aftermath of war will be easy. In the run-up to the Iraq War, however, senior political and military leaders made overly optimistic assumptions about the duration of the receptivity of the Iraqi people to the coalition presence and goals. American leaders then took too long to understand that these assumptions were not valid, and then to recalibrate resources, doctrine and strategy to compensate for the situation in which we found ourselves.

In the decade following the dissolution of the Soviet Union, the US armed forces had become enamored of a so-called "revolution in military affairs" centering on guided munitions coupled with high-tech intelligence, surveillance and reconnaissance systems that would presumably eliminate the fog and friction of battle. Although there was much to this concept in terms of war-fighting, it tended to overlook the equally important role of military forces in stabilizing countries after military operations have terminated. In Iraq we learned – or rather relearned – that shock and awe based on high-tech forces is not a substitute for troop numbers in counter-insurgency operations or for the proper employment of those troops. Iraq did not stabilize until Multi-National Force–Iraq and its Iraqi counterparts and tribal allies reached 600,000 personnel (which, as noted earlier, meant roughly one security person for every fifty Iraqi civilians), and deployed those forces properly. In counter-insurgency operations, numbers matter – as do the overarching big ideas that guide the employment of those forces.

The biggest of the big ideas is that security in a counter-insurgency campaign is job number one. If the population cannot be controlled and protected, all other initiatives will come to naught. Population security requires trained and capable military and police forces operating under thoughtfully crafted rules of engagement that enable them to accomplish their security tasks while protecting the population from excessive violence. This is not an easy balance, but carefully metering security and violence is essential to the successful outcome of these types of conflicts.

Iraq also reminded us of the imperative of truly understanding, in a very granular manner, a country in which we are going to operate. Without a deep appreciation of the ethnic, sectarian, tribal and political elements, as well as how the country is supposed to function *and* how it really does function, it is very hard to govern it. Here it is important to remember that, as occupation force leaders, military commanders were, during the first year or more, exercising the powers of the executive, legislative and judicial authorities in the provinces they oversaw. The same was true of the overall authority in Baghdad. And it became very clear early on that our understanding of Iraq was inadequate.

Beyond that, the experience of the first eighteen months also underscored the importance of using existing organizations and institutions (e.g. a State Department embassy, the US Army's Corps of Engineers, the Defense Contracting Agency, etc.) rather than ad hoc organizations of rotating personnel (the Coalition Provisional Authority and its one-off engineering and contracting elements). Moreover, the experiences of the first year – and beyond – highlighted the importance of not approving policies or operations that will create more bad guys than they take off the streets by their implementation, e.g. overly expansive de-Ba'athification without an agreed reconciliation process, and firing the members of the Iraqi military without an agreed plan for their support. Those policies were exceedingly counter-productive and set our effort in Iraq back for a number of years.

President Bush hoped to see Iraq as the South Korea of the Middle East – a regional US ally with a democratic government and a strong, free-market economy. He understood that this transition would likely require US forces to remain in Iraq for the long-term, just as US forces have remained on the Korean peninsula since their first deployment in

1950. US military forces there provide needed leverage to back diplomacy and encourage the stability that provides the foundation for economic growth. Provided US forces do not take casualties, the American people have proven willing to underwrite long-term overseas deployments in the greater interests of US national security. Bush was willing to underwrite this commitment, but the follow-on Obama administration was not. How to attain a bipartisan consensus on such a crucial national security issue is a question worth considering in the run-up to war, rather than assuming that the natural end state of war will result in the rapid redeployment of US forces from the conflict zone.

Finally, if history has shown us anything, it is that there will be other counter-insurgency conflicts in the future in which the US military will be engaged. US armed forces might not want to prepare for or conduct counter-insurgency operations, but they must be prepared to do so or risk failure when inevitably confronted with such conflicts again. The US Army required more than three years in Iraq to regain the competencies so unwisely jettisoned after the end of the Vietnam War. Keeping counter-insurgency doctrine current and educating mid-grade and senior leaders about it (while understandably focusing more on major combat operations given the developments of the past decade) will be a small price to pay to ensure that when the United States is again confronted with a messy, irregular conflict the leaders of its armed forces are intellectually prepared to cope with it.

Appendix A

Security Incidents in Iraq

January 2004–August 2008

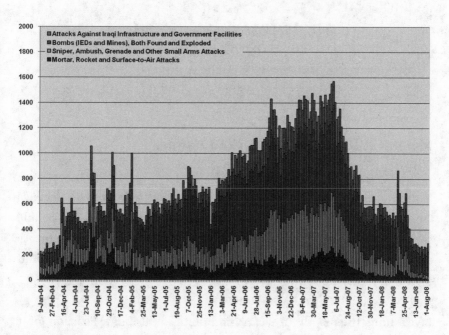

Legend:
- Attacks Against Iraqi Infrastructure and Government Facilities
- Bombs (IEDs and Mines), Both Found and Exploded
- Sniper, Ambush, Grenade and Other Small Arms Attacks
- Mortar, Rocket and Surface-to-Air Attacks

Source: Multinational Force–Iraq, August 2008.

Appendix B

Headquarters

MULTI-NATIONAL FORCE – IRAQ
BAGHDAD, IRAQ
APO AE 09342–1400
15 July 2008

Multi-National Force-Iraq Commander's Counterinsurgency Guidance

Secure and serve the population. The Iraqi people are the decisive "terrain." Together with our Iraqi partners, work to provide the people security, to give them respect, to gain their support, and to facilitate establishment of local governance, restoration of basic services, and revival of local economies.

Live among the people. You can't commute to this fight. Position Joint Security Stations, Combat Outposts, and Patrol Bases in the neighborhoods we intend to secure. Living among the people is essential to securing them and defeating the insurgents.

Hold areas that have been secured. Once we clear an area, we must retain it. Develop the plan for holding an area before starting to clear it.

The people need to know that we and our Iraqi partners will not abandon them. When reducing forces, gradually thin our presence rather than handing off or withdrawing completely. Ensure situational awareness even after transfer of responsibility to Iraqi forces.

Pursue the enemy relentlessly. Identify and pursue Al Qaeda-Iraq and other extremist elements tenaciously. Do not let them retain support areas or sanctuaries. Force the enemy to respond to us. Deny the enemy the ability to plan and conduct deliberate operations.

Employ all assets to isolate and defeat the terrorists and insurgents. Counter-terrorist forces alone cannot defeat Al-Qaeda and the other extremists. Success requires a comprehensive approach that employs all forces and all means at our disposal – non-kinetic as well as kinetic. Employ Coalition and Iraqi conventional and special operations forces, Sons of Iraq, and all other available non-military multipliers in accordance with the attached "Anaconda Strategy."

Generate unity of effort. Coordinate operations and initiatives with our embassy and interagency partners, our Iraqi counterparts, local governmental leaders, and non-governmental organizations to ensure all are working to achieve a common purpose.

Promote reconciliation. We cannot kill our way out of this endeavor. We and our Iraqi partners must identify and separate the "irreconcilables" from the "reconcilables" through thorough intelligence work, population control measures, information operations, kinetic operations, and political initiatives. We must strive to make the reconcilables part of the solution, even as we identify, pursue, and kill, capture, or drive out the irreconcilables.

Defeat the network, not just the attack. Focus to the "left" of the explosion. Employ intelligence assets to identify the network behind an attack, and go after its leaders, explosives experts, financiers, suppliers, and operators.

Foster Iraqi legitimacy. Encourage Iraqi leadership and initiative; recognize that their success is our success. Partner in all that we do and support local involvement in security, governance, economic revival, and provision of basic services. Find the right balance between Coalition Forces leading and the Iraqis exercising their leadership and initiative, and encourage the latter. Legitimacy of Iraqi actions in the eyes of the Iraqi people is essential to overall success.

Punch above your weight class. Strive to be "bigger than you actually are." Partner in operations with Iraqi units and police, and employ "Sons of Iraq," contractors, and local Iraqis to perform routine tasks in and around Forward Operating Bases, Patrol Bases, and Joint Security Stations, thereby freeing up our troopers to focus on tasks "outside the wire."

Employ money as a weapon system. Money can be "ammunition" as the security situation improves. Use a targeting board process to ensure the greatest effect for each "round" expended and to ensure that each engagement using money contributes to the achievement of the unit's overall objectives. Ensure contracting activities support the security effort, employing locals wherever possible. Employ a "matching fund" concept when feasible in order to ensure Iraqi involvement and commitment.

Fight for intelligence. A nuanced understanding of the situation is everything. Analyze the intelligence that is gathered, share it, and fight for more. Every patrol should have tasks designed to augment understanding of the area of operations and the enemy. Operate on a "need to share" rather than a "need to know" basis. Disseminate intelligence as soon as possible to all who can benefit from it.

Walk. Move mounted, work dismounted. Stop by, don't drive by. Patrol on foot and engage the population. Situational awareness can only be gained by interacting with the people face-to-face, not separated by ballistic glass.

Rebels of the Farabundo Martí National Liberation Front (FMLN) in La Anchila, Usulutan, El Salvador, 1983.

Members of the US military carrying their weapons and equipment on a road in Greenville during the invasion of Grenada in 1983.

USAF aircraft of the 4th Fighter Wing fly over Kuwaiti oil fires ignited by the retreating Iraqi Army during Operation Desert Storm, 1991.

Members of Saddam Hussein's "suicide Fedayeen" volunteers parade in the streets of Baghdad, November 1997.

Baghdad ablaze during the allied bombing on the first night of the "shock and awe" campaign, 2003.

A soldier from the Royal Scots Dragoon Guards loads ammunition, 2003.

American Marine Expeditionary Unit helicopters embarking for Iraq, 2004.

Lieutenant General Ray Odierno and General David Petraeus, the top US commander in Iraq, greet US President George W. Bush as he arrives at Al Asad Air Base in Al-Anbar Province, Iraq, 3 September 2007.

General Petraeus explains the security improvements made to Sadr City to Senator Barack Obama, July 2008.

A US Marine and an Afghan solider react as dust blankets the area after an IED exploded while they were under enemy fire in Mian Poshteh, Afghanistan, 17 July 2009.

Coalition forces aid the evacuation at Hamid Karzai International Airport, Kabul, Afghanistan, August 2021.

Major General Christopher T. Donahue, the last US soldier to leave Afghanistan, at Hamid Karzai International Airport in Kabul, 31 August 2021.

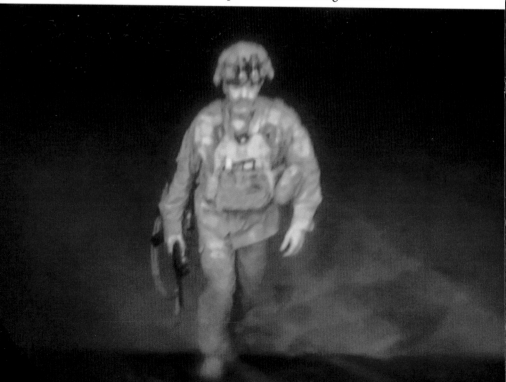

Donetsk People's Republic soldiers pose for a photo in central Mariupol, April 2022.

Ukrainian soldiers of the 3rd Separate Assault Brigade in the Donetsk Oblast reposition a 105mm howitzer to fire towards Bakhmut on 1 July 2023.

A drone pilot in the Ukrainian Army learning to fly unmanned aircraft, Donbas, February 2023.

An Explosive Ordnance Disposal technician observes a member of the Royal Saudi Naval Forces connecting an explosive tool to a PackBot.

Understand the neighborhood. Map the human terrain and study it in detail. Understand the local culture and history. Learn about the tribes, formal and informal leaders, governmental structures, religious elements, and local security forces. Understand how local systems and structures – including governance, provision of basic services, maintenance of infrastructure, and economic elements – are supposed to function and how they really function.

Build relationships. Relationships are a critical component of counterinsurgency operations. Together with our Iraqi counterparts, strive to establish productive links with local leaders, tribal sheikhs, governmental officials, religious leaders, and interagency partners.

Look for Sustainable Solutions. Build mechanisms by which the Iraqi Security Forces, Iraqi community leaders, and local Iraqis under the control of governmental institutions can continue to secure local areas and sustain governance and economic gains in their communities as the Coalition Force presence is reduced. Figure out the Iraqi systems and help Iraqis make them work.

Maintain continuity and tempo through transitions. Start to build the information you'll provide to your successors on the day you take over. Allow those who will follow you to "virtually look over your shoulder" while they're still at home station by giving them access to your daily updates and other items on SIPRNET. Deploy planners and intel analysts ahead of time. Encourage extra time on the ground during transition periods, and strive to maintain operational tempo and local relationships to avoid giving the enemy respite.

Manage expectations. Be cautious and measured in announcing progress. Note what has been accomplished, but also acknowledge what still needs to be done. Avoid premature declarations of success. Ensure our troopers and our partners are aware of our assessments and recognize that any counterinsurgency operation has innumerable challenges, that enemies get a vote, and that progress is likely to be slow.

Be first with the truth. Get accurate information of significant activities to the chain of command, to Iraqi leaders, and to the press as soon as is possible. Beat the insurgents, extremists, and criminals to the headlines, and pre-empt rumors. Integrity is critical to this fight. Don't put lipstick on pigs. Acknowledge setbacks and failures, and then state what we've learned and how we'll respond. Hold the press (and ourselves) accountable for accuracy, characterization, and context. Avoid spin and let facts speak for themselves. Challenge enemy disinformation. Turn our enemies' bankrupt messages, extremist ideologies, oppressive practices, and indiscriminate violence against them.

Fight the information war relentlessly. Realize that we are in a struggle for legitimacy that will be won or lost in the perception of the Iraqi people. Every action taken by the enemy and our forces has implications in the public arena. Develop and sustain a narrative that works and continually drive the themes home through all forms of media.

Live our values. Do not hesitate to kill or capture the enemy, but stay true to the values we hold dear. Living our values distinguishes us from our enemies. There is no tougher endeavor than the one in which we are engaged. It is often brutal, physically demanding, and frustrating. All of us experience moments of anger, but we can neither give in to dark impulses nor tolerate unacceptable actions by others.

Exercise initiative. In the absence of guidance or orders, determine what they should be and execute aggressively. Higher level leaders will provide a broad vision and paint "white lines on the road," but it will be up to those at tactical levels to turn "big ideas" into specific actions.

Empower subordinates. Resource to enable decentralized action. Push assets and authorities down to those who most need them and can actually use them. Flatten reporting chains. Identify the level to which you would naturally plan and resource, and go one further – generally looking three levels down, vice the two levels down that is traditional in major combat operations.

Prepare for and exploit opportunities. "Luck is what happens when preparation meets opportunity" (Seneca the Younger). Develop concepts (such as that of "reconcilables" and "irreconcilables") in anticipation of possible opportunities, and be prepared to take risk as necessary to take advantage of them.

Learn and adapt. Continually assess the situation and adjust tactics, policies, and programs as required. Share good ideas. Avoid mental or physical complacency. Never forget that what works in an area today may not work there tomorrow, and that what works in one area may not work in another. Strive to ensure that our units are learning organizations. In counterinsurgency, the side that learns and adapts the fastest gains important advantages.

DAVID H. PETRAEUS
General, United States Army
Commanding

Appendix C

Anaconda Strategy versus al-Qaeda in Iraq

September 2008

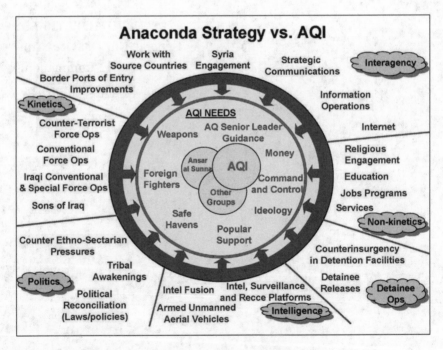

Source: Multinational Force–Iraq, September 2008.

Appendix D
The Battle of Sadr City
March–April 2008

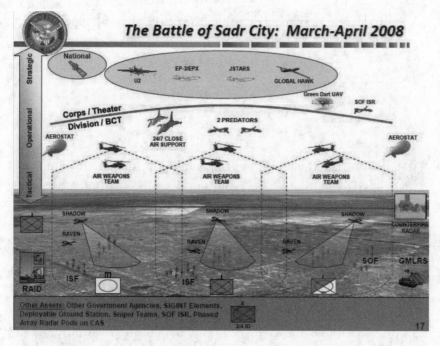

Source: US Central Command, 2009.

Nine

Vladimir Putin's Existential War against Ukraine

2022–

In war more than anywhere in the world, things happen
differently to what we expected, and look differently when
near to what they did at a distance.

Carl von Clausewitz, *On War*, 1832[1]

In January 2013, General Valery Gerasimov, the Russian Chief of Staff, delivered an address on the evolution of warfare to the Academy of Military Science in Moscow, on what he called "non-linear" conflict. "In the twenty-first century we have seen a tendency towards blurring the lines between the state of war and peace," he argued. "Wars are no longer declared and, having begun, proceed according to an unfamiliar template."[2] He went on to describe how "a perfectly thriving state can, in a matter of months and even days ... sink into a web of chaos, humanitarian catastrophe and civil war." This new form of warfare was no longer about "frontal engagements of large formations of force," but instead depended on the use of special forces and fifth columnists "as well as informational actions, devices and means."

While most at the time interpreted these remarks as relating to the conflicts in places that President Vladimir Putin's Russia had attacked earlier, such as South Ossetia and Abkhazia, hindsight suggests that Gerasimov might have had Ukraine in mind, a country the size of Texas and then home to 44 million people. (Ukraine's capital Kyiv covers more than 320 square miles, larger than New York City in area though with less than half the population.)

On 18 March 2014, Putin ordered Russian forces to take control of the Crimean peninsula, an action the Ukrainian government decided not to resist militarily. Crimea had been part of Ukraine since 1954, but historically many of the inhabitants were pro-Russian there. Putin had deterred Western intervention with the less than subtle words: "Thank God, I think no one is thinking of unleashing a large-scale conflict with Russia. I want to remind you that Russia is one of the leading nuclear powers."[3]

In accord with Gerasimov's remarks about special forces and fifth columnists, the *coup de main* was undertaken not by Russian regular soldiers but by operatives whose uniforms carried no labels or markings and who were quickly dubbed "little green men," as though they had landed from Mars. That same month also saw the attempted takeover of south-eastern Ukraine by Russia-controlled proxy forces, which seized administrative buildings and set up two "People's Republics" of Luhansk and Donetsk Provinces, and were soon reinforced by several thousand Russian soldiers posing as "holidaymakers" who had simultaneously decided to vacation there.[4] This use of anonymity for political deniability was not so much an evolution of warfare as a reversion to the past, as when Russian MiG pilots in the Korean War flew without insignia on their uniforms. Nor was there anything innovative in the nature of the fighting in the Donbas region of east Ukraine between the Ukrainian Regular Army and the Russian-backed separatists, which soon developed into First World War-era trench warfare, a so-called "frozen conflict" that was nonetheless to cost thousands of lives.

Ukraine, the USA and Western Europe responded with a degree of timidity to the action in Crimea. The West did impose sanctions against Russian politicians, though they had limited impact.[5] Developments in 2015 bolstered Putin's confidence in murky, grey-zone operations:

Russia intervened in Syria with airpower, special operations forces and mercenaries to assist the Bashar al-Assad regime and tip the balance of the long-running civil war in Assad's favor.

In doing so, Russia turned what had previously existed as an important regional struggle into a strategic one with vast geopolitical implications. Advanced air defense systems based on S-300 and S-400 missiles were swiftly established at the Latakia airbase and Tartus naval base in Syria.[6] Russian troops gained valuable combat experience, and Putin won further sway over an embattled but useful ally in a region of crucial interest, with air and naval bases in the Mediterranean.

Once again, the Russian intervention made use of "anonymized" Russian and pro-Russian soldiers who wore no identifying uniforms and who could therefore be disavowed in the event of failure. Some of these came from the Wagner Group, a shadowy private army that operated as an arm's-length branch of the Russian state – essentially a mercenary force owned by Yevgeny Prigozhin, an oligarch known as "Putin's Chef" and – at least initially – one of his closest henchmen. Though Wagner mercenaries wear no recognized badges or insignia, they do occasionally sport "Grim Reaper" ones bearing slogans such as "Death is our business – and business is good" and "I don't believe in anything. I'm here for the violence."[7]

While the Wagner Group mercenaries themselves demonstrated a nihilistic belief in nothing, their ultimate employer Vladimir Putin believed in a revanchist Russia operating as a great power. He employed the Wagner Group ruthlessly in Syria, although not without occasional significant setbacks. On 7 February 2018, for example, the Wagner Group assaulted pro-Western rebel positions in the Deir ez-Zor province of Syria, which resulted in the Americans not only firing rocket artillery at the attackers, but also unleashing B-52 bombers, AC-130 gunships, F-22 Raptor stealth fighters and AH-64 Apache attack helicopters, killing around 200 Wagner mercenaries while sustaining no US casualties.[8] Moscow decided not to react; to do so would be to acknowledge the Wagner Group as Kremlin-aligned. The unofficial status of such groups as Wagner therefore cuts both ways.

The Ukrainian Army's passivity in Crimea during the 2014 *coup de main* (though it resisted elsewhere) encouraged Russian expectations that the same might transpire in the event of a similar coup against Kyiv, but developments over the next eight years changed everything. Ukrainian humiliation over the loss of Crimea, the fighting in the Donbas, subsequent extensive training and professionalization of the Ukrainian Army by NATO, the inspirational leadership of President Volodymyr Zelensky and the fact that Ukrainians would be defending their own homes and independence were all factors that Russian planners failed sufficiently to take into account in 2022.

The scene was thus set for Putin's colossal miscalculation on Thursday, 24 February 2022. The Russian President hoped his invasion of Ukraine would be a swift, overwhelming, Blitzkrieg attack by special forces, the regular Army and fifth columnists against what appeared to be a weak Ukrainian government and inadequately equipped Ukrainian forces. This would be supported by nuclear threats to discourage Western intervention, followed up by oppressive measures to neutralize anti-Russian elements. One year after the invasion, the Royal United Services Institute published a study which concluded that Russia had planned to invade Ukraine over ten days, before occupying the country to enable full annexation by August 2022.[9] As so often in the past, however, Putin had failed to grasp how warfare had evolved since the days of Blitzkrieg, and how the advantage in recent years had shifted decisively from the offense to the defense.

With history's iron law of unintended consequences, instead of weakening the Ukrainian government, Putin's actions in 2014 had had the exact opposite effect. A fresh spirit of patriotism had been kindled within a country that had existed as a sovereign state for only thirty years, yet which had no intention of seeing its independent existence terminated. And Zelensky proved in the moment to be a truly Churchillian leader, performing each of the key tasks of a strategic leader superbly. (The journalist Jonathan Freedland has described Zelensky as "Churchill with an iPhone.")[10]

* * *

Russia has not been particularly successful when fighting beyond Russian-speaking borders over the past 120 years, something that has transcended politics because it has been true in its Tsarist, Bolshevik and post-Soviet iterations. Its forces attacked Japan in 1904–5, Prussia and Austria in 1914–17, the Baltic States in 1918–20, Poland in 1920–2, Finland in 1939–40 and Afghanistan in 1979–89. Every single time they received a bloody nose, for much the same reasons that they have so far done so badly in Ukraine: poor logistics, badly integrated arms, low morale and so on.[11] Occasionally ground was made up by suicidally costly human-wave attacks. In the Second World War, the Red Army did capture Berlin and destroy Nazism, albeit at the enormous cost of 800,000 killed and wounded in that operation alone. In that conflict, the Soviet Navy rarely ventured out of port.

"For the past half century, European states have not fought one another," the distinguished military historian Professor Sir Michael Howard observed in 2009, "and are highly unlikely to do so again."[12] Even the Yugoslavian Civil War did not see actual state-on-state warfare, so much as state-on-enclave. The extent to which Putin's invasion of Ukraine is thus a remarkable exception to the post-war European rule can hardly be exaggerated. Fortunately, the West swiftly recognized that profound principles were at stake. As the historian Timothy Snyder wrote, it was essentially a struggle between democracy and nihilism, "a test of whether a tyranny that claims to be a democracy can triumph and thereby spread its logical and ethical vacuum. Those who took democracy for granted were sleepwalking toward tyranny. The Ukrainian resistance is the wake-up call."[13]

From the time of Russia's extensive army maneuvers in Belarus and Russia, on Ukraine's northern and eastern borders, in the autumn of 2021, both the CIA and Britain's intelligence service MI6 strongly suspected that Putin was about to invade Ukraine. "We don't have clarity into Moscow's intentions, but we do know its playbook," Antony Blinken, the US Secretary of State, stated in November.[14] The two intelligence organizations then took the unusual but commendable decision to share with other countries large quantities of classified information (with details about the sources and methods of that intelligence removed) about the impending attack, including the way that Russia would be using the breakaway republics as a pretext for invasion.

Warfare was evolving in ways rarely seen before. The sharing of such intelligence was controversial – there is always the fear of sources being compromised – but it was very quickly revealed to be more than justified. On 26 January 2022, US Deputy Secretary of State Wendy Sherman went so far as to publicly predict an attack "in mid-February."[15] As well as satellite imagery of troop, armor, artillery and equipment movements, the information released included myriad other telltale details, such as Russia's deployment of scarce blood-plasma stocks to field hospitals. It was increasingly apparent to the CIA and MI6 that this was no normal field exercise.

In response, Ukraine relocated its air defense systems and took large deliveries of advanced British and then American anti-tank and anti-aircraft weaponry at Boryspil Airport. "Putin rolled the iron dice," Professor Stephen Kotkin, the leading American expert on Russia, later wrote. "Somehow the Russian president's extensively telegraphed invasion stunned almost all Western capitals, Asian capitals, much of the American political establishment, and many members of Russia's loyalist establishment. He did not shock US and UK Intelligence."[16] Russia, of course, denounced the CIA and MI6's "provocations," claiming that troop build-ups on the Belarusian–Ukrainian border were part of normal, if extended, military maneuvers.

Although the US-led coalition had stationed a large army on the borders of Iraq before both the Gulf and Iraq Wars, those deployments were a genuine attempt to force Saddam to negotiate in good faith. Russia's decision to place 190,000 men on the Ukrainian frontier for several weeks prior to its 2022 invasion was therefore somewhat anomalous. More often, attackers try to employ the element of surprise, as we have seen earlier with the Korean War, the Six Day War, the Tet Offensive, the Falklands, Saddam Hussein's invasion of Kuwait and the 9/11 attacks. That would not be the case in this instance.

On 21 February, Putin further advertised his intentions by recognizing the Donetsk and Luhansk People's Republics as independent states to which Russia would send "peacekeeping troops."[17] Two days later, the

44-year-old President Volodymyr Zelensky of Ukraine told President Andrzej Duda of Poland, who was on a visit to Kyiv, that the Russians would invade within hours, adding, "The Russians think we won't fight back, but they are gravely mistaken. Andrzej, this might be the last time we see one another."[18]

An important aspect of General Gerasimov's non-linear warfare was cyber warfare. Russia's malware attack on Ukraine prior to the seizure of Crimea in 2014 had stalled a fifth of the world's shipping capacity, and US Generals John Allen and Ben Hodges believe it might even have been a trial run for a possibly devastating cyberattack in any future war against NATO.[19] On 23 February 2022, just as Russian forces began moving towards the Ukrainian border, parts of the Ukrainian Defense, Foreign and Interior Ministries were subjected to a massive cyberattack, the second such assault in a week.

In the early hours of 24 February, Ukrainian air defense radar was jammed across all frequency bands. "We thought we were going to be denied the entire electromagnetic spectrum around Kyiv," noted a Ukrainian electronic warfare unit commander.[20] Ukrainian banking websites such as Privatbank were temporarily forced offline in an obvious attempt to provoke financial panic.[21] Russia denied it was in any way involved, but Ukrainian military hackers responded by taking down the Kremlin website briefly, while hacking groups such as Anonymous weighed in heavily on Ukraine's side.[22]

At midnight on the 23rd, the Ukrainian parliament declared a state of emergency, giving Zelensky plenipotentiary powers to fight the coming onslaught. Civilians were granted the legal right to carry firearms and 200,000 reservists were mobilized. "Our army is ready," declared Oleksiy Danilov, the Secretary of the National Security and Defense Council.[23]

The assumption of the General Staff in Moscow seems to have been that a dawn airborne assault on Kyiv's Hostomel Airport, located in a suburb of the capital 18 miles north-east of the Presidential Palace in Bankova Street, would open the way for an influx of reinforcements that would then take over the city. Simultaneously, a huge armored attack from Belarus would move on the capital to invest it within three days. Zelensky and his government would be captured or killed by around 300 soldiers from Russia's special forces, Wagner Group operatives and fifth columnists, some of whom had been infiltrated into Kyiv in the days prior to the invasion. After the elimination of the Ukrainian national leadership and replacement of Zelensky with a pro-Russian figure, repressive internal measures would establish control over the rest of the country by Russia's Victory Day on 9 May, the anniversary of Nazi Germany's defeat in 1945.[24]

The Russian airborne units that were to spearhead the invasion had wargamed what they intended to do as recently as December, in conjunction with the 9th Directorate of the Fifth Service of the Federal Security Service (FSB). Unfortunately for them, however, the paratroop officers had spent no fewer than seventy-two hours discussing their orders over lines that the Ukrainian secret service had tapped, and so when they landed at Hostomel Airport the Ukrainian Army was waiting in force, and with artillery.

Hostomel functioned as a center of massive airborne trade, housed the largest plane in the world and operated as a base for the Ukrainian Air Force. Capturing it would thus be strategically consequential, allowing follow-on Russian forces and weapon systems to be airlanded just 6 miles from the Ukrainian capital. At dawn on 24 February 2022, thirty-four Russian helicopters flew hundreds of soldiers there, where they met stiff Ukrainian resistance. The airport changed hands several times during the fighting, and Ukrainian forces rendered it useless for large-scale landings, meaning that the fleet of Russian air transports carrying more heavily armed and armored reinforcements would be unable to land and had to turn back.

The sheer hubris of the Russian battle plan was exposed by the *New York Times* in December 2022, which obtained a copy of it. Airborne troops were ordered to set off from Belarus at 1:33 a.m. on 24 February and reach the outskirts of Kyiv, 140 miles away, at 2:55 p.m. that same day. One unit of the 26th Tank Regiment was expected to get to a point on the other side of the Dnipro River 250 miles away within twenty-four hours, with orders that stated, "There are no forces or equipment for reinforcements."[25] Russian National Guard officers were dispatched in ordinary police trucks for crowd control once Kyiv had fallen, troops were given only five days' provisions for the campaign and were ordered to pack the dress uniforms they would need for the victory parade.[26]

Retired US Major John Spencer has explained how Russian forces "needed to get into the middle of Kyiv as quickly as possible and raise the Russian flag over a government building," because "At that point you've won the war. Yes, you may start the greatest insurgency in history. But you've won the war."[27] Russia's failure to do so swiftly was a disaster for their forces, and the tenuous control of Hostomel rattled the confidence of leaders in Moscow. When Russians did finally take the airport, the prospect of a sudden *coup de main* was already dead. Nonetheless, some Russians were able to head into Kyiv city center in small assassination groups, looking for Zelensky.

Only months earlier, the President of Afghanistan had fled his country prior to the Taliban taking Kabul, allegedly taking suitcases of dollars. It very soon became clear that Volodymyr Zelensky was made of sterner stuff. While he was receiving intelligence and offers of evacuation from Washington and London, Zelensky made the remark that left no one in any doubt about what manner of leader he was: "The fight is here. I need ammunition, not a ride."[28] He was not to know when uttering it that he had coined a ringing battle cry of freedom for the twenty-first century.

Armed skirmishes broke out around Bankova Street, and the Presidential Palace blocked its windows and doors with makeshift barricades, doused its lights, armed its male and female staff – many of whom had never seen an AK-47 before – and defended against at least two Russian attempts to take the building. Although Russian disinformation stated that Zelensky had disappeared in an attempt to flee Kyiv, such reports were soon proven false as the President began to appear regularly in front of well-known Kyiv landmarks, including the Presidential Palace, and commenced what were to become daily phone-linked broadcasts that urged the Ukrainian people to resist the invasion. Perhaps more than anything else, this leadership based on raw personal courage made people throughout the world realize that – like the day the Berlin Wall fell, and on 9/11 – something of great historical import was taking place.

Some officials and government figures did flee Kyiv on the opening days of the attack, and a year later Zelensky remarked contemptuously, "Who disappointed me? All those who left Kyiv. All those who were supposed to fight for the country."[29]

Although he had been elected in April 2019 with nearly three-quarters of the vote, Zelensky's background was as an actor and comedian who had played the role of the Ukrainian president on television, which perhaps explains why Putin underestimated him as fatally as he did. In May 2019, Putin derided Zelensky (whose name he has never once uttered in public) saying, "As far as I can tell, he's amazing at what he does, he's a marvellous actor. But seriously: it's one thing to play a person, but quite another thing to be that person."[30] Before Zelensky's third year in office was complete, Putin would discover that the Ukrainian President was not merely play-acting the role, but was proving that, however much warfare evolves in terms of technology, leadership qualities will always continue to play a vital part.

At 12:30 on the morning of the attack, Zelensky filmed a video message on his phone which he posted on Facebook and other social media. He told the Ukrainian people that he had imposed a state of emergency and then dramatically switched his language to Russian to say:

We fight because this is *our* land. This is *our* history. What will *you* fight for? ... Hear us. The people of Ukraine want peace. The Ukrainian authorities want peace. We want it, and we will do everything we can to build it. We are not alone. Many countries support Ukraine. Why? Because we are not talking about peace at any cost. We are talking about both peace and principles. We are talking about justice and about international law. About the right to self-determination, the right to decide our own future, the right to security and the right to live without being threatened ... If someone tries to take away our country, our freedom, our lives, the lives of our children – we will defend ourselves. And as you attack us, you will see our faces, not our backs.[31]

The role that Zelensky's oratory played in stiffening Ukrainian fortitude for the war was central.

Later that same morning, Zelensky gave a formal speech from a lectern in the Presidential Palace. "Russia has attacked Ukraine in a cowardly and suicidal way, like Nazi Germany did," he said. "We, the citizens of Ukraine, have been determining our own future since 1991. But now, what is being decided is not only our country's future, but also the future of how Europe will live."[32] Zelensky had been only minutes away from being killed or captured in the opening moments of the war, and the first thing he did once the two armed assaults on the Presidential Palace had been driven off was to announce that no military-aged Ukrainian males should leave the country. The response was overwhelmingly supportive. He also called on world leaders to honor the Budapest Memorandum of 1994 that had pledged to defend Ukraine's security when it voluntarily gave up the nuclear weapons that had been stationed there during the Cold War.

Meanwhile, President Putin addressed the Russian people, telling them that the fighting in the Donbas had brought him to "execute a special military operation" in order to "demilitarize" and "de-Nazify" Ukraine and "bring to justice" the Ukrainian high command, whose members he deemed war criminals.[33] The euphemism "special military operation" to describe what the rest of the world recognized was a war was widely ridiculed in Ukraine and the West, as was the claim that Zelensky's government was neo-Nazi – not least because Zelensky and Ukraine's Prime Minister were both Jewish. In Russia, however, the media was banned from using the terms "war" or "invasion" in relation to the Russo-Ukrainian conflict on penalty of fifteen years' imprisonment.

On the first morning of the invasion, Russian missiles struck Kyiv and cities across Ukraine from Kharkiv in the east to Ivano-Frankivsk near the Polish border. Yet history does not tend to favor aggressors, for as the historian Adam Tooze has pointed out, "Other than wars of national liberation, one is hard pressed to name a single war of aggression since

1914 that has yielded clearly positive results for the first mover."[34] Furthermore, as we have already seen time and again in this book, soldiers defending their own homes and families are far more motivated than invaders. In both the wars in which Russia did well – those against Napoleon and Hitler – it was as defender of the motherland, not as aggressor.[35]

The invasion had immediate and serious effects on the world economy. By 6 p.m. on the first day, for example, European stock markets had tumbled and the price of oil had risen above $105 a barrel for the first time in nearly eight years. Meeting at an emergency summit in Brussels on the evening of 24 February, European Union leaders put together an unprecedented sanctions package targeting Russian banks, oligarchs, high-tech exports and the aircraft and oil sectors. Most seriously in the short term, $650 billion worth of Russian assets in financial institutions were immediately frozen. Export controls that partially denied Russia access to microchips crucial to the success of their arms industry forced it to acquire drones from Iran and shells from North Korea, and reduced it to trying to buy back munitions it had already sold abroad.[36] (The Iranian Shahed–136 drone costs only $20,000 each, but can be produced in large quantities and can do significant damage.)

Yet despite the sanctions being the strongest imposed since the Second World War, nations comprising 76 percent of the world's population, such as China and India, did not join them. Furthermore, on 6 April Josep Borrell, the EU's foreign policy chief, admitted that EU countries had undermined their own sanctions towards Russia by having handed over €35 billion to Russia in energy payments since the start of the war, compared to the €1 billion given to Ukraine in arms and weapons. Western leaders could do little about the Russian population being more than treble that of Ukraine's and its territory twenty-eight times the size, but they were in a position to try to reduce Russia's GDP, which was almost ten times that of Ukraine's.[37] More than 1,200 Western companies such as Renault, McDonald's, Formula One and Johnson & Johnson closed or curtailed their activities in Russia over and above what the letter of the sanctions demanded.[38]

Although economics have long played a crucial role in military conflicts, these actions were nonetheless a significant shift in the evolution of warfare in an increasingly globalized world economy, and the Ukraine War led directly to political unrest in South America, high fuel costs and inflation in Europe and near starvation in parts of Africa – though these challenges have been reduced in 2023 and Russian GDP only fell by around 3 percent in 2022 thanks to increased energy prices. One of the reasons that sanctions hit the Russian economy so hard initially was that high-ranking ministers were entirely unprepared for how to deal with them, because Putin had not informed his cabinet that the invasion was going to take place.

Another area in which the Russo-Ukrainian conflict saw an evolution in the nature of warfare was through consumer activism, whereby mainly Western shoppers punished brands that they considered were acting selfishly or in a way that supported Russia. Drinkers boycotted Russian vodka, for example, and when in early March the clothing retailer Uniqlo justified staying in Russia because, as it put it, "Clothing is a necessity of life. The people of Russia have the same right to live as we do," the resulting #BoycottUniqlo social media campaign forced it to reverse its decision only days later. Nestlé was similarly punished. "Commerce's golden age of neutrality is over," observed Elisabeth Braw of the American Enterprise Institute, who also noted that although the equivalent of shareholder political activism had existed since the anti-apartheid campaigns of the 1980s, after Russia's invasion of Ukraine it had exploded: "People know they can't necessarily change government decisions but they sense their power to influence global affairs by pressurising companies."[39]

The desire to punish Russia extended to its citizens and culture: Valery Gergiev, a Putin ally and the chief conductor at the Munich Philharmonic Orchestra, was fired for not denouncing the invasion, film festivals refused to screen Russian films, and the British dinnertime staple Chicken Kiev dish was widely renamed "Chicken Kyiv" in supermarkets. It is remarkable that the Russians should have been taken by surprise by all this, as it was General Valery Gerasimov himself who had observed that "The role of non-military means of achieving political and strategic goals has grown, and, in many cases, they have exceeded the power of force of weapons in their effectiveness."[40]

As we have seen, the first principle of strategic leadership is to get the overall strategy right, yet Putin, Gerasimov, Russian Defense Minister Sergei Shoigu and their staffs all failed miserably in this when they approved the plans to attack Ukraine on no fewer than seven different axes, rather than to launch a diversionary attack but then concentrate on taking Kyiv, the decisive and main effort. The multiple axes were far too diffuse for a war that was intended to be over quickly and decisively, and in none of them were Russian forces in the three-to-one force ratio that history suggests is necessary for success. (That ratio should be even higher against a defense conducted from urban areas, as would be the case around the sprawling metropolitan outskirts of Kyiv.)

The Russians also made a number of basic tactical errors in the opening days of the invasion that severely damaged their chances of success, which is extraordinary considering that they had been on full-scale military maneuvers since the previous autumn, organizing and training for precisely this operation. Although two large columns moved south from Belarus on 24 February, because they had only received twenty-four hours' notice, they were not supported by enough artillery for such a major operation, and Ukrainian units managed seriously to stymie the movement of Russian motor rifle units and the paramilitary security force Rosgvardia troops. These were quickly restricted to travelling just on roads, given the inability of most vehicles to move across country, while Ukrainian special forces were able to travel on all-terrain vehicles along the flanks of the Russian columns.

The topography quickly became a problem, with main supply routes surrounded by dense forests and marshland, which have historically been almost impregnable to armored attack. Ambushes by intrepid and skillful Ukrainians, fighting with the "home field advantage," were common, and the local population used mobile phones to report the Russians' whereabouts. Ukrainian special forces pinpointed areas of congestion on the roads, whereupon artillery was used to destroy Russian armor. "Anti-tank missiles slowed the Russians down,"

General Valeriy Zaluzhnyi, the Ukrainian Commander-in-Chief, stated later, "but what killed them was our artillery. That was what broke their units."[41]

Because the Ukrainians knew they could never match Moscow in terms of sheer numbers of aircraft, ships and combat vehicles, they instead devastated the Russian invasion force and its vulnerable supply lines through relentless, large-scale application of short- and long-range precision firepower. This included anti-tank guided missile systems, guided multiple launch rocket system rounds, precision artillery munitions, suicide drones, guided anti-aircraft missiles and anti-ship missiles such as the Neptune cruise missile.[42] Throughout history, military struggles have been decided less by the balance of material resources than by the creativity and determination with which they have been employed. By making its units both dispersed and highly mobile – firing and then rapidly repositioning themselves – the Ukrainians carried out relentless, withering strikes on Russian targets while evading counter-fire.[43]

"A twentieth-century invasion," commented the historian Niall Ferguson, "has been checked by a twenty-first-century defence."[44] The data analytics company Palantir supplied software that translates up-to-the-second information from satellites, drones, spies' mobile phones and sensors into a model that allowed highly accurate rocket and artillery strikes. The Ukrainians had state-of-the-art night-vision goggles, cutting-edge anti-tank guided missiles which lock on to a target and attack tanks from the top (where they are most vulnerable), and counter-fire radar which detects an incoming artillery projectile and immediately locates from where it was fired.

Meanwhile, the SBU, Ukraine's security service, created a chatbot – "Stop Russian War" – that allowed users of the Telegram social media app to upload the location of Russian vehicles with videos and photographs in real time, while another was added to the government's official e-portal Diia.[45] Once this crowd-sourced information had been verified by drone, the coordinates were passed on to Ukrainian Air Force or artillery units. Ukrainian innovation, determination, agility and sheer courage were all very much on display, with the entire population engaged in defending against the Russian forces. The Ukrainians showed exceptional innovation in finding solutions for a variety of problems – whether adapting Western missiles for use on MiG-29 fighter aircraft or repairing battle-damaged armored vehicles left on the battlefield by the Russians (after recovery by the "tractor army") – or in jamming Russian communications.[46]

It soon became clear that the Russians' precision munitions were not particularly precise, which was one of the reasons that they were unable to make the Ukrainian Air Force's runways inoperable in the early hours of the war, in the way the US-led coalition had in Iraq in 2003.[47] The Russians also failed to provide close air support beyond their ground forces' front lines, in part due to the lethality of the US- and UK-provided shoulder-launched air defense missiles.

On 25 February, Putin broadcast to the Ukrainian Army: "Do not allow neo-Nazis to use your children, women and elderly as human shields. Take matters into your own hands. I believe that we can come to an agreement with you more easily than with that mob of junkies and neo-Nazis that has taken control in Kyiv, and who are holding the entire Ukrainian people hostage."[48] By contrast, Zelensky also gave a speech that evening. Dressed in the military olive-green T-shirt that was to become his trademark, he spoke by videophone from outside the Presidential Palace in the heart of Kyiv surrounded by cabinet ministers and senior generals, with the streetlights for illumination. "Good evening, everybody," he said. "The President is here. Our armed forces are here, along with our whole society. We will defend our independence, our nation. Long live Ukraine!"[49] It was conspicuously brave of Zelensky to have done this in the street, since no one knew how many Wagner Group assassins were still there. Zelensky's approval rating among Ukrainians has remained at more than 90 percent from the invasion onwards.[50]

On 25 February, Petro Poroshenko, the opposition leader and Zelensky's predecessor as president, who had been under house arrest facing treason charges, visited the palace to offer his total support and bury their past differences. National unity is important in any existential struggle, and Ukrainians increasingly began seeing this struggle as a war of independence, in the same way that Americans view 1776 or Israelis the war of 1948. Meanwhile, queues formed across Ukraine as tens of thousands of AK-47s were handed out to ordinary citizens ready to defend their country; barricades went up in towns and villages, and the local populace began making Molotov cocktails from bottles, petrol and rags.

Although the Russians continued to wage cyberwar for months after the invasion, with occasional successes, Ukraine's improvements in cybersecurity were clearly evident (helped reportedly by US and other Western agencies). Moreover, after the Russians had partially destroyed the phone system and jammed it, Ukraine also remained connected to the internet in part due to the Starlink mobile satellite terminals that were made available by the businessman Elon Musk's company SpaceX and paid for by the US government.[51] It effectively gave every Ukrainian military unit with access to a terminal a complete picture of the situation on the ground in real time. The processing of targeting could be completed around ten times faster than in previous conflicts, and more accurately. Warfare was evolving fast, hour by hour.

It is rare in history for a single businessman significantly to affect a conflict in this way. Alfred Krupp's Kruppstahl steel cannon's performance over the French bronze guns in the Franco-Prussian War was one such example. With Musk's satellite internet service aiding Ukraine in everything from facilitating Zelensky's video calls with foreign parliaments and his statements via social media to the military's coordination of artillery missions and airstrikes, Starlink became a lifeline.[52] Unlike Krupp, however, Musk has involved himself in ways that would have surprised the Kaiser, ordaining that Starlink cannot be used in the Crimea and tweeting speculations about a future peace treaty that were later denounced by Zelensky.[53] With individual tycoons such as Elon Musk, Mark Zuckerberg and Jeff Bezos wielding such extraordinary power, wars of the future will have to take their influence into account.

*   *   *

Despite Russian officers telling their troops that they were on a liberating mission that would see many ordinary Ukrainians welcome them, it soon became very clear that that was not the case. At 10 a.m. on the day of the invasion, the Russian ship Moskva, which was armed with cruise missiles, ordered the eighty Ukrainian defenders of Snake Island in the Black Sea to lay down their weapons, in the belief that they might be willing to swap sides. "They were promising us jobs, money, a career in Russia," Major Bohdan Hotskyi recalled. "No one was ready to accept their offer." The exchange that followed became famous enough to be the subject of a commemorative stamp:

WARSHIP: "Snake Island. I, Russian warship, repeat the offer: lay down your arms and surrender, or you will be bombed. Have you understood me? Do you copy?"

UKRAINIAN BORDER GUARD: "Russian warship, go fuck yourself."[54]

Although the Russians attacked the island in force that evening and captured the Ukrainian defenders, the message of defiance had gone out. (Ukrainian forces would later retake the island.) That episode was symbolic of the wider attitude of the Ukrainian Army and population. Over time, it was becoming clear that no Ukrainian political leader had done remotely as much to promote Ukrainian nationalism since Ukrainian independence in 1991 as Putin had by launching the invasion.

Only two days into the war it was evident that fierce resistance on the part of the Ukrainians had already slowed the Russian advance on Kyiv, which was under curfew.[55] The next day, Russian forces attempted to storm Ukraine's second largest city of Kharkiv, but were pushed back to its suburbs after a day of tough street fighting in which citizens joined reservists and regular troops in repelling the assault. "The city is being completely cleansed of the enemy," announced Oleh Synyebuhov, Kharkiv's Governor, soon afterwards. "Control over Kharkiv is completely ours!"[56] There was an important political element to this victory: 2014 had seen pro-Russian separatist demonstrations in the city, which now was almost unanimously fighting back against the invader. The ill-conceived and woefully executed invasion had alienated even those hitherto sympathetic towards Ukrainian–Russian identity.

In a further demonstration of history's law of unintended conse-quences, on 27 February Olaf Scholz, the new German Chancellor, announced "a new era" in which his country would now spend an emer-gency $110 billion on defense and increase general defense spending to meet NATO's goal of 2 percent of GDP. Angela Merkel, who had just left office after sixteen years as chancellor, had been barely able to get German defense spending to nearly 1.5 percent of GDP. "There could be no other answer to Putin's aggression," Scholz said.[57] Moreover, Germany would eventually send military aid, including some lethal systems, to Ukraine, something that the US and Britain had already been doing for months but that was unprecedented for Germany since the Second World War.

The end of Germany's longstanding near-pacifist stance dealt another blow which Putin had almost certainly not considered in his calculations about the invasion. That day, the White House also announced that select Russian banks would be expelled from the Swift global payment system, in the hope that the action would "further isolate Russia from the international financial system and our econo-mies."[58] Over time, it would become increasingly obvious that in setting out to make Russia great again, Putin was actually making NATO great again, fostering the most comprehensive unity there since the end of the Cold War and prompting historically neutral Finland and Sweden to seek membership.

In the first four days of the fighting, over half a million Ukrainians crossed into neighboring countries to the west, the first of what would become around 6 million refugees forced out by the war. Because the Ukrainian government encouraged the evacuation of women, children and the elderly, it was done with remarkably little panic, aided by several generous Western countries allowing Ukrainian refugees to move with ease. It was believed that Ukrainian men would fight better if they knew that their wives and families were safe, and there was virtually no evasion of the call-up. Once they had waved their families off from railway stations across the country, most went to the front with a fierce determination to punish the Russians for the misery caused.

Ukrainian determination was redoubled once it became very clear that, contrary to all the laws and mores of war ancient and modern, the Russian Army was deliberately targeting civilians in order to create the maximum amount of terror. Salvoes of Grad rockets were fired on 28 February at Kharkiv's residential districts, far from any Ukrainian troops, in a pattern that was to escalate to a terrible degree over the coming weeks and months.[59] The multiple 300mm Smerch rocket salvoes against civilian areas, which consisted of high-explosive and cluster-munition-dispensing warheads, constituted a war crime, one of no fewer than 1,400 such incidents by October. The fact that precision weaponry can today be programmed to hit precise windows in a building makes it obvious that the Russians were specifically choosing civilian residential buildings, hospitals, civic centers, markets, electricity substations, museums and art galleries and other entirely non-military targets.

The Russian terror-bombing of Ukraine is one area where, despite much technological innovation since 1945, warfare has not evolved. Boris Johnson, the British Prime Minister, had been stalwart in his support of Ukraine from even before the invasion. On 28 February he condemned the "barbaric Russian airstrikes against innocent civilians, including children," and pledged more military aid.[60] In early March the International Criminal Court opened an investigation into multiple war crimes committed by Russian forces, most probably aided and abetted by the Kremlin.[61] As the Russian ground advance stalled in early March, the rocket and artillery attacks on "soft" targets and innocent civilians increased.

On 1 March, Zelensky delivered a powerful video-linked speech to the parliament of the European Union, which won him a standing ovation. "We have proven our strength, so prove that you are with us," he challenged the politicians. "Prove that you will not abandon us. Prove that you are indeed Europeans. And then life will win over death, and light will win over darkness."[62] If sheer charisma and oratory could win wars, Zelensky would have defeated Putin in days. The EU promised €500 million to help with the refugee crisis, on top of €450 million for military aircraft and arms, which represented a new departure for the organization.

In the UN General Assembly the next day, a vote deploring the invasion and demanding an immediate Russian withdrawal was passed by 141 to 5 – the latter comprising the dictatorships Russia, Syria, Belarus, North Korea and Eritrea. Thirty-one countries, including China, India and Pakistan, abstained, with the Chinese stating that the resolution "does not take into consideration the history and the complexity of the current crisis."[63] Whereas Britain and other countries ended imports of Russian coal, oil, oil products and gas to deprive Putin of funding, by November 2022 India was importing 908,000 barrels of oil per day from Russia at heavily discounted rates.[64]

The Russian armored convoy moving towards Kyiv stalled in early March and created a traffic jam that stretched for 40 miles and was estimated to comprise 15,000 troops. Yet logistics and communications problems born of bad planning started to assail it long before it reached the Kyiv *oblast*. More serious assailants turned out to be unjammable drones and American-made Javelins and British-made NLAWs (Next-generation Light Anti-tank Weapons). With a range of up to 3 miles, Javelins are "fire and forget" weapons, a missile that locks on to its target and allows the soldier who launches it to pull the trigger and leave the area. They are therefore a world apart from the Sagger anti-tank weapon of the Yom Kippur War, which was wire-guided and thus dependent upon an infantryman keeping the reticle on the target and thereby staying exposed until the missile impacts.[65] In 1973, the IDF could load their tanks with anti-personnel rounds and use forward artillery barrages to suppress infantry with those weapons and protect their tanks, but in 2022 not even the Russian Army had enough artillery rounds to stop Javelin operators.

The corruption rife in Russian military procurement and supply also badly damaged that Army's performance in Ukraine. Nothing is more corrosive of trust in a regime than endemic corruption, as seen with the Guomindang in the Chinese Civil War, Fulgencio Batista's anti-communist forces in the Cuban Revolution, and the South Vietnamese Army in the Vietnam War, among others. The corruption that was endemic in Russia's political and military leadership meant that, as the military historian Ralph Peters puts it, "They had trouble getting some of their tanks through the parades in Moscow, let alone fighting in Ukraine. Much of the money spent on military reform went elsewhere, into the pockets of contractors and generals."[66]

Embezzlement, bribery and the purchase of substandard equipment were rife in the Russian Army, which affected its maintenance and supply logistics, with, as one commentator explains, money earmarked for maintenance instead "lining the pockets of officers in charge of the conscripts who would be servicing the trucks."[67] The unprofessional conduct of Russian forces was further compounded by the lack of adequate and rigorous training for conscripts, who made up a quarter of those forces and served only for one-year, lackluster tours, gaining little meaningful skill.[68]

Air superiority, as we have seen in the preceding chapters, has been a vital aspect of conflict since 1945. With it, anything is possible; without it, everything is dangerous. On occasion, as in Kosovo, domination of the skies can even deliver an almost bloodless victory. When Putin's war of choice in Ukraine began, Russia had 1,172 combat-capable aircraft, including ninety-seven Su-35S Flankers, while Ukraine had only 124, including thirty-six MiG-29 Fulcrums.[69] Yet, despite that numerical difference, the Ukrainian Air Force managed to prevent Russia from achieving dominance in the air and thus enabling victory on the ground. This was partly because part of the Russian Air Force had to stay in Syria and north-west Russia, but also because in the first weeks of the war, at least according to Ukrainian estimates, 121 Russian helicopters and ninety-seven fixed-wing aircraft were shot down.[70] Within days, the Russians had to switch to night operations to reduce losses.

Russia's attempts to degrade or destroy Ukraine's Air Force and air defense assets by firing some 400 cruise and ballistic missiles on the first day also largely failed, leaving its planes vulnerable to the admittedly ageing Ukrainian aircraft and air defense systems.[71] Ukraine's practice of moving its planes to random locations also stymied Russian hopes for a Six Day War-style victory. A major problem for Ukraine was that because many of its planes were Russian-made, they could not get spare parts, which reduced its airworthy fighting strength by about a third. "So far," the Ukrainian Defense Ministry stated on 3 March, "the enemy predominates in the number and quality of the aircraft and weaponry, but not the skills."[72] After two months, Western air experts were estimating that Ukraine was down to around twenty operational jets, whereas the Russians had 200 interceptor and strike aircraft available for missions on any given day.[73] Yet the Ukrainians were still able to prevent Russia from turning its air superiority into air dominance.

Russia's inability was in part down to poor training and lack of flight hours, but also the result of NATO sharing intelligence with the Ukrainians that enabled them to target Russian aircraft effectively.[74] The Ukrainian pilots may have been the underdogs, but they often displayed superior skill, situational awareness and tactics. "They are not warriors," a 29-year-old Ukrainian MiG-29 pilot with the callsign "Juice" told The Times in early May in reference to his Russian enemy:

> They're criminals, especially their bomber pilots, because they just bomb our citizens, they bomb our public; they bomb peaceful cities; they bomb targets unrelated to military infrastructure; they bomb hospitals. And they know the co-ordinates of their targets and what it is they are bombing, because they'll check them on the map before their missions and they know what they will be bombing. I have no respect for them at all.[75]

The ferocious motivation of pilots like Juice is reminiscent of those brave Polish and Czech pilots who fought in the Royal Air Force in the Second World War and whose hatred of their opponents was fuelled by what the Nazis had done in invading their homelands.

Pleas by Vadym Prystaiko, Ukraine's Ambassador to London, for NATO to establish a no-fly zone over Ukraine, such as was imposed over parts of Iraq in 1991, Bosnia in 1992 and Libya in 2011, were rejected. Western leaders understandably considered that if it led to a NATO warplane shooting down a Russian, or having to take out Russian air defenses, it might escalate the regional conflict into a world war. "If you have a no-fly zone," Ben Wallace, the British Defence Secretary, pointed out, "it will favor the force with the massive armed forces which will be Russia because they will be able to carry on regardless. They also have lots of long-range missiles. So they can substitute air attack with missiles attacks. So from a military point of view, it wouldn't help Ukraine."[76]

What did help Ukraine, especially in its all-important attack on the extended Kyiv-bound armored column, were drones. Although the Russo-Ukrainian War was not the first conflict in which drones played a part, their use in Ukraine represents another significant evolution in warfare because it was the first conflict in which drones had a central role in ground operations, with many important indications for the future. This may well be history's first full-scale drone war. The Turkish-built Bayraktar TB2 drone had already been deployed successfully in conflicts such as that over Nagorno-Karabakh, for example, but came into its own in Ukraine. The Bayraktar flies at 136mph, has a maximum altitude of 18,000 feet and can be armed with four air-to-ground missiles. Images of their destruction of Russian surface-to-air missile systems, fuel depots and tanks soon went globally viral. Ukraine had only a limited number of them, but they ensured that Russia did not control the skies[77] and also forced the Russians to be much more cautious in their movements.

Smaller than the Bayraktar and carrying only 3 kilos of explosives, but much more numerous, were the Punisher drones with their 7½-foot wing-span and capacity to loiter at 1,300 feet for hours before striking. At less than $100,000 each, they proved a cost-effective way of taking out high-value Russian stationary targets such as fuel and ammunition dumps, electronic and counter-electronic warfare stations and anti-aircraft systems. The number of different types of drones expanded hugely during the conflict. The Americans provided Switchblade drones, which can fit into a rucksack and take off from a portable tube stuck into the ground. They can then loiter above an area for as long as forty minutes, before a target is designated by the operator, and the drone then hits it at 115mph with armor-piercing, tank-destroying warheads. Other drones such as Orlans, Kamikazes, Phoenix Ghosts, Coyote Interceptors and more abounded.[78] Several Ukrainian brigades were well supplied with Mavic 3s, cheap photographer's drones that can be bought from Amazon, and many also have the A1-CM Furia military drones, which have much longer range and endurance.[79] Such weapons make it easy to see how warfare is almost bound to evolve in the future.

The speed with which the very media-savvy Ukrainian Information Ministry and front-line soldiers were able to disseminate videos both of Russian atrocities and of successful Ukrainian drone strikes made an enormous difference to the way the TikTok generation viewed the war, and ensured a high retention of sympathy from foreign observers. While the advent of Twitter had brought images of violent conflicts and twenty-first-century insurgencies to the feeds of journalists and observers alike in previous years, swift and polished videos posted by Ukrainian units and citizens and Kyiv's Ministry of Defense to a relatively nascent TikTok enabled effective engagement of an entirely new generation.

Whereas in earlier conflicts videos could often take hours or days to be widely disseminated, self-generated TikTok videos of the Ukraine War reached screens globally within seconds. With one in three Britons using TikTok, and 80 million monthly active users in the United States, the horror of what was happening to ordinary Ukrainians, and the courageous resistance they were showing, had a profound effect on the way the war was seen. These viral videos won hearts and minds for Ukraine in a way that would probably have been incomprehensible to Field Marshal Templer or General Walter Walker from earlier chapters, but that are increasingly important as warfare evolves.

For all that decent people worldwide could help Ukrainian refugees, demonstrate outside Russian embassies, display Ukrainian flags on their buildings and persons and attach hashtags on Twitter, there was little that global public opinion could do to turn back the gigantic armored column that was still trundling towards Kyiv, threatening to lay siege to the city and crush Zelensky's hopes of saving Ukraine. One defense analyst described it as being more like "a medieval siege train" than a recognized tactic in modern warfare.[80] The column was tracked by US public and private satellites, and was seen to have reached Hostomel Airport by 1 March.

As well as troops numbering a division, towed and self-propelled artillery pieces, main battle tanks and BMP3 armored carriers, the column also contained the pro-Russian quisling government that Putin expected to install in Kyiv once victory was won. Although the next three most important cities of Kharkiv, Mariupol and Kherson were still in Ukrainian hands on the sixth day of the invasion, all three were being encircled. (The strategic importance of Kherson to Russia was to control the water supply to the Crimea, and Mariupol was needed in order to link Crimea to the south-eastern region of Donetsk and the Sea of Azov.) For all that the column was only 17 miles from the outskirts of Kyiv, Zelensky stated that he would not be leaving the city; he was going to fight in his capital, if necessary to the death. The way that the column, once in position outside Kyiv, would be able to devastate the city by indiscriminate shelling was already evident in several other cities. "Kharkiv today is the Stalingrad of the twenty-first century," said Oleksiy Arestovich, Zelensky's adviser.[81]

When the official history of the war comes to be written, honor will be done to the heroic partisans who helped to stall the Russian column. Helped by the accuracy of their information, drone strikes and airstrikes destroyed vehicles in the convoy. Wildly outnumbered and outgunned, well-hidden Ukrainian tanks and especially artillery nevertheless ambushed the convoy and slowed its progress in its attempted envelopment of Kyiv. The Ukrainians blew up bridges, creating bottlenecks and congesting traffic. The Russians suffered from logistical problems with shortages of fuel and food. Once the tanks tried to deploy off-road to the east of the city, they were slowed down badly by mud.[82] It was finally stopped at Makariv, 34 miles from Kyiv's city center.

There are several reasons why the column was unable to encircle Kyiv and finally had to withdraw back to Belarus, some of which were summed up by the American analyst Mark Helprin, who wrote of how, "single-file and bunched together [it was] hit front and rear, the vehicles in between able to move neither left nor right because of obstacles, mud, lack of fuel, tires shredded from sitting too long in depot or due to inferior material substituted by officers who pocketed the difference between the monies allocated and those spent."[83]

By 3 March, less than two weeks into the war, when Putin delivered a speech stating that the special operation was going according to plan, it was already clear that it was not. By that point, some 480 Russian missiles had hit Ukrainian cities, killing hundreds of civilians, destroying infrastructure, cutting off water, electricity and gas supplies and destroying tower blocks and government buildings. Ukraine's military stated that of the 117 Russian battalion tactical groups of over a thousand men each that had been stationed on the border, all but six had entered Ukraine.[84] Russian troops had captured Kherson, and Mariupol had been surrounded and bombarded for thirty-six hours. Nonetheless, the all-important armored column was still 17 miles from Kyiv, having hardly moved in three days and experiencing severe difficulties.[85] There was understandable nervousness in the Russian high command about committing so much armor before they controlled the airspace above, but they took that decision.

Beyond the shortcomings in the air, there were also stories of Russian soldiers sabotaging their own vehicles' gas tanks in order to avoid reaching the front line, a classic example of the demoralization of young conscripts who had been told they would be greeted as liberators but were instead experiencing almost universal hostility and loathing. As we have often seen in earlier chapters, Clausewitz was correct when he wrote that "Moral forces are among the most important subjects in war."[86] Once morale erodes, it is notoriously hard to regain. Some Russian troops were told they were going on a training exercise, others that the limited targets for the special operation were Ukrainian Nazi sympathizers. Few realized on mobilization that they were about to fight a full-blown war. "We have definitely seen cases of very quick surrenders by Russian forces," Ben Wallace told the BBC on 2 March, "we've seen lots of abandonment of incredible pieces of equipment and that is what happens when your military leaders don't prepare you, lie to you, deceive you, and also you base a military plan on the arrogant assumption you are a liberator."[87]

The effects of the widespread demoralization in the Russian Army were plain, even within a week of the invasion. Ukrainian farmers were filmed driving tractors pulling abandoned BMD-4M amphibious infantry fighting vehicles and a Tor-M2 surface-to-air missile system worth $25 million. On the way to Chernihiv in the north, several Russian units gave up their vehicles and simply fled back to Russia. The Ukrainian secret service intercepted many messages from soldiers to their families back home bewailing their fate. Sergiy Kyslytsya, the Ukrainian Ambassador to the UN, read out a text message sent from a Russian soldier to his mother. "We were told we would be greeted with open arms but they call us fascists," it read. When the woman asked her son if he was still out for military drills, he responded: "The only thing I want right now is to kill myself."[88]

In December, members of Russia's 155th Separate Guards Marine Brigade sent a public unsigned open letter to Oleg Kozhemyako, Governor of the Primorsky Krai region, denouncing an offensive in which they had taken part. "As a result of the 'carefully' planned offensive by the 'great generals,'" it stated online, "we lost about three hundred people killed, wounded and missing, as well as half the equipment in four days."[89] Even to send such a letter amounts to mutiny. By that stage of the war, the Ukrainians were discovering the corpses of Russian officers who had been shot in the back by their own men. As with the (admittedly isolated) incidents of "fragging" in the US Army in Vietnam, there can be no clearer evidence of demoralization. Nevertheless, on 2 March Putin told President Macron of France – to whom he spoke no fewer than seventeen times before Macron finally recognized that his mediation was doing no good – that Russia would fight on until the entire territory of Ukraine was in his hands.[90]

The lengths to which Ukraine was willing to go in its defense were illustrated on 4 March when it decided to scuttle its flagship, *Hetman Sahaidachnyi*, a Krivak III-class frigate, in the port of Mykolaiv, to prevent her capture. "It's hard to imagine a more difficult decision for a brave warrior and the entire team," wrote Oleksiy Reznikov, the Ukrainian Defense Minister, on Facebook, "but we're building a new fleet ... The main thing now is to stand up."[91]

On 3 March, the Russians admitted that Major General Andrei Sukhovetsky, a Spetsnaz (special forces) commander and the Deputy Commander of the 41st Combined Arms Army, had been killed on the front line, the highest-ranking officer to die so far.[92] Because Ukrainian special forces operations discouraged the Russians from placing their high-value electronic warfare and air defense systems deep into Ukraine for fear of capture, and because of ineptitude and deteriorating morale at the front, the Russian attack had stalled along five of their seven invasion axes. Furthermore, the Russian communications system (single channel and unencrypted) had proved to be insecure and prone to Ukrainian eavesdropping and jamming. Generals were forced to go forward themselves in order to determine the nature of delays and chastise local commanders. Thus exposed near the front line, they were successfully targeted by Ukrainian snipers, drones and artillery. As the war progressed, an increasing number of Russian generals were killed, a development that sent shockwaves through the Russian high command in Moscow.[93] Nor was it only generals; on 9 August Lieutenant Colonel Vitaly Tsikul became the 100th Russian colonel or lieutenant colonel to die in the "special operation."

One way in which the Ukrainians were able to locate senior Russian officers was through their own data harvesting. Russian mobile phones in Ukraine emitted a roaming signal that connected to Ukraine's cellular network, which allowed the intelligence services to triangulate their position using the three closest cell towers. "It is not hard to geo-locate someone on a phone talking in the clear," General Ben Hodges, the former Commander of US Army Europe, told the *New York Times*.[94] Russians gave away their positions even when using stolen Ukrainian iPhones, because Ukrainian intelligence used the "Find My iPhone" app, which worked even when the phones were switched off. Russian forces did not destroy the Ukrainian telecommunications network because they needed its 3G and 4G for their own communications, and although their special forces had encrypted communications, ordinary Russian troops did not. This – and the presence of smart phones and social media posts from the battlefield – is why this has been described as the first open-source war.

America lost twelve generals during the whole course of the Vietnam War, yet only one month into the invasion of Ukraine, Russia had lost fifteen generals and colonels.[95] The deaths of so many commanders might not have been quite so serious for another army, differently constituted, but the Russian Army has long been a top-heavy organization, with officers controlling almost everything and with little power and responsibility entrusted to non-commissioned officers (NCOs), who are not comparable in professional development or authority to their counterparts in most Western armies. Control over the means and methods of completing tactical objectives is entirely held by the officers in the Russian Army, which severely hampered its effectiveness in Ukraine. "Russian NCOs do not dictate the tactical decisions of their units, nor discipline or motivate the troops," writes the military historian Katherine Bayford. "Instead, higher-ranking officers assume control of everything. The result is that units are inflexible, chains of command vulnerable, and troop morale low."[96]

A terrifying new aspect to the war was unleashed on 4 March, when Russian artillery shelled the Zaporizhzhya nuclear plant, with its six reactors the largest in Europe, thus threatening a radiation leak. A fire was started in one of the plant's buildings, but no leak occurred. Later in the year, shelling knocked out power to one of the reactors, forcing engineers to use back-up generators and prompting the Ukrainian authorities to issue iodine tablets to local residents to be taken in the case of a leak.[97] The truly breathtaking irresponsibility of shelling anywhere near a place where a nuclear catastrophe could occur was denounced by Rosemary DiCarlo of the United Nations at an emergency meeting of the Security Council as contrary to international humanitarian law, but to no avail.[98] Russia was the first to use cluster munitions, forcing Ukraine to reciprocate much later on.[99]

The Russians have also used white phosphorus bombs – which are banned from use in built-up areas by international convention – and the illegal anti-personnel POM-3 Medallion landmines, which land via parachutes and are designed to detect approaching footsteps and obliterate everything within a 16-yard radius.[100] Russia has also admitted using thermobaric weapons in Ukraine. A vacuum bomb, the TOS-1 Alpha has two charges. The first disperses fuel into the air and the second ignites it, sucking all the oxygen out of people's lungs. Its twenty-four unguided rockets can be fired in six seconds from a multiple launch rocket system mounted on the chassis of a T-72 tank, and have a range of 2 miles. Its use was not specifically illegal, but it nonetheless represented a significant escalation in Russian barbarity, especially against civilians trapped in enclosed spaces.[101]

Warfare had evolved so much by 2022 elsewhere in the world that smart weaponry ought to have minimized civilian casualties and collateral damage to infrastructure. During the Iraq and Afghan Wars, military operations were regularly called off by the US-led coalition if it was felt by commanders that too many innocent civilians would become casualties. By total contrast, the Russians have deliberately targeted civilians in order to create terror, as well as to punish ordinary Ukrainians for not welcoming them as liberators. Putin has chosen to fight the war using the methods of earlier, less sophisticated and far more brutal periods of history.[102]

Since 1945, a series of laws and conventions have been established over what constitutes war crimes, yet Russia has casually cast these aside, and when upbraided for it adopts the old Soviet *dezinformatsiya* (disinformation) tactic of accusing the other side of whatever it was that Russia was doing or planning to do. A new low of barbarity was reached on 4 March when the Russians bombarded Red Cross evacuation routes out of the cities of Mariupol and Volnovakha, ones which they themselves had agreed to as part of a ceasefire only hours earlier.[103] The Kremlin later claimed, without offering evidence, that it had been Ukrainian nationalists who had stopped people from leaving the cities.

When the history of the siege of Mariupol comes to be written, it will rank for courage of the defenders alongside the great sieges of history, such as those of Gibraltar and Leningrad. Mariupol had been denuded of artillery, air defenses, reserves and anti-tank weaponry earlier in the war because of the decision of the Ukrainian General Staff that Kyiv must not fall. Strategically, its position on the coast of the Sea of Azov, separating Russian elements in the Donbas from those north of Crimea, meant that the Russians needed to capture it in order to complete the land bridge that connected those places. However, Mariupol's system of enormous Cold War-era tunnels and bunkers, especially under the massive Azovstal iron- and steelworks, made it supremely defensible. The Russian response was essentially to flatten Mariupol, but, just as with Stalingrad, Monte Cassino, Caen and other Second World War urban battlefields, it is often easier to defend rubble than untouched buildings.

Mariupol was home to 450,000 people before the war, and it became the ultimate symbol of Ukraine's remarkable determination, skill, resourcefulness and courage. It also demonstrated how costly clearing an urban area can be, as had been seen in very different circumstances in Iraq at Ramadi, Fallujah, Baqubah, Mosul and large parts of Baghdad. After the water and electricity had been cut off, Mariupol ran out of medical supplies, especially the morphine that is so important in relieving pain for the wounded. Additionally, the shelling of refugees attempting to escape the city within previously agreed deadlines produced death and panic.[104] Russia also only provided safe corridors towards its own territory, thereby effectively turning Ukrainian citizens into Russian prisoners, a ploy that James Cleverly, the future British Foreign Minister, described as "cynical beyond belief."[105]

As Mariupol continued to hold out heroically, buying time for the Ukrainians to realign their forces in the Donbas, heavy fighting took place around the armored column which had reached Hostomel outside Kyiv, where the Russians experienced problems performing the most basic of soldier tasks, such as dispersing and seeking cover when they stopped for any period of time. A vehicle column should never close up and halt in the open on a major highway where it can be spotted by drones and hit by artillery, yet this 40-mile convoy/traffic jam did just that, and repeatedly paid the price.[106]

A major problem was that, although General Alexandr Dvornikov had been in nominal command of the invasion, it was Moscow that controlled every aspect of it, which had predictably bad consequences for on-the-spot decision-making. Nor did Dvornikov's writ include all seven axes of attack. Unity of command is an essential principle of war, because commanders have to be in overall charge when coordinating battles, directing logistics, committing reserves, adjusting concentration on different axes and so on.[107] That might seem like an obvious prerequisite for a full-scale war of this nature, but Putin and his war minister Sergei Shoigu had not grasped it. The Russian leadership not only failed to design a proper campaign to achieve its objectives in Ukraine, it also failed to establish the proper organizational architecture and authorities, not unlike some of the coalition shortcomings of the early months – and years – in Afghanistan after the Taliban were toppled.

On 8 March, President Zelensky spoke to the British parliament over video link from Kyiv. He channelled his inner Winston Churchill into the speech, telling a packed meeting of parliamentarians that in expelling the Russian invader Ukrainians would fight "in the forests, in the fields, on the shores, in the streets ... We will fight till the end. At sea. In the air. We will continue fighting for our land whatever the cost."[108] It was of course a conscious paraphrase of Churchill's great 4 June 1940 speech after the Dunkirk evacuation, in which he explained to the British people how they would expel the Nazi invader. Zelensky's leadership drew the adjective "Churchillian" from all over the world, underlining both how brave and eloquent Zelensky was, and also how relevant Winston Churchill remains as the benchmark for inspirational wartime leadership even more than half a century after his death.

Zelensky would receive stalwart support from President Joe Biden in the US and Prime Minister Boris Johnson in the UK – though it was not unconditional. In March, the US rejected Poland's offer to send twenty-eight MiG-29s to Ukraine via an American airbase in Ramstein,

Germany. Citing "serious concerns for the entire NATO alliance," the Biden administration feared that such a powerful addition to the Ukrainian Air Force would be interpreted as an act of aggression by Putin, who might escalate the conflict outside Ukraine.[109] After the Canadians and Germans had also opposed Poland's offer, Zelensky remarked in a video, "Listen, we have a war. We do not have time for all these signals. This is not ping-pong – this is human lives. We ask once again: solve it faster."[110] It later emerged that the Biden administration had vetoed the planes being sent in exchange for a promise from President Xi of China to put pressure on Putin not to use tactical nuclear weapons in Ukraine.[111]

On 10 March, Ben Wallace stated that Britain would send the portable, laser-guided Starstreak missile system to Ukraine, to shoot down Russian helicopters and warplanes.[112] A high-velocity weapon that uses a system of three dart-like projectiles, Starstreak carries out multiple hits on low-flying fighter jets and helicopters. The operator points a laser at the target and the missile flies along its beam, allowing the pilot little time to take defensive measures. The British also sent the Martlet, a laser-guided, lightweight, multi-role missile not unlike Starstreak. The US had earlier sent Stinger anti-aircraft missiles, a lightweight, self-contained air defense system that can be rapidly deployed by ground troops against anything flying below 11,000 feet. Its accuracy derives from an infrared seeker which locks on to the heat from the enemy's exhaust. This was turning into a proxy war between Russia and the West of the kind seen so often in the Cold War chapters, though on a larger scale than most earlier ones.

Also shooting down Russian planes was the so-called "Ghost of Kyiv," an anonymous Ukrainian fighter ace who had allegedly accounted for no fewer than ten opponents. "Hello, occupier, I'm coming for your soul!" the Ghost was reported telling Russian pilots.[113] Whether he was a real person, urban legend or figment of Ukrainian propaganda, he provided a substantial boost to Ukrainian morale. High morale, which essentially stems from a belief in ultimate victory however long and hard the road may be, is a vital factor in warfare and an essential prerequisite for victory, and is easier to engender in a war of independence such as this one, rather than a war of occupation such as Russia's. With high morale, soldiers are willing to undergo extraordinary dangers and privations; without it they simply are not.

High morale was evident in Kyiv at this time, where bunkers, trenches and barricades were being built to contest the city, if necessary street by street. Russian tanks had reached Irpin, 12 miles to the north-west, and Brovary on the eastern bank of the Dnipro, but they were unable to surround the capital. His Blitzkrieg might have faltered, but Putin probably assumed (and still seems to believe) that Russia could simply "out-suffer" the Ukrainians, Europeans and Americans, in the same way that Russians had out-suffered Napoleon's and Hitler's invasions. Yet once again the key difference that eluded him was that while Napoleon and the Nazis were the aggressors invading Mother Russia, this time it was the Russians themselves who were the foreign aggressors.

Outside Kyiv and in the Donbas and elsewhere, deep and well-dug trenches proved their efficacy, even though they suffered from the same problems as they had in the First World War, such as rats, rainwater and mortar fire. Trenches provide the inestimable advantage of protecting soldiers from the direct fire of machine guns that can fire up to a thousand rounds a minute at ranges of up to a mile. Similarly, they provide solid-earth defenses against artillery shells which can be fired from 15 miles away, and whose fragments flying at 3,000 feet per second are lethal to anyone caught within 50 yards of the blast. The fire from six such guns can cover every square foot of an area the size of a football field, so trenches are necessary for anyone caught outside concrete fortifications or armored vehicles.[114] For all that war is constantly evolving, some aspects of it stay remarkably static.

When Russia's initial failures in Ukraine became apparent by the end of the first week of the war, its General Staff and the FSB started to blame each other, with the former arguing that poor intelligence over Ukrainian resistance was the problem, while the latter pointed out that combat units had failed because the General Staff had overestimated their capabilities. Casting around for scapegoats for what was ultimately his own error, Putin settled on the intelligence services, and on

11 March General Sergei Beseda of the FSB and his deputy Anatoly Bolyukh were put under house arrest.

The intelligence regarding the likely Ukrainian response to invasion had of course been catastrophically poor, but as a former KGB agent himself Putin ought to have viewed it through a far more skeptical lens than he had. In his book *Ukraine and the Art of Strategy*, Lawrence Freedman points out that "Russian strategic thought is often thought to be influenced by chess, a game in which it is necessary to think several moves ahead of the opponent, yet in Putin's case his strategic thought appears to be more influenced by judo, a game which he played (earning a black belt) and follows closely."[115] Judo is not about long-term strategy, but instead about trying to catch your opponent momentarily off-balance through speed and cunning, to be the last person standing. By mid-March 2022, it was clear that Putin did not have Zelensky on the mat, and the war would much more closely approximate chess rather than judo.

Just as foreign fighters had joined International Brigades to fight fascism during the Spanish Civil War, so the Ukrainian ranks were swelled by some 20,000 volunteers from fifty-two countries who joined Ukraine's International Legion to frustrate Putin's dream of annexing Ukraine.[116] Americans made up the largest single contingent, but volunteers came from all over the world. Alongside the actual warriors were flocks of foreigners who helped take medical supplies over the frontiers and act as medical orderlies. Internet donors inside and outside Ukraine provided medical supplies, and also weapons. People across the globe were raising funds to equip volunteers with sniper rifles, flak jackets and essential equipment. This was the first crowd-funded war, and warfare was evolving in ways that Putin and Gerasimov had entirely failed to foresee.

Furthermore, NATO was already helping Ukraine significantly in providing intelligence assessments and real-time information, efforts that would continue to expand as the war progressed. Although they had strict orders never to cross into Ukrainian airspace, MQ-9 Reaper drones above Ukraine's neighbors monitored Russian activity in Ukraine, and Russian communications were eavesdropped by Boeing

RC-135 and AWACS (airborne warning and control system) planes, for the benefit of Ukraine's defense – although the intelligence is reportedly not passed to Ukrainian intelligence in real time, in order to prevent the Russians claiming that the US is aiding Ukrainian targeting.[117]

Shortly after 10 a.m. on 16 March, the eyes of the civilized world focused on the Donetsk Regional Academic Drama Theatre in Mariupol, particularly its air-raid shelter where hundreds of civilians were staying after their homes had been destroyed. Despite the word "Дети" – Russian for "Children" – having been painted in giant Cyrillic letters on the forecourts on either side of the building, which would have been clearly visible to Russian pilots and also on satellite imagery, two Russian warplanes dropped two 500-kilo bombs on the theatre. As Amnesty International concluded after an investigation, it "was a hub for the distribution of medicine, food and water, and a designated gathering point for people hoping to be evacuated via humanitarian corridors. It was clearly recognizable as a civilian object, perhaps more so than any other location in the city."[118]

No one knows the exact number of civilians who were killed in that particularly egregious war crime, but Associated Press estimated around 600. The Russian government initially claimed that the Mariupol theatre had been a military target, but after that was comprehensively disproven they claimed it had been deliberately destroyed by the ultranationalist Ukrainian Azov Battalion defending Mariupol, an equally ludicrous suggestion. Of all the many war crimes of this conflict, the Mariupol theatre was undoubtedly one of the worst. At least, so far.

The notion that the presence of children might have deterred the Mariupol theatre atrocity was chillingly corrected on 8 April, when, after a Tochka-U missile attack on refugees at Kramatorsk railway station which killed fifty-two people and wounded dozens, it was discovered that the words "For children" had been spray-painted on the side of the bomb.[119] There is a long tradition of airmen and artillerymen chalking humorous remarks on shells – "For Kaiser Bill," for example, or "For Adolf" – but writing "For children" on a missile that was ultimately destined for a civilian target, where five children were indeed killed, indicates a truly depraved psychology.

On 19 March, the Ukrainian General Staff announced that it had lost all access to the Sea of Azov, although Mariupol was somehow still holding out despite many thousand mainly civilian deaths. Several Ukrainian units were still defending the city center and the Azovstal iron- and steelworks. The latter plant was one of the largest in Europe, with its extensive series of massive Cold War-era bunkers deep underneath the plant. Although around 40,000 civilians had been evacuated, some 300,000 remained, facing an ever-worsening humanitarian catastrophe.[120] On occasion corpses had to be buried alongside their car numberplates in order to be identified for later reburial.

It had already become clear by mid-March that as well as alienating Ukrainian hearts and minds by attacking civilians, being surprisingly poor at performing basic tasks such as employing armor, infantry, engineers, artillery and mortars to achieve combined-arms effects and inadequately maintaining their vehicles and weapons systems, the Russian Army had another serious weakness.[121] Over one-quarter of Russia's troops consisted of conscripts serving fixed, year-long conscription terms, largely performing logistical tasks such as driving trucks, fuel tankers and maintenance vehicles. These conscripts had suffered harsh if limited training, resulting in endemic demoralization and limited leadership or technical qualities.[122] Historic instances of brutal hazing that had haunted the immediate post-Soviet Army – which had included merciless beatings, torture and even the prostitution of soldiers by their own officers – were not entirely eradicated: in 2019, one conscript shot eight of his fellow soldiers, claiming that they were planning to rape him.[123]

Low Russian morale was highlighted on 23 March when the Ukrainian intelligence services broadcast a report made by a Russian officer that it had intercepted. Propaganda such as this is all the more powerful when it can be seen to be genuine. Because some officers in the Russian Army relied on mobile phones and analogue walkie-talkies, they were highly vulnerable to being overheard. The officer, who was besieging Mykolaiv, complained of a lack of body armor, noted that half his men were suffering from frostbite because of a lack of tents and stoves and carped about repeated and costly friendly-fire incidents. "In short, it's shit here," he said, reporting that a Russian Grad rocket had been fired at his unit only the previous day. "We asked if it was ours, but no one could tell us."[124] The lack of supplies might have been because Lieutenant General Yakov Rezantsev, the commander of the 49th Combined Arms Army, had predicted on Day 4 of the war that they were "hours" from victory. (Rezantsev was killed near Kherson on 25 March.)

With morgues in Mazyr in Belarus full of Russians, yet no credible casualty numbers being released by the Kremlin, this low morale was having an effect on recruitment, with some young men in Krasnodar in southern Russia refusing to join the National Guard if it was ordered to serve outside the motherland. A further indicator of rock-bottom morale came when Colonel Yury Medvedev, the commander of Russia's 37th Motor Rifle Brigade, which had been fighting on the outskirts of Kyiv, was deliberately run over by his own troops after half the brigade had been lost.[125] Another came in late March when a Russian soldier named "Misha" handed over a T-72B3 main battle tank for a £7,500 reward and the opportunity of Ukrainian citizenship, after his two crewmates had fled the battlefield.[126]

In late March, the village of Chornobaivka near Kherson airport became a byword for Russian military incompetence when the same tactics were employed in attempts to capture it no fewer than ten times over several days, each with the same bloody result. Russian troops and heavy equipment were landed on the airfield on ten separate occasions hours apart, and each time were hit by Ukrainian artillery. Memes flooded Ukraine featuring the Bill Murray movie *Groundhog Day*, before the Russians finally adopted new tactics.[127] Chornobaivka represented a startling indication of the notorious inflexibility of Russian command techniques.

It was to be the uncovering the following month of the atrocities committed in the town of Bucha, to the north-west of Kyiv just beyond Irpin, that came to symbolize Russian war crimes. The word Bucha now takes its place alongside Guernica, Lidice, Oradour and Srebenica as a ghastly reminder of man's inhumanity to man. As the Ukrainian Army liberated the town, its vehicles were forced to swerve to avoid the corpses of civilians littering the streets.[128] As mass graves were exhumed, it soon became clear that hundreds of citizens had been summarily executed there.[129]

The Russians had hubristically assumed that this was territory from which they would never have to retreat, so their war crimes would go undetected and thus unpunished, but now satellite imagery was used to identify where they had hastily buried bodies.[130] Many of the corpses in the main mass grave in Bucha had their hands tied behind their backs as well as bullets in their chests or heads, indicating execution. Evidence of torture was also abundant. On 8 April, Iryna Venediktova, Ukraine's Prosecutor-General, announced that 650 bodies had been found – forty of them children – and the search was still ongoing.[131]

When the Russian propaganda machine inevitably started to allege that the corpses had been placed in Bucha by the Ukrainians themselves, their claims were comprehensively disproven by a team at Maxar Technologies, a company that used satellite technology to determine how long the bodies had been there, as well as the exact date when a mass grave had been dug in a church cemetery.[132] Drone and satellite before-and-after footage of places such as Bucha, Irpin and Mariupol easily disproved Russian disinformation about how the deaths had come about from Ukrainian friendly fire.

Meanwhile, the Russian people were receiving nothing but positive propaganda from the state, as well as absurdly low casualty figures. In mid-March, the government closed down all non-government social media and opposition news outlets, made those it termed "defeatists" liable to a fifteen-year prison sentence and continued its ban on the word "war," even though there were also orders to liken it to the Great Patriotic War of 1941–5. Moreover, even though Russia was under no strategic threat whatever from Ukraine, local authorities were ordered to form local defense detachments and instruct the population about artillery shelling and airstrikes, and businesses were told to prepare anti-tank defenses as a way of heightening a nationalistic response. Instructions were given to the media to present the "special operation" as a struggle for Russia's very survival at the hands of an imperialistic

NATO.[133] The letter Z – which had been used to mark Russian tanks during the invasion – was popularized.

Yet no amount of propaganda could alter the realities of the war on the ground. Because the Russians rarely paused to collect their dead from battlefields during their retreat, Ukrainian soldiers and officials were able to apply Clearview AI facial-recognition technology to identify some 852 Russian corpses, before sending photographs of the corpses to the victims' mothers back in Russia. The practice was intended to stoke up domestic opposition to the war, and is illustrative of yet another way in which this conflict saw warfare evolve.[134] The Ukrainians even set up a Russian-language Telegram page where ordinary Ukrainians could post photos and identify killed, wounded or captured Russian soldiers.[135] There was an undeniably gruesome aspect to history's first open-source war.

At 8 p.m. on 13 April, Ukrainian intelligence made a positive sighting of the guided-missile cruiser *Moskva*, the 12,490-ton flagship of the Russian Black Sea fleet. She had a crew of 510 and had spent seven weeks bombarding Ukrainian targets such as Mariupol and Snake Island, had launched Kalibr cruise missiles against Ukrainian cities and was the most powerful and best-equipped vessel in Russia's Black Sea fleet. The United States, which had a P-8A Poseidon maritime surveillance aircraft in the area, confirmed the sighting, whereupon the Ukrainians launched two land-based Neptune cruise missiles at subsonic speeds at her. The Neptunes ought to have been shot down by the *Moskva* but were not, and the ship became the largest vessel to be sunk in conflict since the Second World War, even larger than the *General Belgrano*.[136] She had a relic of the True Cross aboard, but even that did not save her.[137]

Despite the Russian Defense Ministry's ridiculous claim that the ship had sunk as the result of a mysterious fire, it was a signal humiliation for President Putin, who had proudly shown President Al-Sisi of Egypt around it some years earlier. "The notion of Ukraine's military, the veritable underdog in the fight, bringing down the *Moskva*, a warship named for Russia's capital city, was not only a strategic victory but a symbolic one," noted the American military website *Breaking Defense*.[138] The race between weapons being invented that can sink ships and those that can protect them is an ancient and ongoing one, with big prizes going to the winner. As warfare evolves, so too do debates about the immense costs of aircraft carriers, for example, vis-à-vis the weapons, such as drone swarms (where large numbers of less sophisticated, and thus less expensive, drones converge en masse and the first to reach the target emits a signal which the others latch on), that might one day be able to sink them.

For all their success over the *Moskva*, mid-April saw the Ukrainians facing a difficult task in redeploying forces to the Donbas and elsewhere, and they used every means of transportation they could find. They did not have enormous expeditionary logistics, but employed "fixed-base logistics" instead. This involves forces in the field being supported from largely fixed supply depots that provide fuel, ammunition, food, water, spare parts and so on, rather than by large mobile supply elements that travel with the army as it moves (as is the case for US units, which have several times the logistics force structure of comparable Russian units and can thus enable rapid advances such as the race to Baghdad).

Fixed-base logistics were appropriate for the Ukrainians since they were fighting on their own soil and had the great advantage of working on interior lines. However, the lack of expeditionary logistics in the Russian Army, once it had left the Russian rail system on which it is heavily reliant, resulted in serious shortcomings. These were issues, along with combined-arms operations, that ought to have been ironed out during the long military maneuvers in Belarus and Russia before the invasion but were not. "It's very clear to me that they were camping, not training," observed General David Petraeus.[139]

The shortcomings of Russia's military were not just strategic and logistical but systemic, and coming under increasing pressure. Russia could not employ its entire arsenal of more accurate missiles without undermining the stockpiles needed for any possible war with NATO, and the supply chain to produce more was threatened because it was so heavily dependent upon critical specialist components that were not manufactured domestically.[140] The 9M727 cruise missile fired from the Iskander-K, for instance, has considerable precision but carries a specialist computer which has no fewer than seven socket attachment points that allows data to be moved through its heat shield. Yet six of the seven were made by US companies, as were the circuit boards.[141]

Many of Russia's weapons systems had the same vulnerability; the 9M949 guided 300mm rocket used an American-made fiber-optic gyroscope for navigation, for example. The TOR-M2 air defense system used a British-designed oscillator in its radar computer. The communications for the Il-76 transport plane had no fewer than eighty components that could not be produced in Russia. Russian military radios contained critical components imported from the US, Britain, Germany, Japan and the Netherlands. Western export controls would, over time, target not just the microchips needed in virtually all modern armaments but also other necessary components, demonstrating yet another area where NATO was able to help Ukraine without putting so much as a single boot on Ukrainian territory. The whole area of proxy war, evident in every chapter of this book, was being taken to a new level in Ukraine.

Western sanctions thus posed a military as well as an economic threat to Russia's war-making capacity, at least after the existing supplies run out. Of course, Russia took active steps to counteract this, and in March set up an interdepartmental committee under Deputy Defense Minister Alexei Krivoruchko to try to produce equipment either domestically or in friendly countries which do not enforce Western sanctions, particularly Serbia, Armenia, Kazakhstan, Turkey, India, Pakistan and China, although there has been little evidence of China violating the export controls established.[142] Russian munitions were increasingly built with semiconductors repurposed from household machines and gadgets, because ever since the Western sanctions of 2014 it had been running low on military-grade electronics such as transistors, transformers and microchips.[143]

In November 2022 it was discovered that Russia was buying fridges, dishwashers, electric breast pumps and wargame consoles in order to cannibalize them for microchips for its arms industry, a sure sign of desperation.[144] Meanwhile, Western intellectual property rights were entirely ignored.

The transparent nature of this open-source war became particularly obvious after Russia's major spring offensive in the Donbas in late April 2022, when satellite and signals intelligence gave early warning of unfolding operations. As General Sir Richard Barrons points out, "Data provided by both expensive, capable geo-stationary military satellites and cheap, commercial low-Earth-orbit satellites and drones mean that the battlefield is transparent."[145] There was even a website, Oryx,* that carefully documented every Russian weapons system and vehicle that was destroyed. Platforms such as UAWeapons[†] and JominiW[‡] on Twitter and the Enforcer Channel[§] and Institute for the Study of War[¶] allowed people around the world to follow the war in depth in a way that had never happened before.

As the Russian offensive in the Donbas developed, long-range artillery kept up a ferocious bombardment in scenes reminiscent of the First World War or the battle of Grozny in Chechnya in 1994–5. Pisky, a suburb of Donetsk, was shelled daily and Russian tank advances were supported elsewhere in the Donbas, such as the fronts south of Izyum and west of Popasna. Fears that large parts of the Ukrainian Army might be surrounded and forced to capitulate proved unfounded, however, as its high command proved flexible in ordering strategic withdrawals whenever that danger arose, and then proved equally nimble in counter-attacking. The sharp counter-attack is one of the most important tactics in warfare, capable of knocking an opponent off-balance, and the Ukrainians excelled at it.

---

* https://www.oryxspioenkop.com/2022/02/attack-on-europe-documenting-equipment.html

† https://twitter.com/UAWeapons

‡ https://mobile.twitter.com/JominiW

§ https://www.bing.com/search?q=the+enforcer+channel&PC=U316&FORM=CHROMN

¶ https://www.understandingwar.org

In May, President Biden signed into law the Ukraine Democracy Defense Lend-Lease Act of 2022, the title a reference to the military aid with which the United States helped Great Britain during the Second World War. Biden's package was eventually to comprise well over $30 billion in arms, ammunition and other military assistance over the next year and exceed $41 billion by July 2023. The US State Department also provided tens of billions of dollars in financial and humanitarian assistance, with the total equal to one-third of Ukraine's entire pre-war GDP.[146] By July 2023, the Biden administration had used its drawdown authority no fewer than forty-two times. Without this enormous contribution, underlining the United States' status as the indispensable nation, it would have been near-impossible for Ukraine to have continued the fight after its own supplies of weaponry and ammunition had been exhausted.[147] Ammunition was consumed at rates not seen since the Second World War, with the Ukrainians regularly firing 6,000 artillery shells a day, and the Russians even more. At one point during the struggle for the Donbas area, Russia employed more ammunition in three days than the entire stock of the British military.[148] Without supplies from the United States, and to a much lesser extent from European countries, Ukraine could simply not have fought on, at least conventionally.

The second largest national contribution after America – of £2.3 billion in military aid and £220 million in humanitarian assistance in the first twelve months – came from Britain, where Boris Johnson's government and its successors continued their proselytizing support for Zelensky and the Ukrainian cause. When Putin invaded, Britain immediately sent 4,000 NLAWs, more than 200 Javelin missiles and a further 6,000 anti-tank missiles. Other high-explosive weapons were supplied in a second package in March. After that came large numbers of Starstreak air defense missiles, hundreds of armored vehicles such as the Bulldog and 13.5-ton Stormer, a fourteen-strong squadron of Challenger 2 tanks, AS-90 guns with 100,000 artillery rounds, minefield-breaching machines, sophisticated missiles including GMLRS rockets, Harpoon anti-ship missiles, armored cars, body armor, night-vision goggles, helmets and boots.[149]

Britain provided training on Salisbury Plain and elsewhere for 10,000 Ukrainian soldiers, taught by instructors from Britain, Canada, Denmark, Finland, Sweden, Norway, New Zealand, Lithuania, the Netherlands and Australia.[150] Britain also sent troops and the Sky Sabre medium-range anti-air missile system to Poland.[151] The dispatch of tanks by NATO was further confirmation of the fact that they still have an important role on the modern battlefield, even in the era of NLAWs and Javelins. "Tanks are like dinner jackets," once quipped the Australian Major General Kathryn Toohey, "you don't need them very often, but when you do, nothing else will do."

The practical application of British help was proven on 8 May when GIS Arta, a computer system developed by Ukrainian programmers in collaboration with British digital mapping companies, cut targeting times from twenty minutes to one, and led to the destruction of more than seventy Russian tanks, armored fighting vehicles and personnel carriers in two days of coordinated shelling and airstrikes as they attempted to cross a bridge over the Siverskyi Donets River in eastern Ukraine. What the press nicknamed "Uber-style technology" used GPS and NATO-donated radars, along with reconnaissance drones, rangefinders and smart phones, to pinpoint enemy positions, whereupon the GIS Arta "shooting calculator" software worked out which weapons were most suitable to carry out the strike.[152]

Similarly, equipment that Antony Blinken, the US Secretary of State, had promised at a meeting of forty countries in Ramstein in late April began to arrive on the battlefield in May, including seventy-two 155mm howitzers, which came with 144,000 artillery rounds and had three times the range of Ukraine's artillery; further prodigious quantities of US artillery ammunition would follow, eventually exceeding two million rounds by summer 2023. Equally welcome were the ten AN/TPQ-36 counter-artillery radars.[153] Yet at the time NATO refused to provide strategic weapons to Ukraine such as attack aircraft, helicopters, NATO-standard tanks, cruise missiles and long-range missile systems, both for fear of antagonizing Russia and in return for China's opposition to Russia's use of nuclear weapons.[154]

* * *

By mid-May, the Russian advance in the Donbas had been slowed to half a mile a day, though it finally captured the town of Rubizhne after a three-week offensive. Putin did receive some good news when on Monday, 16 May, after eighty-two days of siege, Mariupol finally fell to Russia, and 260 Ukrainian defenders of the Azovstal iron- and steel-works, fifty-three of whom were seriously wounded, surrendered and were taken into captivity. The defense had been heroic, though, and it had moreover denied Putin the breakthrough of which he wanted to boast at the time of Russia's Victory Day parade of 9 May. It was esti-mated that the Mariupol defenders had prevented no fewer than seventeen Russian battle groups, amounting to many thousands of troops, from being redeployed to the Donbas. With the fall of Mariupol, however, the land link between Crimea and the separatist republics was established and more Russian troops could now be freed up for the Donbas offensive.

Ukrainian forces withdrew from Severodonetsk, a city with a pre-war population of 100,000 people, on 23 June. At the same time, the first American-supplied HIMARs (High Mobility Artillery Rocket System) started to arrive in the Donbas. Given the range of the precision rockets of nearly 50 miles, these proved invaluable in taking out Russian head-quarters, ammunitions dumps, fuel depots, rear-area barracks, bridges and other transport nodes, thus impeding the delivery of food, fuel, artillery, replacements and ammunition to front-line Russian units and slowing the bombardment of Ukrainian positions.

The HIMARs were not enough, however, to stop the heavy barrage behind which the Russians advanced to capture Lysychansk, the last city under Ukrainian control in the Luhansk province, on 3 July. Once again, the Ukrainian high command sensibly ordered its troops to withdraw, to avoid encirclement. Just as counter-attacks can be invalu-able, so can strategic withdrawal, and there are few armies that do not have to make use of it at least occasionally. Zelensky stated that earlier delivery of Western rocket systems such as the HIMARs might have saved Severodonetsk and Lysychansk, and requested more help, espe-cially from France and Germany, whose earlier promises of support were not being translated into significant arms deliveries on the ground, while additional American deliveries were arriving several times a month.

On 25 August, Putin took the decision to raise an extra 137,000 soldiers for the war, increasing the Russian Army's target to 1.15 million by January 2023. He did not order the general mobilization that was widely expected, however, and made sure that citizens of Moscow and St. Petersburg were largely excused from the new draft, since he did not want anti-mobilization sentiment building in those politically sensitive cities. Nonetheless, protests across Russia the next month resulted in more than 1,300 people being arrested and there was a flight of tens of thousands of military-aged men and their families from the country, which blocked several of the major exit routes for days.[155] It is possible that more Russians left the country than reported to the conscription stations.

The Ukrainians seemed sanguine about the massive new call-up of Russians. Valeriy Zaluzhnyi, now Commander-in-Chief of the Ukrainian Army, pithily tweeted of the new effort to mobilize additional recruits, "We finished off the Russian professional army, now it's time to finish the amateur one." The defeats around Kharkiv and the partial mobilization in September were the first indications for Russians that Russia might lose the war – or at least not win it – and no soldier wants to die in a struggle for a messy diplomatic compromise.

A long-awaited Ukrainian southern offensive began on 29 August, which attempted to push Russian forces back from the Dnipro River towards Kherson and press on to Zaporizhzhya, east of Crimea, on the way to the Sea of Azov. As well as HIMARs reaching well behind Russian front lines, Ukraine also had the signal advantage of being able to conduct a partisan campaign in Russia's rear area, with every Ukrainian with a mobile phone a potential artillery spotter or intelligence collector. Here, too, conflict entered a new phase of its history. The advancing Ukrainians issued Russian-language leaflets, complete with toll-free numbers for Russian soldiers to call in order to surrender, along with QR codes. Before the invasion, each Ukrainian special operations brigade had created and trained a "resistance company" recruited from the local population in precisely the areas such as

Kherson, Zaporizhzhya and the Donbas that were most likely to be attacked.

Just as the Russians redeployed large numbers of troops southwards in order to defend Kherson, the Ukrainian high command unleashed a surprise offensive in the Kharkiv region in the east, one that had been planned well in advance and wargamed at the same simulation center in Germany where the invasion of Iraq had been planned in early 2003. The Ukrainian offensive was a masterstroke that caught the Russians unawares and off-balance, and early September saw a string of Ukrainian victories around Kharkiv, as the morale of the Russian Army broke and a panicked retreat became general across the eastern theatre of operations, in much the same manner as that witnessed during the Afghan Army's collapse in the summer of 2021. "The speed of advance has been impressive," noted Lawrence Freedman at the time, "as tens of square kilometres turn into hundreds and then thousands, and from a handful of villages and towns liberated to dozens."[156]

In a week of heavy fighting to the east and south-east of Kharkiv from 5 and 11 September, the Ukrainian Army routed the elite 4th Guards Tank Division, part of the 1st Guards Tank Army, Russia's best armored formation. A hundred tanks were destroyed or captured in scenes reminiscent of the Israeli counter-attack in the Yom Kippur War. So much Russian military equipment was abandoned that Ukrainians joked that Russia had now overtaken America as their largest supplier of military hardware.

Despite Ukraine's GDP contracting by 15.1 percent year on year in the first quarter of 2022, and by 37 percent in the second quarter, its morale remained remarkably high. In August, an opinion survey found that 98 percent of Ukrainians believed they would win the war, and that around 90 percent of them either strongly or somewhat approved of President Zelensky, while 88 percent strongly approved of the Ukrainian Army.[157] These are the kind of approval ratings that Winston Churchill enjoyed during the Second World War, but are otherwise almost unknown in the history of political opinion polling. A country that has persuaded itself that it will eventually win is capable of undergoing unimaginable hardships in order to see that day dawn.

On the morning of 21 September, President Putin gave a televised address in which he made the thinly veiled threat against NATO that "Those who try to blackmail us with nuclear weapons should know that the prevailing winds can turn in their direction," despite NATO having made no nuclear threats against Russia whatever.[158] "The danger is serious, real," Sergei Lavrov had already stated of nuclear war, "and we must not underestimate it. NATO, in essence, is engaged in a war with Russia through a proxy and is arming that proxy. War means war."[159] Putin's and Lavrov's blatant nuclear saber-rattling saw the first instance since the Cuban missile crisis in 1962 that Western leaders had seriously to consider the heightened likelihood of nuclear weapons being used by a great power. There had been standoffs on the Indian subcontinent in the 1970s and in 2008, and more recently with North Korea, but otherwise the use of nuclear weapons had not seriously impinged on decision-making for the sixty years before Putin put his nuclear forces – which comprise 6,000 warheads – on heightened alert. In February 2023, he withdrew Russia from the New START nuclear agreement of 2010 with the United States.

The Russian stance was particularly worrying because there is a qualitative difference between the way the Russians regard tactical nuclear weapons and the way the West does, with the former concentrating on the word "tactical," while the latter tends to focus on the word "nuclear." Russian wargames routinely envisage the use of tactical nuclear weapons, whereas ever since Douglas MacArthur, Western leaders have reacted with horror to the idea that a conflict might ever "go nuclear."[160] This dichotomy exists despite the glaring historical irony that only the USA (in consultation with Britain) has ever actually used nuclear weapons. With Western officials threatening "catastrophic consequences" for the use of any nuclear weapons, and with both China's and India's leaders warning Putin as well, it appears that the Kremlin has been dissuaded from employing them.

By early November some 4.5 million Ukraine households had no electricity, including nearly half a million in Kyiv, and the situation worsened when in revenge for the loss of Kherson the Russians unleashed a campaign known as "weaponizing winter," aimed at freezing the Ukrainian people to death in their homes, barracks and foxholes, by destroying power stations, water treatment plants and electrical grids. Sub-zero temperatures were to be used as a weapon of war, which the Russians hoped might force the Ukrainians into a humiliating armistice. Yet, once again, the Russians would achieve the opposite of what was intended, and Ukrainian resolve hardened still further. Moreover, the international community stepped up, supplying 1.23 million winter supply items, including cold-weather clothing, heavy-duty sleeping bags and insulated tents to Ukraine. The United Kingdom alone sent 900 generators.[161] The United States pledged $53 million to start to repair Ukraine's badly damaged electrical grid.

General Mark A. Milley, the US Chairman of the Joint Chiefs of Staff, considered that the Russians' new strategy would be to defend what they held, continue to destroy Ukraine's energy-generating plants and other infrastructure and hope that the weather would change the course of the war.[162] The Russians had depended on snow, ice and the wind-chill factor in the past – notably in the Retreat from Moscow in 1812 and during the Second World War – although neither Napoleon's French nor Hitler's Germans were as used to extremely cold winters as were the Ukrainians.

Looting of artworks and historical artifacts by nation states, as well as by individual soldiers, has also long been a phenomenon of war, and the liberation of Kherson revealed that the Russo-Ukrainian War was no exception. It is hard to escape the conclusion that by looting some museums and art galleries, and deliberately targeting others for destruction, the Russians were hoping to destroy Ukraine's sense of cultural and historical identity. In late October, shortly before Kherson was recaptured, Russian cultural officials arrived at the Kherson regional art museum and directed the removal of all the most important artifacts by seventy laborers. "They took swords, guns, our entire collection of weapons," recalled Elena Yeremenko, the museum's secretary. "They took coins, icons, gold necklaces, 18th and 19th century furniture. This was not just looting – this was theft by the Russian state."[163]

The museum had 15,000 paintings comprising a rich collection of Western European art. All were stolen by Russian soldiers in trucks and taken to Simferopol in Crimea. Even the coffin containing the bones of Prince Grigory Potemkin was removed from Kherson Cathedral. "This shows what Russia means when they're fighting against our identity," said Oleksandr Tkachenko, the Ukrainian Culture Minister. "If they steal our heritage, they believe that we won't continue to live and to create. But we will."[164]

The next major stage of the war would once again demonstrate how the Ukrainians were clinging on tenaciously, while still Putin was not evolving Russian strategy, and Russian tactics were not evolving. In the seven-month battle for Bakhmut, the Russians by March 2023 (especially the Wagner Group pushing westwards) took heavy losses, impaling themselves on the town.[165] It was reported by the *Economist* that the Russians had lost 1,200 men in a single day there.[166] Meanwhile the Ukrainians were training multiple new brigades in Britain, France, Germany, Poland and elsewhere for a much anticipated summer offensive, with an emphasis on combined-arms operations incorporating Western weapons systems.[167]

There are many historical examples of particular strongpoints, both in the offensive and the defensive, so fixating commanders to the point that they lose perspective on the bigger picture, as the Russians did in their focus on Bakhmut. Napoleon devoted far too many men to the (failed) attempt to capture the Hougoumont chateau during the battle of Waterloo, for example, and, on a far greater scale, taking the city of Stalingrad obsessed both Hitler and Stalin far beyond its actual strategic value. At Guadalcanal in 1942–3, the Japanese kept pouring irreplaceable forces into an exhausting six-month battle over control of a place that held no essential importance, while the story offstage was that of the United States mobilizing its continent-wide resources that would ultimately overwhelm Japan. Strategists always need to retain a sense of perspective.

One year after the invasion, neither side was giving casualty figures, but British Ministry of Defence officials estimated that 180,000 Russians had been killed or wounded, and between 80,000 and 100,000 Ukrainians.[168] Other estimates give 65,000 Russian dead, which would be consistent with the overall casualty numbers.[169] Russia had bombed Ukrainian targets 41,500 times and destroyed 152,000 residential buildings in the country.[170]

Volodymyr Zelensky's performance of his leadership role could hardly be more different from that of President Putin. When Putin got the big overall strategic picture catastrophically wrong in thinking that he could take over a vast modern country by *force majeure* in seventy-two hours, Zelensky understood that with courageous, determined and intelligent resistance – and with huge support from the US and Western nations – Ukraine could retain its sovereignty.[171]

Putin told his generals of the invasion only hours before it happened, and has consistently lied to his people about the nature of it; he also believed he could communicate effectively by giving orders from the end of a ludicrously long white table and making occasional television broadcasts. Zelensky, by contrast, spoke to his people nightly from his capital by video link, sometimes from its streets, and addressed world forums using powerful language understood by all. Clearly, just as it had been for Ronald Reagan, Zelensky's training as an actor was a great asset. Moreover, he spoke truthfully and bluntly and largely eschewed propaganda. As US forces had tried to do in Iraq during the Surge, he sought to "be first with the truth" and avoid spin.

Another critical task of strategic leadership is, of course, to drive the implementation of the big strategic ideas, and here too the two leaders diverged remarkably. Putin sacked generals, ministers and especially intelligence chiefs – there were three commanders-in-chief in Ukraine within twelve months, and more since – yet still the strategy could not be implemented and major offensives failed because it was fundamentally flawed. Zelensky, who has also dismissed a few of his underperforming lieutenants, has reinforced success and learned on the job, and pulled off some extraordinary strategic coups such as luring the Russians southwards before delivering the killer blow around Kharkiv in the north. Moreover, he has provided an admirable example of personal courage, in addition to his impressive energy and indomitable will.

Whereas the Russians have signally failed to revise or refine their strategy or tactics effectively – on occasion even adopting the same methods of attack ten times in succession despite repeated setbacks, as in Chornobaivka – Zelensky and his team have ensured that Ukrainian tactics have been flexible, smart, nimble and unpredictable. The Ukrainian government and its armed forces have been "learning organizations" – and in prolonged wars the side that learns and adapts the fastest often prevails. (One major area where the Ukrainian Air Force will be forced to adapt further is in transitioning from Eastern Bloc aircraft such as MiG-29s to Western ones such as F-16s, not least because the supply of MiGs in Europe is limited.)[172]

There have been innumerable reasons why the once-feared Russian Army has so consistently underperformed in Ukraine since 24 February 2022. Any list of them must include a vast overestimation of Russian capabilities and considerable underestimation of Ukrainian ones; the lack of unity of command; a campaign design so ambitious that it exceeded the ability of the theatre commanders to follow it; failure to achieve combined-arms effects (employing armor, infantry, engineers, mortars and artillery together); wholly inadequate training (what were they *doing* for all those months when they were deployed on the northern, eastern and southern borders of Ukraine?); the employment of massed twentieth-century-style armored formations inadequately supported by other arms; the lack of a professional non-commissioned officer corps, with, instead, a top-down command system that does not promote initiative at lower levels; a vastly inadequate logistics structure to support the forces deployed once the rail system was not available (a major organizational design shortcoming); poor equipment (exemplified by turrets blowing off tanks when fires ignite in them); a culture that condones war crimes and the abuse of local populations; poor command and control by headquarters at many levels; a seeming lack of standards for performance of the most basic of tasks (such as a failure to stay dispersed when moving and halted); using analogue, unencrypted, single-channel easily jammed communication systems (a major reason why so many generals and colonels were killed); relying on conscripts for 20–25 percent of the forces; inadequate equipment (night-vision goggles, armored vehicles, close combat optics, internal communications and so on), and a pervasive corruption that undermines every aspect of their forces.[173] That list might read as if it is exhaustive, but it is not.

Of course there could be plenty more offensives to come in this terrible war, in which the Russian Army in Ukraine, since January 2023 led by General Gerasimov himself, might learn from their multiple failings and errors.[174] Yet, although Vladimir Putin has had thousands of accomplices, he alone is ultimately to blame for launching the most consequential war of recent times, the first year of which will be taught in staff colleges around the world for decades to come as providing a masterclass in how *not* to fight a war. Instructors in those colleges might well conclude that, in the words of the retired American Army colonel Joel Rayburn, "A bad army was ordered to do something stupid."[175]

Ten

The Wars of the Future

Wars are not just contests of weapons and will; they are also laboratories of a sort.

Peter Warren Singer, 2023[1]

Trying to predict the future of warfare is notoriously difficult. General Giulio Douhet commanded the Italian Air Force in the First World War, an experience that led him to predict in the 1920s that armies and navies would be rendered obsolete, because all future wars would be fought and won by aerial bombing alone. Needless to say, that assessment proved wrong, even though airpower did come to play an increasingly important role.

Although the historian Margaret MacMillan has likened predicting the outcome of future wars to "betting on horses or guessing where new technology is going," there have of course been innumerable attempts at glimpsing at least the outlying contours of forthcoming conflict.[2]

Few have proved accurate, and Major General Bob Scales, a respected thinker and writer who served as commandant of the US Army War College, once described such attempts as "the least successful enterprise in Washington DC."[3] Despite that caveat, having examined the last seven decades of warfare in these pages, we feel compelled to venture some observations and reflections on some of the potential developments that lie ahead.

Technological and strategic developments in warfare are ever evolving: the previous nine chapters of this book bear eloquent witness to that, especially the last. As with so much else in human affairs, change is also accelerating exponentially. Arsenals are always acquiring new weapons and moving into new dimensions.[4] By the end of the First World War, warfare spanned all three of the earthly elements, and by the end of the Second World War capabilities in the air and under the sea were proving decisive in widely different theatres. Today, emerging technology leaves us reasonably confident that future wars will be fought in even more dimensions – space and cyberspace among them – some of which are in their infancy but are now becoming increasingly important and, indeed, central to modern conflict.

"The lines between war and peace can blur into near-irrelevance," political scientist Mark Galeotti has observed, "and 'victory' just means today was a good day, with no guarantees for what may happen tomorrow."[5] Economic sanctions, targeted assassinations and political-influence campaigns are becoming more common, more alluring and, in some cases, less detectable than actions on the usual battlegrounds of land, sea, air and space. Perpetual low-level conflict masked by plausible deniability both de-intensifies war and makes it more ubiquitous. Disconnected from traditional battlefields, such actions comprise a form of warfare where computers, currencies and public opinion become primary battlegrounds, with little definitive indication of final success or failure. This has been termed "the weaponization of everything," and we can assume it to be a growing phenomenon.[6]

Conflict is thus likely to diminish in clear, geographical scope – though that will still be the major focus of the great powers – while simultaneously reaching further than ever before. In a world in which everyone has a computer in their pocket, almost all of which are controlled by just a few companies, espionage, corporate sabotage, subversion, cyberattacks and online disinformation can each become highly effective methods of future struggle.[7] Yet none of this means that advanced states will not continue to need well-funded, sizeable militaries with cutting-edge technology.

Encouraged by swift and significant developments in civilian technology, militaries have found themselves testing a whole range of technologies previously seen as belonging to the realm of science fiction. Generals John R. Allen and F. Ben Hodges note that these currently include:

> Artificial Intelligence, deep learning, natural language processing, computer vision, and other related characteristics: super-computing, eventually quantum-computing, and nano and bio technologies; advanced big-data analytics; and other emerging technologies are beginning to offer an entirely new way of war, and at command speeds hitherto unimaginable. The revolution in sensor and command and control technologies is matched and enabled by developments in long-range, hypersonic "intelligent" weaponry and new swarms of killing machines allied to a range of directed-energy weapons. Such a potentially revolutionary change in the character and conduct of war must necessarily impose entirely new ways of defense.[8]

In wars of the twentieth century, militaries pioneered technological advances. Areas such as computing, cryptography, radar, plastic surgery and the internet would have taken far longer to emerge had it not been for early military funding. Today, something of the reverse is happening. Having spent so much time on traditional forms of warfare, and sinking vast amounts of money into them, Western armies now find themselves ever more reliant on civilian developments in robotics and increasingly autonomous systems to illuminate their blind spots and provide the means for identifying and responding to such threats.

In particular, data analytics and now artificial intelligence enable the processing and analysis of all types of intelligence, as well as their fusion, and also enable increasingly lethal, precise and capable weapons systems and autonomous capabilities. A related, enabling development is the advent of more and more powerful big-data analytics, computational capability (using cloud computing), quantum computing and other capabilities that enable the rapid analysis of enormous, dissimilar data sets and can pull needles – such as locations of high-value targets, major weapons systems and platforms – out of digital haystacks.

This is particularly significant given the proliferation of surveillance capabilities of all types in all domains. So reliant is the military on advances in the civilian-consumer sphere that Allen and Hodges call for "a new strategic public–private partnership in which technology plays an ever-larger role in driving security and defense policy and strategy." They are particularly blunt about how "all the strategic and political assumptions underlying Europe's current defense effort must change."[9]

In 2017, the US Army prioritized six new conventional capabilities for future large-scale combat operations; namely long-range precision fires, the next-generation combat vehicle, future vertical lift, the Army network, air and missile defense, and initiatives to increase the lethality of individual soldiers.[10] In August 2018, a new four-star Futures Command was established in Austin, Texas, to oversee the Army's acquisitions and modernization program, just one of many such initiatives in the US military and beyond intended to deepen the relationships between the military services and those in industries on the cutting edge of capabilities with military potential and relevance. More can be anticipated as each US military service strives to foster innovation and relationships with advanced technology companies.

Hybrid warfare

The term "hybrid warfare" has many soubriquets: grey-zone war, asymmetric warfare, non-linear warfare – the terminology changes almost as much as the means to fight it. While it may be a recent and somewhat ill-defined term, the concept is nothing new. It means the use of unconventional or asymmetric methods of war that stretch across subjects, arenas and domains of warfare, and we have seen this in earlier chapters in theatres as far removed as Kashmir, Borneo, Dhofar, El Salvador and pre-2022 Ukraine. Each side aims to disrupt its opponent's efforts while maintaining an air of deniability, existing just below open hostilities and within the "grey zone" of covert action. Proxies, economic warfare, targeted disinformation, political manipulation and covert military action – all fall short of traditional forms of conflict, and come under the umbrella terminology of "hybrid war."

The phrase "hybrid war" was popularized in a 2005 paper in the *Naval Institute Proceedings*, authored by then-Lieutenant General James Mattis and Lieutenant Colonel Frank Hoffman.[11] It included a discussion of the concept of a "three-block war" that had been put forward by US Marine Corps Commandant General Charles C. Krulak in 1997, in which US forces in the future would have to conduct military operations against an enemy, conduct peacekeeping operations and also provide humanitarian assistance to refugees "all on the same day" and "all within three city blocks."[12] In other words, Marines and other forces would need to possess versatility, adaptability and varied operational capabilities.

Writing eight years later, in the light of the Afghan and Iraq campaigns, Mattis and Hoffman added a discussion of "the psychological or information aspects" which turned it into a "four-block war." The term "hybrid war" soon came to include the full spectrum, comprising terrorism, insurgency, criminality, conventional operations and disinformation campaigns.[13] To take just one example of this, the criminal violence in Mexico today counts as a conflict, as it seriously undermines the elements of the rule of law: the police, security forces, judiciary and prison system. And the criminal activity in Mexico is waged in virtually all domains of warfare, including the employment of increasingly heavy weaponry.

In 2013, eight years after Mattis and Hoffman's article, Valery Gerasimov, the Russian Chief of the General Staff, set out what Mark Galeotti has dubbed the "Gerasimov Doctrine," which emphasized the way that political, economic, humanitarian and informational warfare could work in tandem with military power.[14] "The role of non-military means of achieving political and strategic goals has grown, and, in many cases, they have exceeded the power of force of weapons in their effectiveness," stated Gerasimov, who at the time of writing is Commander-in-Chief of the Russian forces in Ukraine, as well as Chief of the General Staff.[15]

This "non-linear" form of psychological subversion ("psyops") was seen during the annexation of Crimea in 2014, with the aim of reducing Ukraine to a "web of chaos, humanitarian disaster and civil war."[16] As we have seen so graphically, however, hybrid cold war of the Gerasimov

Doctrine can easily turn into the hot war that has been fought there since 24 February 2022, yet again justifying the need for well-equipped militaries. The results of the Russian invasion of Ukraine suggest that Gerasimov was not quite the infallible military genius he had been assessed to be after Crimea and Russia's brutal support for the Bashar al-Assad regime in Syria.

Hybrid warfare particularly appeals to China and Russia, since they are much more able to control the information their populaces receive than are their Western adversaries. A 1999 book, *Unrestricted Warfare*, written by two People's Liberation Army colonels suggests that militarily, technologically and economically weaker states can use unorthodox forms of warfare to defeat a materially superior enemy – and clearly they had the United States and NATO in mind. Rather than focusing on direct military confrontation, the weaker state might succeed against the dominant opponent by shifting the arena of conflict into economic, terrorist and even legal avenues as leverage to be used to undercut more traditional means of warfare. The subtitle of their book, *Two Air Force Senior Colonels on Scenarios for War and the Operational Art in an Era of Globalization*, notes a core truth of the early twenty-first century: an increasingly globalized world deepens reliance upon, and the interdependence of, nations, which in turn can be used as leverage to exploit, undermine and sabotage a dominant power.

The two colonels might not be happy with the lesson their book teaches Westerners, which is that no superpower can afford to be isolationist. One way to keep America great, therefore, is to stay firmly plugged into – and leading – the international system, as it has generally done impressively in leading the Western world's response to the invasion of Ukraine. The siren voices of American isolationism inevitably lead to a weaker United States.

Another practitioner of hybrid warfare theory is Iran. In the late twentieth century, Iran chose to pursue an active form of hybrid warfare against the United States and its allies. It recognized that it could not hope to prevail against the West using traditional means; indeed, in the 1990s it had seen the US-led coalition demolish Iraqi forces that had themselves largely defeated Iran three years earlier.[17] Yet the Iranian leadership also recognized that the United States had imposed specific limits on when it could and would use outright military force.

Iran's asymmetric strategy thus focused on attacking America in those areas that would not trigger the United States to engage in a direct military response against it: primarily diplomatic, propaganda and financial subversion, and support for terrorist proxies.[18] Through a wide range of non-traditional means, Iran tried to straddle a delicate line, inflicting enough punishment to encourage the United States to pull its forces out of the Arabian Gulf, but never so much as to invite direct retaliation.[19] It was happy to wound, but careful not to strike too overtly.

In the final years of the twentieth century and at the beginning of the twenty-first, Iran, Russia and China learned from each other's asymmetric, hybrid war theories and practices to disrupt and weaken conventional Western powers. China has done this while rapidly modernizing its military in recent decades, forcing American planners to keep developing new conventional concepts and capabilities, and to spend huge amounts of money on weaponry and high-end platforms such as high-technology carrier strike groups, fifth-generation fighter aircraft and stealth drones – and, now, unmanned platforms in all domains. This must be done even though, as counter-insurgency expert David Kilcullen argues,

> the decisive action may be taking place elsewhere, in a domain that we do not consider to be warfare, through the manipulation of technology transfer, the leveraging of cyberwarfare by civilian actors, control of key mineral resources, or the purchase of strategic real estate. Simultaneously, out on the fringes, China has made swift and significant advances in fields (such as 5G mobile telecommunications, cyber operations, nanotechnology, artificial intelligence, robotics, human performance enhancement, quantum computing, genomics and biotech political warfare, as well as financial manipulation) that lie outside the ken of Western war-fighters and thus invoke limited direct military competition.[20]

Robotics and Artificial Intelligence (AI)

For over fourteen years, Israel had wanted to assassinate Mohsen Fakhrizadeh. In 2004, the Israeli government had issued Mossad with carte blanche to find ways of preventing Iran from developing its own nuclear arsenal; following that, Israel's intelligence service had reportedly been diligently thinning out the number of Iran's best nuclear scientists. Mossad believed, with good reason, that the gentle-looking Fakhrizadeh, acknowledged in public simply as a professor of nuclear physics, was in fact the lynchpin of Iran's efforts to create its own nuclear weapons, which the mullahs who rule Iran had often and unmistakably boasted would one day be used to turn the Jewish state into a pile of radioactive rubble.

Fakhrizadeh had survived previous assassination attempts, but in late 2020 the time had come for Mossad to test out a new technology. On the night of 27 November, as Fakhrizadeh travelled back to Teheran from his holiday home near the Caspian Sea protected by armed guards, his chauffeur-driven car passed a parked pickup truck on an empty stretch of road. The truck had a tarpaulin cover that hid a machine gun. As Fakhrizadeh's car drove past, the gun was able to achieve a good "visual" on the physicist, ascertain his identity and fire thirteen rounds, mortally wounding him.

News reports over the following days revealed that Fakhrizadeh had been killed by a remotely controlled weapon. The tarpaulin had hidden a gun, but no human being. The equipment needed for the operation weighed around a ton, necessitating each of its components to be broken down and smuggled into the country. After killing Fakhrizadeh, the truck, which was packed with explosives, was then detonated, but enough equipment survived for the Iranian Revolutionary Guard to piece together the fact that Iran's top nuclear scientist had been killed by a remote-controlled robotic machine gun.[21]

The ability of increasingly sophisticated computers to assess vast amounts of data means that AI and robotic-directed assassinations may well proliferate over the course of this century. Human beings are able to use AI to comb through photographic data in order to locate anything from individual park benches to nuclear facilities. AI also increases the power of states to, in the words of Henry Kissinger, "deploy machines and systems employing rapid logic and emergent and evolving behavior to attack, defend, survey, [and] spread disinformation," but also to "identify and disable one another's AI."[22] In an area where warfare will almost certainly evolve beyond all recognition, the United States and its allies must be in the forefront. AI has already been used in the Russo-Ukrainian War in areas as wide as facial-recognition software, making military supply chains more efficient and producing "deepfake" videos. We can be certain that it will continue to evolve in these areas and others.[23]

The great majority of robotics and AI research is civilian. From NASA to domestic vacuum cleaners, non-military organizations lead the technological development; however, it was inevitable that their technology would be adapted to the battlefield as well. Peter Warren Singer, who worked in the Pentagon, State Department and CIA, wrote in his 2009 book *Wired for War* that:

> Robotics in Iraq and Afghanistan today are sketching out the contours of what bodes to be a historic revolution in warfare. The wars of the future will feature robots of a wide variety of size, design, capabilities, autonomy, and intelligence. The plans, strategies, and tactics used in these future conflicts will be built from new doctrines that are just now being created, potentially involving everything from robotic motherships and swarms of autonomous drones to cubicle warriors managing war from a distance ... In these battles, machines will take on greater roles, not just in executing missions, but also in planning them ... Our robotic creations are creating new dimensions and dynamics for our human wars and politics that we are only just beginning to fathom.[24]

Wired for War prompted the military historian Max Boot to ask, "Will wars someday be fought with Terminator-like machines?" At present, it certainly looks that way. Unmanned ground vehicles are already efficient and helpful tools; they are used for de-mining and searching for and destroying roadside IEDs. The American V60 Q-UGV is a robot that can cross terrain and sense, scan and communicate a wide range of environmental factors. AI can, in Jeremy Black's words, "link, sense, scan and fire technologies and systems at a speed and sophistication that humans are incapable of matching."[25] This is a particularly significant observation, as it has considerable implications for keeping humans in the loop in future tactical operations.

Somewhat paradoxically, unmanned remote-controlled systems require heavy manning. The US Air Force estimates that keeping a single Predator UAV orbit in the air non-stop needs a crew of 168 military personnel to maintain, arm, fuel, fix and fly the vehicle on station and flying to and from the location of the "unblinking eye," as well as to process, exploit, analyze, store and disseminate the intelligence it gathers. The larger Reaper 180 and Global Hawk unmanned aerial vehicles require as many as 300 people per orbit. "The number one manning problem in [the US] Air Force," states one of its generals, "is manning our unmanned platforms."[26]

Yet this would not be the case with autonomous robots, where humans would be "on the loop" monitoring the robot, but not "in the loop" controlling it, though many tasks beyond flying and payload operations would still need to be performed by humans. Over time, the humans in the loop are likely to become those who write the code and algorithms to guide the system. The humans on the loop monitoring its operations are those who permit it to take lethal action when certain conditions are met, with the determination of those conditions employing other advanced capabilities enabled by machine learning and AI.

The "PackBot," an agile and hardy robot first developed in the late 1990s, was used in the aftermath of the 9/11 attacks to help rescue teams at Ground Zero. Weighing 42 pounds and the size of a lawnmower, it would soon be employed in the wars spawned by that atrocity. The first battlefield robots to be used by the US Army were employed in scouting out booby-trapped cave complexes in Afghanistan, before being used to defuse IEDs in Afghanistan and Iraq. In 2006 there were 2,500 IED attacks in Iraq per month, and as PackBots began to proliferate, the insurgents began offering bounties of $50,000 for the destruction of a PackBot.[27]

Depending on how much explosive has been packed into an IED, a soldier on foot has to be roughly 50 yards away to escape death from its detonation. The PackBot was a much easier loss to bear. "When a robot dies," one bomb disposal chief observed, "you don't have to write a letter to its mother."[28] It was fitted with cameras and sensors, and moved on four tank treads known as "flippers" due to their ability to flip up and down to climb stairs. By 2011 there were no fewer than twenty-two different robot systems operating on the ground in the US Army.[29]

Many more have been introduced in the intervening decade. Current robots include the Marcbot (multifunction agile remote-controlled robot) – a device able to search for enemies and bombs under cars – and Swords (special weapons observation reconnaissance detection system) on which a variety of weaponry can be mounted. American Special Forces have already performed tests in which they have worked alongside Robotic Combat Vehicles to best understand how robots and humans can operate together. In the autumn of 2016, the US Army conducted its first force-on-force training cycle that featured a robot. The University of Manchester and the Ministry of Defence in the United Kingdom recently experimented with a "Project Origin" vehicle to perform primarily defensive actions, such as obstructing a helicopter landing zone.[30]

An industry cliché has it that humans employ robots to do the three Ds – work that is dull, dirty and dangerous. Over a decade ago, robots were able to detect mines more accurately than a human about one-fifth of the time; since then, they have improved enormously.[31] Robots can also operate in all weathers, as well as during chemical and biological attacks, and can take cover when their radar detects incoming fire. Moreover, as an official at the Defense Advanced Research Projects Agency has pointed out, because pilots black out at extreme gravitational forces, "the human is becoming the weak link in defense systems."[32]

Once robots fight robots, one US Army colonel predicts, "there won't be any time ... for humans."[33] Peter Warren Singer agrees, arguing that "A soldier who learns French or marksmanship cannot easily pass that knowledge on to other soldiers. Computers have faster learning curves. They not only speak the same language, but can be connected directly via a wire or a network, which means they have sharable intelligence."[34] He envisages that the dawn of the robot soldier represents nothing less than the end of the human monopoly on war.[35]

The Pentagon already distinguishes between what it calls "self-directed" and "self-deciding" systems in its *Unmanned System Integrated Roadmap*: "self-directed" systems are semi-autonomous and pre-programmed to perform actions independently. The jet-powered Global Hawk UAV, which can stay aloft for thirty-five hours and fly as high as 60,000 feet, can already operate independent of human control, albeit within parameters set by humans. By contrast, a "self-deciding" system is a fully autonomous robot which responds on its own to sensors and does not have a pre-programmed response, but rather seeks "optimal solutions in unforeseen situations" by acting like a human brain.[36] Once again the challenges of keeping a human in the loop with such systems, especially if in operations against an adversary's autonomous systems, become obvious.

While we currently witness remote-controlled, robot-on-human warfare, the political scientist Dr. Elinor Sloan predicts that soon we shall see robot-on-robot warfare, with battles fought in the skies between unmanned combat aerial vehicles.[37] Fighter jets piloted by AI already tend to beat human pilots in simulated dogfights, not least because they can near-instantaneously "identify patterns of conduct that even an adversary did not plan or notice, then recommend methods to counteract them."

Any states that want a modern military must embrace AI, and, as Kissinger and Eric Schmidt argue, "The solution to these complexities is neither to despair nor disarm."[38] An arms race in AI between China and the United States is already well under way, with each aiming to develop it for tactical – and even strategic – purposes. The dawn of AI brings the world to what Kissinger and Schmidt call "the threshold of a strategic transformation as consequential as the advent of nuclear weapons – but with effects that will be more diverse, diffuse, and unpredictable."[39]

Dr. Sloan has summed up the ethically positive aspect of robots on the future battlefield, pointing out that they "will not carry out revenge attacks on civilians, commit rape, or panic in the heat of battle. They do not have human emotions like fear, anger, and guilt, which may lead to war crimes, and they are not constrained by desire for self-preservation."[40] It has fallen to Lawrence Freedman to set out a useful corrective, pointing out what robots *cannot* do. In a sober synthesis of robotic potential and its realistic limitations, he argues that:

> Set against a human commander, an AI commander will still have many drawbacks. Machines cannot lead by their brave example, or explain the rationales of their decisions to humans. They have no moral sense or personal fear. They might be able to manage multiple tasks, but would struggle to switch to a wholly new endeavor. Moreover, because AI is about machine learning, situations with many new features might spook the AI commander ... Unlike a human commander, an AI commander might fail to attach significance to a few odd points in the incoming data that would attract human curiosity ... AI might be tactically brilliant but it is strategically banal.[41]

None of that means that AI does not represent the future of war, however. The US Air Force has undertaken plane flights and radar systems operations without human overrides. Unresponsive to human logic, such systems are not affected by long-established norms, signals or misdirections. The AI-flown warplane knows "only its instructions and objectives, not morale or doubt ... When two AI weapons systems are deployed against each other, neither side is likely to have a precise understanding of the results their interaction will generate or their collateral effects."[42] Any future conflict might thus be liable to escalation that is highly unpredictable.

This is, of course, most dangerous in the nuclear sphere. AI strategist Kenneth Payne notes "intriguing parallels" between the developments of artificial intelligence and nuclear strategy. Both are "offset" developments that can make up for the disadvantages in conventional weaponry. In the nineteenth century, Clausewitz believed that war was limited "both by the capabilities available to the belligerents and by their desire to pursue particular goals through violence"; however, as Payne points out, in the twenty-first century, "War involving AI need not be subject to the same limitations. If the AI miscalculates what its human principals want, there is tremendous potential for catastrophic and unwarranted violence."[43] A robot intent solely on defeating an enemy, therefore, might blow up the world in the process.

For all that this might sound like the stuff of science fiction, it is currently evolving, and rapidly. Delegating autonomy to machines saves all-important time in active conflict. Today, more than thirty countries have robot autonomized weapons for situations for which the speed of engagement is simply too fast for the human mind to respond to in real time.[44] Sixteen nations operate armed drones. The Israeli Harpy drone searches areas for enemy radars which it can then destroy without requesting human permission, something which the Chinese have already reverse-engineered.[45] Such systems typically have very clear, machine-verifiable criteria that the machine has to meet before taking action. For over a decade, South Korea has employed a heat-detecting robot sentry gun facing North Korea across the demilitarized zone to dissuade North Korean infiltration.[46]

Attempts to stem the more morally dubious elements of technological development often rest on the tender consciences of individual states, which may not prove substantial – and non-state actors might have even fewer qualms. Russia is developing armed ground robots to lessen its human casualties in any future war against NATO, though their failure to be deployed in Ukraine suggests that they are not yet ready, and General Gerasimov has stated that in the near future a "fully robotized unit will be created, capable of independently conducting military operations."[47] Keeping in mind that one of the few factors limiting Putin in Ukraine is the heavy losses that Russia is currently sustaining, one can only imagine what might happen if it were mainly robots who were bearing the brunt of action instead of flesh-and-blood fathers, brothers and sons.

For the advantages robots enjoy on the battlefield are immense: they can be faster, smaller (or larger), lighter (or heavier) and more dexterous than humans. And they can have far greater resilience, require no sleep, can cross any terrain and can take part in suicide missions that rest on the conscience of neither themselves nor their operators. The idea that they will not be employed by countries out of moral considerations is ludicrous; rather than presenting a moral Rubicon, many nations consider robotics a moral boon. That some states, fearful of the moral compromises involved, may not use or develop robotics is similarly absurd – especially in the light of history, which shows that humans have tended to grasp at the most lethal weapons possible, especially in dire situations. Moreover, in future wars the need to try to jam or "fry" the electronics of enemy robots will mean that the electromagnetic spectrum will be even more contested than it is already. Of course, many of the systems contesting that spectrum will be robotic, as well.[48] On Russia's Knowledge Day in 2017, Vladimir Putin told Russian school-children that "Whoever becomes the leader in [AI] will become the ruler of the world."[49] In this, if little else, he may well be right.

Nuclear weapons

President Kennedy predicted that sixteen states would have achieved nuclear arsenals by the end of the 1960s; however, today there are only nine: the United States, Russia, the United Kingdom, France, China, Israel, India, Pakistan and North Korea. His prediction would have been closer had not South Africa, Belarus, Kazakhstan and Ukraine acquired and then disposed of their own nuclear capabilities.

Nuclear threats interwove the major conflicts of the second half of the twentieth century, from Douglas MacArthur's request to use tacti-cal nuclear weapons during the Korean War to the Cold War nadir of the Cuban missile crisis. More recently, Vladimir Putin has elliptically threatened the use of tactical nuclear weapons in 2014 and 2022. Implied nuclear threats between India and Pakistan have been a contin-ual source of mutual antagonism since 1974, and the advent of tactical nuclear weapons in Pakistan makes that situation even more delicate.

Despite that, easily the most serious present-day threat to the deli-cate balance of nuclear non-proliferation is Iran. In addition to its asymmetric or hybrid approach to warfare, Iran's concurrent response to its own conventional military inferiority has been a longstanding interest in acquiring nuclear weapons. Ayatollah Khomeini had originally been uninterested in Iran's nascent nuclear program and had begun to close it down, but after the devastation of the Iran–Iraq War, his military aides convinced him that developing nuclear weapons would provide security against Iraq, Israel and the United States.[50] Had Iran already possessed the bomb, they argued, it would have prevented the Iran–Iraq War entirely.[51] The United States needs to stay at the forefront of efforts to prevent Iran acquiring a nuclear bomb, which would both pose a threat in itself and serve as an incentive for several of its enemies in the region to acquire one too.

When a possible future invasion of Taiwan by the Chinese is wargamed at high level by the Center for a New American Security, the confrontation between China and the United States escalates into the nuclear sphere very quickly and disturbingly.[52] This seems to reflect Lawrence Freedman's observation that much in future nuclear warfare will depend on "the likely passions raised by the preceding conventional campaigns, in which many would have already died."[53] The desire for vengeance is one of the most basic human emotions and one of the most mutually destructive, but, as held true during the twentieth century, hopefully the sheer ubiquity of the devastation caused by nuclear weaponry might give pause to even the most bloodthirsty of leaders. "Since the dawn of the nuclear age we have progressively acquired the capacity to call the ultimate moment of truth, and we are not gods," stated Father Bryan Hehir, the Dean of Harvard Divinity School, in 1987. "But we must live with what we have created."[54]

Disinformation

There are currently five widely recognized dominions of warfare – land, sea, air, cyber and space – but it appears that a sixth should be added, namely information, which is more important now than ever before. "A great part of the information obtained in war is contradictory, a still greater part is false, and by far the greatest part is uncertain," wrote Carl von Clausewitz in the aftermath of the Napoleonic Wars. Two hundred years later, that observation is still true, and still pertinent. From Stalin's *desinformatsiya* campaigns to Sergei Lavrov's 2022 insistence that Russia "did not attack Ukraine," disinformation has been a crucial component of modern warfare that shows no sign of lessening.[55]

In 1988, just before the fall of the Berlin Wall, the East German Ministry of State Security (the "Stasi") employed 90,000 people directly and had a further 175,000 "unofficial collaborators," or informers, on its books. Some were the teachers, landlords, employers and even parents and spouses of targets. Its files alone cover 70 miles of archival shelves in fourteen locations serviced by 1,400 people. The Stasi was exceptionally harsh, ideologically unbending and thorough. According to one history, it "even collected samples of its enemies' body odour from chairs and sofas on which unsuspecting victims had been sitting."[56] One of its principal tasks was to sow conspiracy theories in the West that were detrimental to its ideological foes.

The methods of promoting disinformation changed over time in Russia as the KGB First Chief Directorate became the SVR.[57] Much like the various terms for hybrid warfare, however, it retained the same core characteristics. The internet was to give the phenomenon of disinformation an entirely new lease of life in the 1990s, as the historian Thomas Rid has observed, although the level of craftsmanship by disinformation specialists fell significantly:

Digital storage made it possible to breach targets remotely and extract vast amounts of compromising material. The internet facilitated acquiring and publishing unprecedented volumes of raw files at distance and anonymously. Automation helped to create and amplify fake personas and content, to destroy data, and to disrupt. Speed meant that operational adaptation and adjustments could take place not over years, months or weeks – but in days, hours, even minutes ... And the darker, more depraved corners of the internet offered teeming petri dishes of vicious, divisive ideas, and guaranteed a permanent supply of fresh conspiracy theories.[58]

Genuine and legitimate leaks can mask those planted by hostile actors, who can also plant disinformation. An FSB hack into the Ukrainian Central Electoral Commission's network in May 2014, for example, managed to invent a chart that suggested that the far right was likely to win the upcoming election, which was then reported on Channel One, Russia's most watched TV station.[59] The FSB hoped that this would confirm in Russian minds the truth of their propaganda that the Ukrainian government was neo-fascist – a line they continued to push to justify their invasion in 2022 as an effort to de-Nazify Ukraine. As Rid writes, such measures are designed "to erode the capacity in an open society for fact-based, sober debate, and thus wear down the norms and institutions that resolve internal conflict peacefully."[60]

With the advent of deepfakes and video footage that can be altered by AI, seeing should no longer equate with believing.[61] It is not hard to create a deepfake; the technology can create original images from the ether.[62] One such was "Katie Jones," a young woman with professional connections to the Center for Strategic and International Studies and the White House. She had impeccable professional standing, ability and competence, but simply did not exist.[63]

China in particular uses networking websites for espionage purposes. "Instead of dispatching spies to some parking garage in the U.S. to recruit a target," states William Evanina, Director of the US National Counterintelligence and Security Center, "it's more efficient to sit behind a computer in Shanghai and send out friend requests to thirty thousand targets."[64] In the spring of 2019, a retired CIA officer was sentenced to twenty years in prison for selling classified information to a Chinese intelligence agent who had contacted him while posing as a recruiter on LinkedIn.[65] So common is this use of deepfake-illustrated profiles used to contact Western targets with access to sensitive defense information that NATO intelligence agencies have publicly warned of the tactic.

As well as videos, audio deepfakes can mirror the tones, accents and intonations of public figures, since most anti-hacking software focuses on traditional fraudulent activity and cannot easily detect computer-generated voice mimicry. Politicians and military leaders might be opening themselves up to audio mimicry every time they speak in public, yet in the words of one cybersecurity expert, "You can't go around and be silent the entire time. You're going to run into situations like this where you expose information that you never thought could be used against you."[66] Efforts are always under way to detect such technology, and there will always be an arms race between detection and generation.[67]

In 2019, some 42 percent of the global population regularly used social media in some form or other, so there is an enormous number of people who are vulnerable to such misrepresentations and disinformation.[68] There is also an information crisis presently haunting democracies; one MIT study has shown how it takes truth about six times as long as falsehood to reach 1,500 people, and that disinformation was 70 percent more likely to be shared on social media than a piece of news that was true.[69] Once again, advances in human nature lag far behind advances in technology.

Operation Denver, a KGB disinformation campaign of the 1980s focused on blaming the AIDS virus on a CIA-funded biological laboratory, was a significant success. In a telegram sent to its allied Bulgarian secret service, the KGB reported that:

> We are conducting a series of measures in connection with the appearance in recent years in the USA of a new and dangerous disease, Acquired Immune Deficiency Syndrome ... and its subsequent, large-scale spread to other countries, including those in Western Europe. The goal of these measures is to create a favorable opinion for us abroad that this disease is the result of secret experiments with a new type of biological weapon by the secret services of the USA and the Pentagon that spun out of control.[70]

The KGB then leaked false information to sympathetic newspapers, claiming the development of the disease occurred through Pentagon operatives travelling to Zaire, Latin America and Nigeria to collect samples of deeply malignant viruses which could be used to create the

HIV supervirus, and which were secretly tested on Haitian and American homeless, addicted and homosexual people. Denver was a massive and costly undertaking that spanned several intelligence agencies and relied on traditional forms of media to pass into the public consciousness. By 1992, it had succeeded in convincing 15 percent of Americans that the virus had indeed been deliberately produced by the CIA in an American laboratory.[71]

Directly comparable is the Chinese operation undertaken to convince the world that Covid-19 was an American bioweapon. This disinformation was tweeted to Western conspiracy theory websites and only required a small fraction of the infrastructure of Operation Denver. Whereas Denver had taken months and years fully to germinate, within days of the Chinese campaign starting in 2020, some 29 percent of the American population believed that Covid had been artificially created in a laboratory outside China.[72] Social media may have revolutionized information-sharing, but it frequently fails adequately to differentiate between fact and fiction, with generally harmful effects for Western democracies.

After 298 lives had been lost when the Malaysian civilian airliner MH17 was shot down by pro-Russian separatists with a Russian-made missile in 2014, no fewer than 260 different explanations were suggested by Russia and its supporters, from the predictable accusation that it was shot down by the Ukrainians to the inventive one that each passenger had been a carefully arranged cadaver, put into a plane and deliberately shot down over Donbas as a provocation.[73] Similarly, the attempted murder of Alexei Navalny through the poisoning of his underwear with a nerve agent in 2020 was blamed on Navalny drinking too much on a flight to Tomsk.

Putin's war in Ukraine in 2022 led to a freshly galvanized series of Russian disinformation campaigns, spearheaded by his intelligence services. Ukrainian soldiers fighting in the Donbas region were targeted with cell-phone messages from supposed comrades urging them to desert or surrender, for example. Russia employed a drone-based Leer-3 electronic warfare system capable of targeting thousands of mobile connections at any given time, hoping that they could capitalize on enemy soldiers' loneliness and fear.[74] Ukrainian expertise with such technologies and operations was also used against Russian forces and citizens.

Moscow's disregard for the truth can backfire, to the point that sometimes Russian statements are disbelieved even when truthful.[75] This disregard is often so blatant that it can be immediately revealed. After an attack at a Ukrainian shopping mall in 2022 that killed around a dozen civilians, Sergei Lavrov explained that a hangar nearby which was used to store European and American ammunition and weapons had been bombed, an action which had unintentionally resulted in the mall catching fire soon afterwards. Unfortunately for this narrative, many witnesses testified that a Russian missile had indeed directly hit the shopping center, in what one Ukrainian politician described as a "predetermined, deliberate attack on civilian infrastructure knowing well enough civilian people, peaceful people would be there."[76]

Open-source intelligence

Open-source intelligence has come into its own during the Russo-Ukrainian War, overwhelmingly on Ukraine's side, where social media, commercial satellite images and cell-phone data have yielded almost as much actionable intelligence and information as the traditional sources. This marks a significant development in the evolution of warfare, reflecting the fact that there is a massive, almost overwhelming, quantity of unclassified material available online or for purchase that includes posts on social media and web forums, as well as commercially available data and imagery collected by technology firms and advertising companies.[77] There are also firms that can track all ships at sea and others that track all aircraft in the skies.

Commercial satellite imagery enabled Western defense and intelligence officials (augmenting classified imagery, but more easily releasable) to monitor and publicly report on Russia's mobilization before February 2022, and has continued to provide strategic and tactical military insight since. This has enormous consequences. It has been estimated that, by the late summer of 2022, at least thirty-two

Russian battalion tactical group command posts had been destroyed by Ukrainian military strikes that were helped in some way by these new technologies.[78] Location-tracking apps installed on Russian cell phones have also offered particularly useful insights. In April 2022, a Ukrainian whose Apple AirPod earbuds had been stolen by a Russian soldier was able to track Russian troop movements by using the "Find My" app.

Drones

Although unmanned aerial vehicles (UAVs, or drones) were originally intended for reconnaissance in the air, they were swiftly developed into lethal systems, and their maritime counterparts are now also being used underwater, combining mine and torpedo technology.[79] By 2009 there were already more than 5,300 drones being used by the US military, and virtually no mission took place without their help.[80] Drones costing as little as $3,000 have also been used to direct artillery strikes, especially effectively by the Ukrainians, and it is likely that in the Russo-Ukrainian War, just as in the First and Second World Wars, artillery will have caused more military casualties than mortars, machine guns, aerial bombing and all other kinds of weapons combined.[81] "All wars have their iconic weapons, from the AK-47 to the I.E.D.," the *New York Times* has commented. "In Ukraine, it's the drone."[82]

In his 1886 poem "Arithmetic on the Frontier," Rudyard Kipling pointed out how cheap were the weapons that could kill a British public-school-educated officer. "Two thousand pounds of education," he wrote, "Drops to a ten-rupee *jezail.*" The heavy, long Afghan *jezail* flintlock rifle proved the financial asymmetry of warfare in the nineteenth century, but the phenomenon is just as true today, when a missile from a $35,000 Next-generation Light Anti-tank Weapon (NLAW) can destroy a $2 million T-72 tank.

There are dozens of different types of drones, some of which will inevitably play a significant part in future conflict. The US Army has near-silent, pocket-sized Black Hornet drones that weigh only 1.1 ounce (32 grams) and survey surroundings to give troops situational awareness through live video transmissions. By contrast, the Predator is a 27-foot-long, $4.5 million, propeller-powered drone that can spend up to twenty-four hours as high as 26,000 feet in the air and is piloted (after locally controlled takeoff) some 7,500 miles away in Nevada. It can be armed with laser-guided Hellfire missiles, a 500 lb. precision bomb and other munitions. In the first year of the war in Afghanistan, Predators took out 115 targets there, a number that grew exponentially over the years.[83] To be sure, the experiences for the operators who fly them from bases in the United States can be jarring. "You are going to war for twelve hours, shooting weapons at targets, directing kills on enemy combatants," says one operator, "and then you get in the car, drive home and within twenty minutes you are sitting at the dinner table talking to your kids about their homework."[84]

Russia has used Iranian-made drones in Ukraine since midsummer 2022, and that October they were used to attack civilian targets in Kyiv.[85] The world's first drone war is undoubtedly a harbinger of the future, although it does not yet feature the extensive drone-on-drone engagements that will probably become ubiquitous in future conflicts.

In the last days of June 2022, a precision munition launched from a drone struck and killed a senior member of a Syrian al-Qaeda-linked group. Rather than employing a traditional explosion that might risk civilian collateral damage, the Hellfire R9X missile instead used a 100 lb. warhead with six extendable blades designed to shred just the intended target. Described by one US official as "an anvil that falls from the sky at full speed," it leaves little but the mangled remains of a car, balcony or, in this case, the motorcycle that the terrorist leader had been riding. One British newspaper reported the attack with the headline: "Motorbike-riding Al Qaeda chief is MINCED by US Hellfire Ninja Missile in Precision Drone Strike on Syrian Dirt Road."[86]

Defense against drones is thus a priority for the US. Among other initiatives, it has developed a new "proximity" airburst to protect its Stryker armored fighting vehicles against incoming drones. This ejects small munitions equipped with radio-frequency sensors able to detect drones and explode. While tests have been successful against small numbers or groups of drones, swarms of drones have a much better chance of getting through, as has been seen in Ukraine, though recent breakthroughs in directed-energy weapons by both Israeli and US companies suggest that even swarms will be vulnerable to high-powered microwave and laser weapons systems.[87]

US armed drones have operated successfully in Bosnia, Iraq, Afghanistan, Yemen, Somalia, Libya, Mali and many other countries during the halcyon period of near-total US technological supremacy. They were invulnerable in those locations in a way that they may well not be against a scientifically advanced nation in the future, especially against swarming. As noted above, it is feared that even aircraft carriers might be vulnerable to this kind of attack, an area which the Iranians are believed to be pursuing. Dr. Elinor Sloan believes that, should this swarming concept be implemented, it would mark a change of direction in the evolution of warfare, by reintroducing the importance of mass and quantity after more than a century of developments focusing on mobility and agility.[88]

Enormous investment is being concentrated on technology aimed at countering such possibilities, from hacking and jamming to directed-energy weapons that can melt or incinerate drones and their electronics.[89] One of the more significant anti-drone developments is the potential use of microwave energy, currently effective at destroying drones but imprecise at distinguishing between friend and foe. One invention, the Leonidas Pod, is a small unit that uses gallium nitride transistors to produce microwaves, and can be mounted on ground platforms or attached to larger drones capable of lifting heavy weights. When a swarm of drones emerges, the combination of a precise, nimble drone or a ground platform and a streamlined high-powered microwave system can destroy it. One engineer notes that the operator can distinguish between friendly and enemy drones, and is able to create a protective sphere around a "friendly system."[90]

In August 2018, there was an assassination attempt on President Nicolás Maduro of Venezuela using two DJI-M600 drones armed with a kilogram of explosives. It failed, wounding nearby soldiers, but future such attempts will likely become more sophisticated, more accurate and more effective. Over time, it is likely that increasing numbers of government facilities and even venues with large crowds will need to be protected with counter-drone systems.

Drones – in all domains – have many uses. They can take supplies to nuclear submarines, for example, thus prolonging their time on station.[91] Unmanned maritime vessels and aircraft give militaries much improved visibility over the oceans. By the summer of 2023, the US Navy had around a hundred small maritime and aerial surveillance drones operating from the Suez Canal to waters off the Iranian coast, feeding information to the US Fifth Fleet. Captain Michael Brasseur believes that in this crucial area the United States is "truly on the cusp of an unmanned technological revolution." Maritime drones can float at sea for up to six months, sending detailed images back to land-based military analysts. Vice Admiral Brad Cooper says that the US Navy uses them "to detect activity that we simply did not know was previously happening."[92] The drone fleet has already detected Chinese naval ships moving through Iranian waters, as well as likely illicit transfers between ships and actions by vessels using electronic trackers to disguise their identities. It is likely that undersea surveillance systems – and, over time, lethal systems – will become increasingly capable and ubiquitous.

As we saw in Chapter 9, drones that are relatively modest in capability compared to a US Reaper have become an essential tool for Ukrainian and Russian forces. Yet the capabilities we are seeing employed on the battlefield in Ukraine are really quite limited in range, surveillance capability, armament and duration compared to what may be coming.

The future will also bring increasingly precise and capable munitions for all forms of weaponry – surface-to-surface, air-to-surface, air-to-air, sub-sea to surface, sub-sea to sub-sea, cyber weaponry and so on. The advanced systems already in use by the United States and others are incomparably more capable, and also integrate imagery and other information collected by overhead systems operated in outer space, with the resulting intelligence enabled considerably by other systems that exploit AI and machine-learning capabilities. "The power of advanced algorithmic warfare systems is now so great that it equates to having tactical nuclear weapons against an adversary with only conventional ones," states one executive of a company producing this technology.

"The general public tends to underestimate this. Our adversaries no longer do."[93]

Sensors and electronic jammers

Sensors are an integral part of most major modern militaries, and they, too, are experiencing revolutionary developments in capability. The so-called "internet of things" – encompassing any objects with sensors and processing software that exchange data with others via communication networks – creates a world of new opportunities for military operations. Today's civilian sensors, which can be used by the military, are often imperfect and vulnerable to manipulation and exploitation, but they contribute an unprecedented amount of data for virtually any given area.[94]

Sensors allow soldiers to detect and hit opponents without revealing their own positions. Increasingly, forces can be detected even without movement, as infrared and thermal imaging devices proliferate on modern battlefields.[95] Soldiers and civilians alike are able to track and be tracked through data pouring in and out of battlefield technology, making it very hard to hide from anyone.

The electronic warfare undertaken in Ukraine by Russia, including the jamming of electronic equipment, has invigorated the response of some NATO armies. Electronic warfare is one of the hardest capabilities to discern on the battlefield, and Russia's electronic efforts in Ukraine imply that the days when troops could sit in a tactical operations center with perfect communications are over.[96] Ukrainians have taken to using unjammable analogue landline phones, and American servicemen have been taught to expect their own electronic communications to be jammed in future combat.

There was a Cold War adage that has even greater relevance today than it did in those times of much less sophisticated intelligence and surveillance systems – as well as much less precise and shorter-range munitions and before the advent of today's armada of unmanned and robotic weapons systems. "What can be seen can be hit," it observed, "and what can be hit can be killed." Given the advances of the past three decades, particularly the proliferation of sophisticated surveillance systems aided by computer analysis, this adage is one that has to be taken very seriously.[97] For today, virtually any significant military platform – ship, plane, fighting vehicle, base, headquarters, logistical site and assemblage of troops – can be seen in a given theatre of war. And if it can be seen, it can be hit and killed with weapons that fly at hypersonic speeds, can overwhelm defenses with capabilities that include swarming munitions and have extraordinary precision. It is this confluence of factors that often leads military leaders to observe that they must dramatically transform their forces and all aspects of them in order to ensure deterrence of the most capable potential adversaries. In future wars, expensive and important weapons systems will have to be protected from inexpensive drones and precision missiles linked to networks of sensors.[98]

Cyber warfare

In 2007, Moscow launched a cyberattack against Estonia, shutting down websites associated with its parliament, ministries and banks, and hoping to spark a run on Estonia's currency. A similar but more sophisticated attack occurred a decade later, when cyberattacks on Maersk brought the shipping conglomerate with seventy-six ports and nearly 800 ships to a halt – and, with it, a fifth of the world's shipping capacity. While the GRU (Russian military intelligence) hired smart young computer science graduates to work for it, the FSB by contrast tended to employ illegal hackers, who escaped jail by being transformed into what it termed "patriotic hackers," and who attacked Estonia in 2007, Georgia in 2008 and Ukraine almost continuously since the Russian invasion of Crimea in 2014.[99]

By the time of the 2022 invasion, Ukraine had long been a testing ground for Russian cyberattacks. With the release of the NotPetya virus on its systems in mid-2017, for example, Ukraine lost an estimated $10 billion. Russia was testing its cyber warfare tactics, with great success and very little blowback. However, those attacks and subsequent Western assistance helped Ukraine to dramatically improve its cyber defenses and resilience, as was evident when Russian cyberattacks during the invasion achieved much less disruption than Moscow had hoped to create.

In 2010, Iran became an unwilling host to the world's first digital weapon. Stuxnet, a tenacious and malicious computer worm, had been smuggled into the country and let loose upon Iran's Natanz nuclear facility. Computer viruses and worms had existed prior to this, but Stuxnet differed in a dramatic way: rather than simply hijacking computers, it also destroyed the equipment that the computers controlled.[100] For two years, the worm operated undetected, sabotaging Iran's uranium enrichment program.[101] Enrichment at Natanz ground to a halt multiple times throughout 2010, and centrifuge operational capacity fell by 30 percent.[102] Though Iran took the brunt of Stuxnet, the worm spread to Indonesia, Pakistan, India and even the United States – which was believed to be one of its two architects, alongside Israel. Ultimately, America took less than 2 percent of the total Stuxnet damage, a bearable amount of collateral damage in light of the greater aims being achieved.[103]

Worms and viruses are not the only means of extracting information from electronic devices. For all that TikTok was useful to the Ukrainians in disseminating information about the war, the platform had disadvantages. Security experts fear that TikTok user data can be harvested by the Chinese company ByteDance and passed on to the Chinese government, which can then track the location of government and military employees, steal intellectual property and build blackmail cases against individuals.

The Russo-Ukrainian War has already provided militaries all over the world with invaluable insights into the many ways that warfare is evolving. It is already being studied as intently as the Yom Kippur War was in the years after 1973. One lesson has been that, as an influential study by the Royal United Services Institute puts it, "There is no sanctuary in modern warfare."[104] Because the Russians have been able to strike deep into Ukraine, in future wars survivability will depend on the wide dispersal of, among other things, ammunitions stocks, command-and-control centers, maintenance areas and, especially, aircraft. In previous wars we have seen air forces destroyed on the ground – in June 1941 some 40 percent of the Soviet bomber force was destroyed before it could take off – but the Ukrainians avoided that through dispersal.

Another major lesson is that large stockpiles of shells are necessary for any modern war lasting more than a few days. Although Kyiv was saved by the massed fires of two artillery brigades, and the Ukrainians kept up rough equality of fires in the first six weeks of the war, by June the Russians were firing ten shells to every one Ukrainian.[105] Without NATO's supply of shells, Ukraine could not have stabilized the line in the Donbas. Yet no NATO country besides the USA has the depth of shell production and stockpiling necessary to fight a general war, especially not the United Kingdom.[106] As in so many other areas, it is the American taxpayer who is ultimately expected to pick up the lion's share of the costs, and the United States that is the sole indispensable nation in defending the rules-based international order from growing threats.[107] But even the US has to increase its defense industrial base to support the consumption rates in Ukraine.

The constant struggle to deny the enemy space in the electromagnetic spectrum, be it in radio, radar or infrared, and thus to be able to disrupt his use of precision weapons, is only likely to grow as warfare evolves. Ukraine has underlined how electronic warfare can defeat precision weaponry and how it will be essential to victory in the wars of the future, but this will require huge and ongoing investment. With Putin set to spend $143 billion on defense, security and policing in 2023, some 30 percent of Russia's tax income, the West must not be left behind in this vital area.[108]

There are profound lessons from the Russo-Ukrainian War about NATO's best strategic posture for the future, which needs to be oriented along its borders adopting the "hedgehog" approach that assumes defense to be dominant, at least for the immediate future, and strategic maneuver extremely difficult.[109] For the moment, the ancient offense-defense balance has tipped towards the latter because of the ubiquity and relative low cost of surveillance and targeting technologies.[110] In any likely future war between advanced powers, the intelligence, surveillance and reconnaissance systems would be incomparably more capable than we have seen in Ukraine, and precision munitions would have vastly greater range, speed and explosive power. There would also be far larger numbers of very much more capable unmanned systems (some remotely piloted, others operating according to algorithms) in every domain – not just in the air, but also at sea, in the sub-sea, on the ground, in outer space and in cyberspace, and operating in swarms not just individually. And every intelligence and strike capability will be integrated and connected by highly advanced command, control, communications and computer systems.[111]

The implications of the Russo-Ukrainian War for China and Taiwan

We can be certain that military lessons from Ukraine are being studied very carefully by the high command of the People's Liberation Army, and that a good deal of intensive staff work is currently being undertaken in Beijing to apply these lessons to the case of Taiwan.[112] As we noted in Chapter 1, Taiwan is the most sensitive and dangerous situation in the world today besides Ukraine itself, because it has the capacity to draw the world's two superpowers into direct conflict in a way that never happened during the Cold War.[113] China may have been emboldened after the chaotic American withdrawal from Afghanistan in August 2021; in fact, the Department of Defense's November 2022 annual report to Congress on national security developments noted that China had "employed multiple diplomatic tools in an attempt to erode U.S. and partner influence," including "highlighting the U.S. withdrawal from Afghanistan."[114]

As well as learning from the many Russian blunders in Ukraine which we list at the end of Chapter 9, Chinese planners might also feel deterred from invading Taiwan by seeing the Russians' challenges, the Ukrainians' resilience, the dangers of committing tanks and aircraft against NLAWs and Stingers, as well as by the West's unity against Russia and its willingness to employ economic sanctions.[115] "The war in Ukraine has in certain respects accelerated and amplified certain elements of great power *competition*," observes Vance Serchuk of the Center for a New American Security; "however, it has also likely reduced the prospects of great power *conflict*."

That is because the Chinese, always astute students of history, have probably seen what has transpired in Ukraine as a cautionary tale, despite the numerous obvious differences between Russia and China. The reverses that Russian forces have experienced in Ukraine will have reminded Chinese planners that large-scale operations are exceedingly complex and no amount of training can truly prepare a force for what can happen once the "iron dice" are rolled. Certainly, a war between China and the United States and its allies would unfold in ways that are impossible to predict and risk catastrophic global consequences.[116]

The Taiwanese are undoubtedly also watching Ukraine closely, and have been advised to pursue what the *War on the Rocks* website has described as "an asymmetric dragon-choking 'porcupine strategy,' prioritizing a large number of small things for its defence."[117] These include ballistic missile defense, air defense, sea-denial fires, shore-denial fires, mine warfare, information warfare, civil defense and resilient critical infrastructure. Taiwan clearly will seek to glean key lessons and apply them to ensure that their defenses are so substantial as to constitute an effective deterrence.

Conclusion

A number of observations emerge from our review of those conflicts of the post-Second World War era that have seen warfare evolve. Firstly, tragically, Plato was right: only the dead have seen the end of war. Contrary to what was widely expected at the start of the present century, the world has not even seen the end of major wars involving large conventional forces, much less the end of so-called small wars – insurgencies, terrorist campaigns and guerrilla warfare. Beyond that, nuclear weapons, clearly still a critical element of superpowers' and major nations' arsenals and deterrence regimes, certainly have not precluded wars; instead they seem in several cases merely to have placed undefined limits on war rather than preventing it.

We have also been reminded about the potential duration of major wars, not just insurgencies. The Russo-Ukrainian War has, in particular, dispelled the notion that major wars will be short and decisive affairs, and thus militaries can get by with only modest stockpiles of munitions and weapons systems. Russia and Ukraine have been expending such vast quantities of munitions that they and their supporters have had to undertake urgent programs to produce more of virtually every major weapons system and related munitions.[118] Any hope that major wars would be resolved as quickly as was the Gulf War has clearly been dispelled.

Ukraine also provides glimpses of what a major war among superpowers might look like, but only glimpses. With a few exceptions such as anti-ship missiles, the Ukrainians' longest range of precision munitions is presently no more than 50 miles (though additional precision munitions being provided by the United States are doubling that), a fraction of that of superpower arsenals. The drones used by both Russia and Ukraine are also fairly limited in range and capability, although Russian use of Iranian drones to swarm Ukrainian air defenses has shown how difficult it might be to defend against such a capability in the future, for such a tactic undoubtedly would feature in war between more advanced militaries, likely in concert with missile strikes.

Air force operations in Ukraine have also been constrained by the air defenses on both sides, and remotely piloted and algorithmically operated air, ground and maritime systems have made only minor appearances beyond the extensive use of relatively short-range aerial drones. Finally, the intelligence, surveillance and reconnaissance assets on each side have been relatively modest, compared with the state of the art of the United States military, for example. The Ukraine War does suggest, however, that actual combat between high-end forces would take to a new level and to vast distances the old Cold War adage noted earlier: "What can be seen can be hit, what can be hit can be killed." Furthermore, combat between superpowers would be carried out with systems that fly many times the speed of sound, are increasingly unmanned and either remotely controlled or autonomous, functioning in accordance with algorithms that instruct the machine, and enabled by extraordinary networks of sensors and computers, all of which are constantly updated by machine learning and Artificial Intelligence.

This world of unmanned machine-on-machine conflict, in every domain of warfare including cyberspace, and driven by algorithms rather than remote operators, enabled by self-healing communications and surveillance networks, might seem futuristic, but it is not that many years from becoming reality. Ukraine has, once again, provided merely a glimpse of what is already possible and only hints at what might be possible in the future.

Beyond the hardware and technology, however, our examination of post-Second World War conflicts also underscores the central importance of leadership. As we have repeatedly highlighted, effective leadership, especially at the strategic level, is absolutely critical – and can, in many cases, determine the outcome of a conflict. The abilities of the leaders at the top, and also those below that level, to understand the context and nature of the conflict and to get the big ideas – the strategy – right; to communicate those big ideas throughout the breadth and depth of a unit, a country, a coalition, the world; to oversee the implementation of the big ideas, providing example, energy, inspiration, determination and solid operational direction; and to determine how to refine and adjust the big ideas so that the leader can perform the four tasks again and again – these abilities are absolutely critical.

The importance of the ability to perform the four tasks of strategic leadership cannot be overestimated. Indeed, we have seen the impact of great leadership in several of the conflicts – Ben Gurion and Yitzhak Rabin in Israel, Ridgway in Korea, Briggs and Templer in Malaya, Giap in Indo-China and Vietnam, Qaboos and Akehurst in Oman, George H. W. Bush and General Thurman in Panama; George H. W. Bush and Schwarzkopf (with Colin Powell in the Pentagon) in Desert Shield and Desert Storm; Volodymyr Zelensky and his senior lieutenants in Ukraine – just as we have seen what happens when such leadership is lacking – as in Vietnam especially, and in Ukraine where Vladimir Putin and his high command have so far repeatedly demonstrated terribly flawed performances of the four tasks of strategic leadership.

The qualities of a successful military strategist will change from person to person, but there are a central few that all of them need. These include, above all else, strategic judgment, but also stamina, interpersonal skills and a feel for people; an ability to energize, inspire and motivate; the ability to communicate effectively orally and in writing; a degree of personal presence and charisma; a sincere love of servicemen and women; an ability to be tough when needed, but also compassionate when that is appropriate; fortitude in the face of adversity and the capacity to stay calm in the midst of chaos; an ability to deal with setbacks, missteps and mistakes; a sense of what leadership style is required to bring out the best in those immediately below, and also for the organization collectively. A great strategic leader also needs to be able to foresee how a conflict will end.

Conflict also reminds us never to underestimate the moral component and the role of morale in war. The recognition by the Ukrainians that they are, in essence, fighting their war of independence has been an element of incalculable importance to them. So has been the resulting mobilization of the entire country to defend Ukraine against the Russian invaders, who clearly lack the motivation, commitment and, increasingly, professional competence and initiative of their Ukrainian counterparts. Napoleon clearly was correct in observing that "The moral is to the physical as three is to one."

The professional expertise and training of a force, as well as the importance of a professional non-commissioned officer corps, must also be highlighted, as it is non-commissioned officers who are the standard bearers, standard enforcers and trainers at the critical small-unit level. Rigorous and demanding training will always be a vital component of combat readiness. So should be the intangible but critical element of initiative, especially of competent junior leaders empowered and encouraged to exercise initiative and acting in accordance with one of the admonitions in the counter-insurgency guidance issued during the Surge in Iraq: "In the absence of orders, figure out what they should have been and execute aggressively."

The war in Ukraine and several of the other conflicts we have reviewed in this book have also demonstrated the importance of being able to adapt quickly and respond to changing circumstances on the battlefield. A military should always strive to be a "learning organization," especially in war. The side that learns and adapts the fastest typically prevails, and commanders have to take explicit actions to promote a culture of learning and adaptation, to institute various practices and engage in individual actions that all contribute to the overall entity being a learning organization. Commanders have to remember that a lesson is not learned when it is identified, only when it has been incorporated in the big ideas, communicated effectively and actually implemented.

The importance of coalitions also deserves highlighting. As Churchill observed during the Second World War, "There is only one thing worse than fighting with allies, and that is fighting without them." The value of the coalition that the United States assembled and led in supporting Ukraine and in imposing financial, economic and personal sanctions and export controls on Russia – and in encouraging major businesses to curtail or halt their operations in Russia – has been enormous. The same has been true of other successful coalitions, particularly those in the Gulf War.

Related to the importance of coalitions is the importance of non-military means to complement and augment battlefield operations. These clearly deserve note: diplomatic initiatives in the United Nations and the broader international community; financial, economic and personal sanctions; export controls; actions by elements of the business community; and information campaigns.[119] All of these are currently contributing significantly to the effort to convey to Vladimir Putin that the war in Ukraine is unsustainable not just on the battlefield but also on the home front in Russia. Such an ultimate recognition is crucial, because until Putin acknowledges that Russia will not be able to out-suffer the Ukrainians, Europeans and Americans he will not make the exceedingly painful decision to suspend or negotiate an end to the war he so rashly began.

The significance of the information domain – striving to "be first with the truth" (not with spin or "putting lipstick on pigs") – has also been shown to be a key element in warfare. The Ukraine War has demonstrated the importance of actions in this realm, particularly the impressive ones taken by the CIA and MI6 before and after the invasion to release information publicly that clearly came from highly classified intelligence assessments without jeopardizing sources and methods.

While much of this chapter has focused on Ukraine, we cannot forget that other forms of conflict persist in addition to conventional warfare. These include terrorism, insurgency (based on ideological, religious or criminal motivations), cyber warfare, guerrilla actions, major criminal activities (such as those in Mexico) and, increasingly, information warfare. The United States and other major powers have to be able competently to craft campaigns to address such conflicts, as well as major combat operations, ideally to help a host-nation partner resolve them, or, if that is not possible, at least keep the problem from spreading. It needs to be recognized that while major conventional warfare will hopefully be infrequent at most, small wars will continue to be fought and will often continue to require Western involvement in some fashion.

In this regard, it is important that the US and its partners keep the lessons of Iraq, Afghanistan and Syria – and other similar endeavors – fresh in mind and continue to refine them as time goes on. It is also critical that the militaries of the United States, the United Kingdom and other countries sustain the various doctrinal, training, organizational, materiel, leader development and weapons system initiatives and "tactics, techniques and procedures" that, in recent years in particular, have enabled our forces to help local partners defeat serious threats without Western forces having to do the front-line fighting.

In northern Iraq and Syria, for example, the US-led counter-Islamic State coalition provided critical assistance to Iraqi security forces and Syrian democratic forces that enabled them to destroy the Islamic State caliphate and defeat the Islamic State as an army, although remnants of ISIS continue to operate as terrorists and insurgents. Coalition forces were only rarely in the lead, typically only in the most demanding operations conducted by our elite counter-terrorism forces against high-value targets. Rather, our elements provided training and equipping and, most importantly, advice, assistance and enablers – the latter including an armada of drones for surveillance, precision air attack, cyber capabilities, counter-drone assets, intelligence fusion and other forms of support.

Over time, the US Army has established Security Forces Assistance Brigades, and the British Army a Ranger Regiment, in order to augment special forces elements in the conduct of "advise, assist and enable" missions, as the demand for such elements has outstripped the capabilities of the special forces units in our militaries.* It is clear that the demand for such units will remain robust in the decades ahead. While the focus will rightly need to be on deterring the most potentially cataclysmic conflicts, there should be no doubt about the need to continue missions to combat extremists, insurgents and the most dangerous of criminal organizations – but to do so by advising, assisting and enabling local partners rather than fighting on the front lines.

Yet when the United States does have to fight on the front line again, it must be ready to do so. A lesson of Ukraine and of the other post-9/11 wars is that the US, in particular, has to maintain the capabilities for all types of conflict and also other missions – not just major combat operations with a peer competitor across a vast theatre, but also various forms of irregular warfare – counter-insurgency, counter-terrorism, "advise, assist and enable" operations in support of local partners – as well as humanitarian support operations and support to civil authorities, and all of them in all domains.

Conflict also underscores the inevitable human toll of war – both for those engaged in combat and for the innocent civilians caught in it. It recalls for us – especially as we observe the ongoing Russian campaign to destroy Ukraine's civilian infrastructure – the terrible toll of war on civilian communities writ large. This accentuates the critical need to deter war whenever possible – keeping in mind that deterrence is a function of two factors: a potential adversary's assessment of our capabilities

---

* Such missions have often been described as "by, with and through" missions; however, beyond that phrase being ungrammatical, it is also not as accurate a description of such efforts as is "advise, assist and enable."

on the one hand, and of our willingness to employ those capabilities on the other. Ensuring there are no doubts about either of those components of deterrence is imperative. Hence the need, for example, for the United States and its Indo-Pacific allies and partners to transform their forces in terms of the reach, resilience, survivability, lethality, speed and intelligence means of modern military weaponry and systems. The amount of money that needs to be spent might seem vast, but historically it has always proven to be a mere fraction of what it costs in blood and treasure when deterrence fails.

ACKNOWLEDGMENTS

I am most grateful to my wife Holly (the consummate editor), as well as to my colleagues at KKR, all of whom supported my embarking on this endeavor. I am also grateful to an old battlefield comrade, retired Colonel (and History Professor) Pete Mansoor, for his assistance with research and helping me craft a number of my contributions to *Conflict*. Sincere thanks also to Jonathan Jao, Arabella Pike and the other members of the great HarperCollins team, as well as to our experienced and enthusiastic agent Georgina Capel.

My thinking benefited substantially from exchanges with many experts on global issues, but I must specifically thank Ambassador Ryan Crocker – my extraordinary partner as U.S. Ambassador to Iraq – and both the Institute for the Study of War and the American Enterprise Institute that have done such exceptional conflict analysis in recent decades, particularly on Iraq, Afghanistan and Ukraine.

I am also grateful for the constructive comments on and critiques of my commentary on Afghanistan and Ukraine from innumerable netizens on LinkedIn.

I dedicate this book to all those who have borne the battle – and their families. To that end, my profits from *Conflict* will be donated to organizations that support our men and women in uniform and our veterans and their family members.

General David Petraeus, US Army (Ret.), Arlington, Virginia

I should like to thank Roger and Martha Mertz for their generosity in funding my visiting fellowship at the Hoover Institution at Stanford, where much of the research for this book took place, along with Lady (Dambisa) Moyo of Knightsbridge and her husband Jared Smith for their hospitality in Palo Alto during it. I am also very grateful to Jeremy Solel, Katherine Bayford and Michael Bishop for their invaluable research assistance, and to Dr. Conrad Crane, General James Mattis, General H. R. McMaster, Captain Larry Robinson, Air Marshal Edward Stringer, and General Sir Peter Wall for their recollections and advice.

I should like to thank my wife Susan for her characteristically superb support and advice, my excellent literary agent Georgina Capel, and Arabella Pike and Jonathan Jao at HarperCollins for their fine editing. Many thanks also to Jerry del Missier, Paul Felton, Larry Robinson and Nataliya Bulgayova who read the manuscript in proof.

This book is dedicated to Simon Sebag Montefiore, my close friend and co-conspirator of forty years.

Andrew Roberts, House of Lords, London, England

MAPS

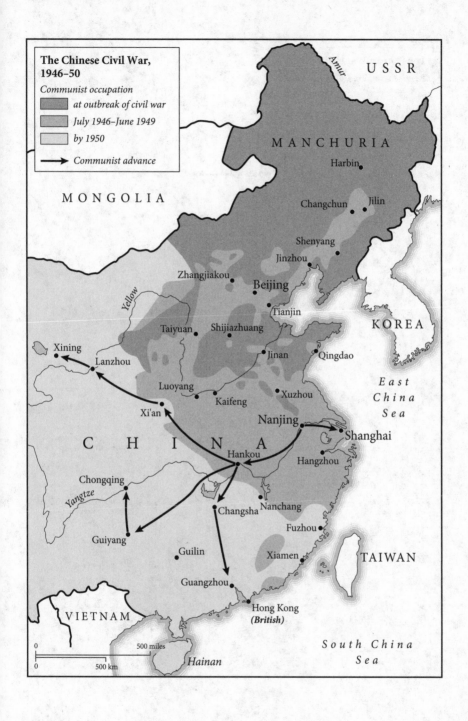

USSR

MONGOLIA

MANCHURIA

Amur

Harbin

Changchun   Jilin

Shenyang

Jinzhou

Zhangjiakou   Beijing

KOREA

Tianjin

Yellow   Taiyuan   Shijiazhuang

Xining   Jinan   Qingdao

Lanzhou

*East
China
Sea*

Luoyang

Kaifeng   Xuzhou

Xi'an

Nanjing

Shanghai

C H I N A   Hankou

Hangzhou

Chongqing

Yangtze   Changsha   Nanchang

Guiyang

Fuzhou

Guilin   Xiamen

TAIWAN

Guangzhou

VIETNAM   Hong Kong
(British)

0 ——— 500 miles

0 ——— 500 km   *Hainan*

*South China
Sea*

The Korean War, 1950–53

- - - - - *Front lines, with date*

Vladivostok

USSR

MANCHURIA

(CHINA)

Tumen

Chongjin

Yalu

UN maximum advance,
Nov. 1950

Chosin
Reservoir

Sinuiju

Osan

Hamhung

Sea of
Japan

NORTH
KOREA

Yongdu

Pyongyang

Wonsan

Imjin

Armistice Line, 27 July 1953

Sariwon

Ichon

Panmunjom

Haeju

Kap Yong

38th Parallel

Seoul

Han

Wonju

Inchon

Chinese & North Korean
maximum advance, Jan. 1951

Osan

SOUTH
KOREA

Andong

North Korean maximum
advance, Sep. 1950

Yellow
Sea

Taejon

Naktong

Taegu

Mokpo

Pusan

Strait

Western Channel

Tsushima

Korea

JAPAN

0   100 miles

Eastern Channel
(Tsushima Strait)

0   100 km

Jejudo

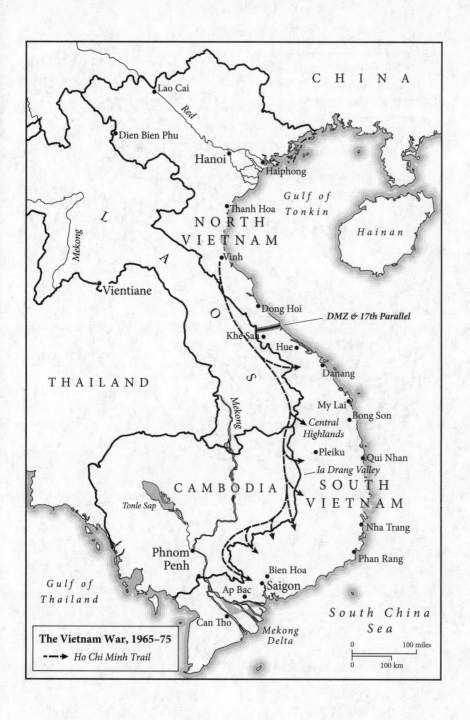

C H I N A

Lao Cai

Red

Dien Bien Phu

Hanoi

Haiphong

Gulf of Tonkin

Thanh Hoa

N O R T H
V I E T N A M

Hainan

Vinh

L
A
O
S

Vientiane

Mekong

Dong Hoi

DMZ & 17th Parallel

Khe Sanh

Hue

T H A I L A N D

Danang

My Lai

Bong Son

Central Highlands

Mekong

Pleiku

Qui Nhan

Ia Drang Valley

C A M B O D I A

S O U T H
V I E T N A M

Tonle Sap

Nha Trang

Phnom Penh

Phan Rang

Bien Hoa

Gulf of Thailand

Ap Bac

Saigon

Can Tho

Mekong Delta

South China Sea

	100 miles
0	
0	100 km

The Vietnam War, 1965–75

- - -▶ *Ho Chi Minh Trail*

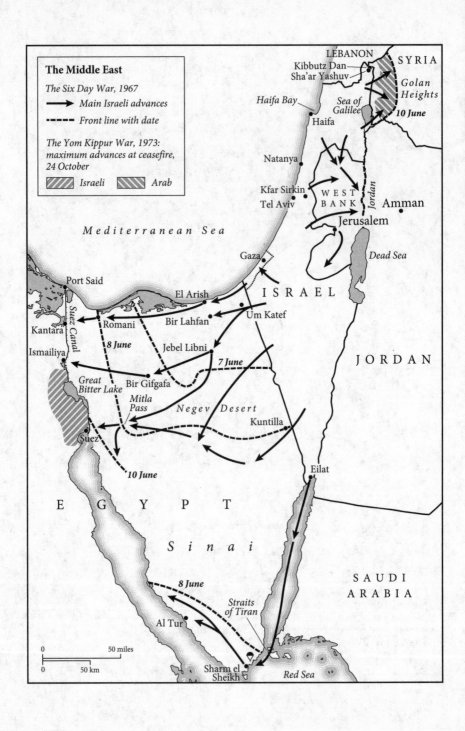

The Middle East

The Six Day War, 1967
→ Main Israeli advances
- - - Front line with date

The Yom Kippur War, 1973:
maximum advances at ceasefire,
24 October

Israeli   Arab

LEBANON
Kibbutz Dan
Sha'ar Yashuv
SYRIA
Golan
Heights
10 June
Haifa Bay
Sea of
Galilee
Haifa
Natanya
Kfar Sirkin
Tel Aviv
WEST
BANK
Jordan
Amman
Jerusalem
Gaza
Dead Sea
Mediterranean Sea
Port Said
El Arish
ISRAEL
Kantara
Suez Canal
Romani
Bir Lahfan
Um Katef
Ismailiya
8 June
Jebel Libni
7 June
JORDAN
Great
Bitter Lake
Bir Gifgafa
Mitla
Pass
Negev Desert
Kuntilla
Suez
10 June
Eilat
E G Y P T
S i n a i
SAUDI
ARABIA
8 June
Straits
of Tiran
Al Tur
0   50 miles
0   50 km
Sharm el
Sheikh
Red Sea

Pebble Island

Douglas

Port San Carlos

San Carlos
Water

West
Falkland

*East
Falkland*

Mt. Longdon
Two Sisters
Mt. Harriet
Bluff Cove
Mt. Tumbledown

Port
Stanley

Darwin

Goose Green

Fitzroy

South Atlantic Ocean

The Falklands War, 1982
→ *Main British advance*
0   20 miles
0   20 km

Karshi-Khanabad Airbase ✈

Amu Darya

TURKMENISTAN

Mashhad

JAWZJAN

Mazai

Shiberghan

Sari Pul

Maymana

FARYAB

SARI PUL

IRAN

BADGHIS

Herat

Chaghcharan

HERAT

GHOR

DAY
KUNDI

Nili

FARAH

URUZGAN

Farah

Tarin Kowt

Helmand

Arghandab

Camp Bastion ✈

Kandahar

Lashkar Gah

Kandahar Airbase ✈

NIMROZ

HELMAND

KANDAHAR

Zaranj

US air assault

Zahedan

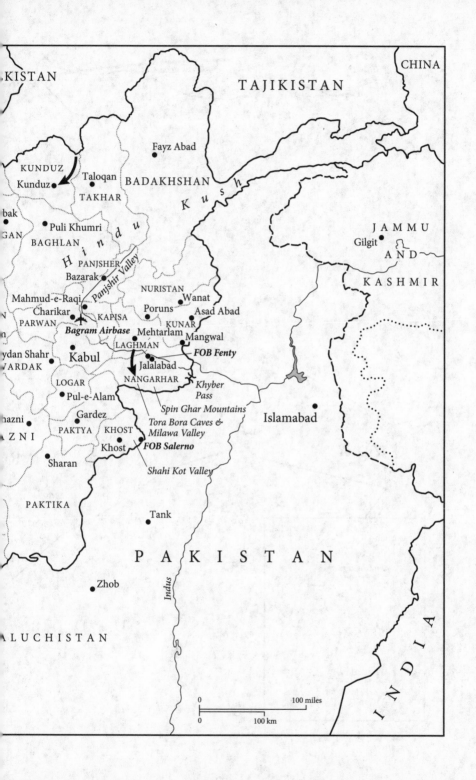

T U R K E Y
airborne assault
DA

Aleppo

Tal Afar
Mos

Euphrates

N I N E V E H

S Y R I A

Homs

SAL

Damascus

Rama

A N B A R
Lake Thar

J O R D A N

I R

Karmah
Taji
D I Y A L A
Fallujah
Sadr City
Abu Ghraib Prison
Baghdad
Euphrates

Lake Tharthar

Tigris

S A U D I

Karbala
B A B I L
Karbala Gap
KARBALA
Hillah

A R A B I A

0   100 miles
0   100 km

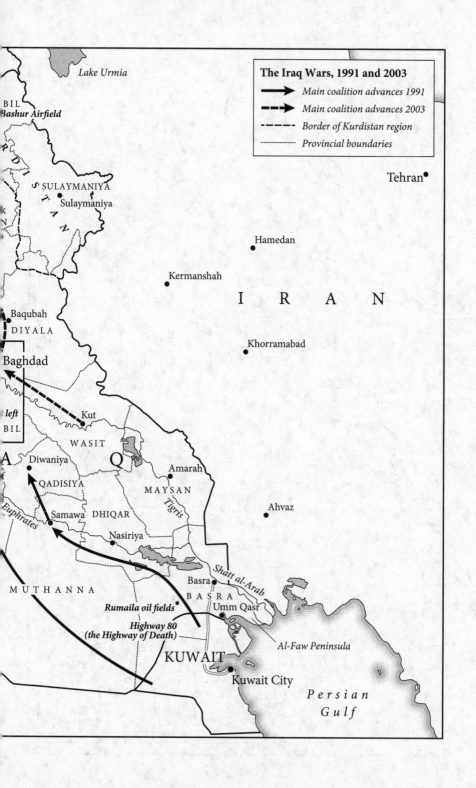

Lake Urmia

BIL
Bashur Airfield

The Iraq Wars, 1991 and 2003
⟶ Main coalition advances 1991
⟶ Main coalition advances 2003
--- Border of Kurdistan region
⋯ Provincial boundaries

Tehran

SULAYMANIYA
Sulaymaniya

Hamedan

Kermanshah

I R A N

Baqubah
DIYALA

Khorramabad

Baghdad

left
BIL

Kut

WASIT

A   Diwaniya   Q
QADISIYA

Amarah

MAYSAN

Ahvaz

Samawa   DHIQAR

Nasiriya

Shatt al-Arab

Basra

MUTHANNA

BASRA

Rumaila oil fields

Umm Qasr

Highway 80
(the Highway of Death)

Al-Faw Peninsula

KUWAIT

Kuwait City

Persian
Gulf

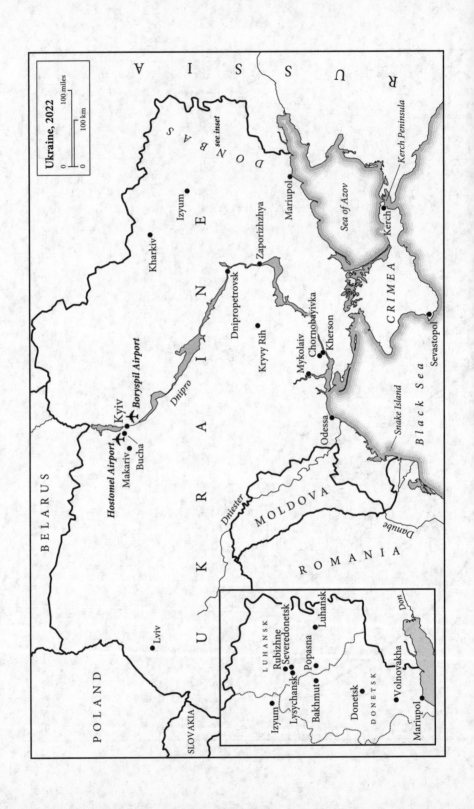

BIBLIOGRAPHY

Books

Akehurst, John, We Won a War: The Campaign in Oman 1965–1975 1982

Albright, Madeleine, Madame Secretary 2013

Allawi, Ali A., The Occupation of Iraq: Winning the War, Losing the Peace 2007

Allen, John R., et al., Future War and the Defence of Europe 2021

Andrew, Christopher, The Secret World: A History of Intelligence 2018

Andrews, Timothy, et al. (eds), The Last Card: Inside George W. Bush's Decision to Surge in Iraq 2019

Archer, Christon I., et al., World History of Warfare 2002

Arkin, William M., The Generals Have No Clothes: The Untold Story of Our Endless Wars 2021

Asmus, Ronald D., A Little War that Shook the World: Georgia, Russia, and the Future of the West 2010

Asselin, Pierre, Vietnam's American War 2018

Atkinson, Rick, Crusade: The Untold Story of the Persian Gulf War 1993

—, In the Company of Soldiers: A Chronicle of Combat 2004

Auerswald, David P., and Saideman, Stephen, NATO in Afghanistan: Fighting Together, Fighting Alone 2014

Barnet, Richard J., Intervention and Revolution: The United States in the Third World 1968

Bellavia, David, House to House 2007

Bergen, Peter, Manhunt: The Ten-Year Search for Bin Laden from 9/11 to Abbottabad 2012

Berman, Larry, Planning A Tragedy: The Americanization of the War in Vietnam 1982

Beschloss, Michael, *Kennedy v Khrushchev* 1991

Bew, John, *Realpolitik: A History* 2016

Biddle, Stephen, *Military Power – Explaining Victory and Defeat in Modern Battle* 2006

Bierbauer, Alec, and Cooter, Col. Mark, with Michael Marks, *Never Mind, We'll Do It Ourselves: The Inside Story of How a Team of Renegades Broke Rules, Shattered Barriers, and Launched a Drone Warfare Revolution* 2021

Black, Jeremy, *War since 1945* 2004

—, *Air Power: A Global History* 2016

—, *A Short History of War* 2021

Blair, Tony, *A Journey: My Political Life* 2010

Blumenthal, Sidney, *The Clinton Wars* 2002

Boot, Max, *Invisible Armies: An Epic History of Guerrilla Warfare from Ancient Times to the Present* 2013

—, *The Road Not Taken: Edward Lansdale and the American Tragedy in Vietnam* 2018

Borger, Julian, *The Butcher's Trail: How the Search for Balkan War Criminals Became the World's Most Successful Manhunt* 2016

Bose, Sumantra, *Kashmir at the Crossroads* 2021

Bowden, Mark, *The Three Battles of Wanat: And Other True Stories* 2016

Braun, William G., et al. (eds), *Robotics and Military Operations* 2018

Bremer III, L. Paul, and McConnell, Malcolm, *My Year in Iraq: The Struggle to Build a Future of Hope* 2006

Bush, George W., *Decision Points* 2010

Carter, Ash, *A Lasting Defeat: The Campaign to Destroy ISIS* 2017

Casey Jr., George W., *Strategic Reflections: Operation Iraqi Freedom, July 2004– February 2007* 2012

Chandrasekaran, Rajiv, *Imperial Life in the Emerald City: Inside Iraq's Green Zone* 2004

Cheney, Dick, *In My Time: A Personal and Political Memoir* 2012

Chinkin, Christine, and Kaldor, Mary, *International Law and New Wars* 2017

Churchill, Winston, *The Story of the Malakand Field Force* 1898

—, *My Early Life: A Roving Commission* 1930

—, *Marlborough: His Life and Times*, vol. III 1936

Clark, General Wesley, *Waging Modern War* 2002

Clausewitz, General Carl von, *On War*, trans. J. J. Graham, 1873

—, *On War*, ed. and trans. Michael Howard and Peter Paret, rev. edn 1984

Clewlow, Ade, *Under a Feathered Sky: The Untold Story of NATO's Role in Newly Independent Kosovo* 2020

Clinton, Bill, *My Life* 2005

Clinton, Hillary Rodham, *Hard Choices* 2015

Clodfelter, Michael, *Warfare and Armed Conflicts* 2008

Colbeck, Graham, *With 3 Para to the Falklands* 2002

Cole, Ronald H., *Operation Urgent Fury: Grenada*, Joint History Office, Office of the Chairman of the Joint Chiefs of Staff, 1997

Coll, Steve, *Ghost Wars: The Secret History of the CIA, Afghanistan, and Bin Laden, from the Soviet Invasion to September 10, 2001* 2004

Collins, N. W., *Grey Wars: A Contemporary History of U.S. Special Relations* 2021

Collins, Richard D., and Hooker, Joseph J., *Lessons Encountered: Learning from the Long War* 2015

Conquest, Robert, *The Dragons of Expectation: Reality and Delusion in the Course of History* 2004

Cordesman, Anthony H., *The Lessons and Non-Lessons of the Air and Missile War in Kosovo* 1999

Cormac, Rory, *How to Stage a Coup: And Ten Other Lessons from the World of Secret Statecraft* 2022

Corrigan, Jim, *Desert Storm Air War: The Aerial Campaign against Saddam's Iraq in the 1991 Gulf War* 2017

Cosmas, Graham A., *MACV: The Joint Command in the Years of Escalation 1962–1967* 2006

Crane, Conrad C., *Cassandra in Oz: Counterinsurgency and Future War* 2016

Currey, Cecil B., *Victory at Any Cost: The Genius of Vietnam's General Vo Nguyen Giap* 1996

Daalder, Ivo H., and O'Hanlon, Michael H., *Winning Ugly: NATO's War to Save Kosovo* 2000

Daddis, Gregory A., *Westmoreland's War: Reassessing American Strategy in Vietnam* 2014

—, *Withdrawal: Reassessing America's Final Years in Vietnam* 2017

Dallaire, Roméo, *Shake Hands with the Devil: The Failure of Humanity in Rwanda* 2003

Degen, Edmund J., and Reardon, Mark J., *Modern War in an Ancient Land: The United States Army in Afghanistan, 2001–2014*, 2 vols 2021

Department of the Army, Field Manual 23-4, *Counterinsurgency* 2006

Derix, Steven, and Shelkunova, Marina, *Zelensky: Ukraine's President and His Country* 2022

De Waal, Thomas, *The Caucasus: An Introduction* 2010

Dikötter, Frank, *The Tragedy of Liberation: A History of the Chinese Revolution 1945–1957* 2013

Donati, Jessica, *Eagle Down: The Last Special Forces Fighting the Forever War* 2021

Donnelly, Thomas, Roth, Margaret, and Baker, Caleb, *Operation Just Cause: The Storming of Panama* 1991

Dorman, Andrew M., *Blair's Successful War: British Military Intervention in Sierra Leone* 2009

Doyle, William, A Soldier's Dream: Captain Travis Patriquin and the Awakening of
    Iraq 2011
Dunstan, Simon, The Six Day War 1967 2009
Ebrey, Patricia Buckley, The Cambridge Illustrated History of China 2022
Eversmann, Matt, and Schilling, Dan (eds), The Battle of Mogadishu: Firsthand
    Accounts from the Men of Task Force Ranger 2005
Fall, Bernard B., Last Reflections on a War 1967
—, Hell in a Very Small Place: The Siege of Dien Bien Phu 2002
—, Street without Joy: The French Debacle in Indochina 2018
Felix, Antonia, Wesley K. Clark: The Biography 2004
Ferguson, Niall, The War of the World: History's Age of Hatred 2006
—, Kissinger 1923–1968: The Idealist 2015
FitzGerald, Frances, Fire in the Lake: The Vietnamese and the Americans in
    Vietnam 2002
Fontenot, Colonel Gregory, Degen, Lt. Col. E. J., and Tohn, Lt. Col. David, On
    Point: The United States Army in Operation Iraqi Freedom 2004
Ford, Harold P., CIA and the Vietnam Policymakers: Three Episodes 1962–1968
    1997
Forsyth, James, The Caucasus: A History 2015
Fowler, William, Operation Barras: The SAS Rescue Mission, Sierra Leone 2000
    2012
Franks, Tommy, and McConnell, Malcolm, American Soldier 2004
Freedman, Lawrence, Atlas of Global Strategy: War and Peace in the Nuclear Age
    1985
—, Kennedy's Wars: Berlin, Cuba, Laos, Vietnam 2000
—, The Cold War: A Military History 2001
—, Strategy 2013
—, The Future of War 2017
—, Ukraine and the Art of Strategy 2019
—, Command: The Politics of Military Operations from Korea to Ukraine 2021
—, (ed.), War 1994
—, (ed.), Strategic Coercion 1998
—, and Gamba-Stonehouse, Virginia, Signals of War: The Falklands Conflict of
    1982 1990
—, and Karsh, Efraim, The Gulf Conflict 1990–1991: Diplomacy and War in the
    New World Order 1993
—, and Michaels, Jeffrey, The Evolution of Nuclear Strategy 2019
Gaddis, John Lewis, We Now Know: Rethinking Cold War History 1998
—, The Cold War 2005
—, On Grand Strategy 2019
Galeotti, Mark, The Weaponisation of Everything 2022
—, Putin's Wars: From Chechnya to Ukraine 2022

Galster, Steve, *Afghanistan: The Making of U.S. Policy 1973–1990* 1990

Galula, David, *Counterinsurgency Warfare: Theory and Practice* 2006

Galvin, John R., *Fighting the Cold War: A Soldier's Memoir* 2015

Gant, Jim, *One Tribe at a Time: The Paper that Changed the War in Afghanistan* 2014

Gates, Robert, *Duty: Memoirs of a Secretary at War* 2014

Gelb, Leslie H., and Betts, Richard K., *The Irony of Vietnam: The System Worked* 1979

Giap, Vo Nguyen, *How We Won the War* 1976

Gilbert, Martin, *Israel: A History* 1999

Giles, Keir, *Russia's War on Everybody and What It Means for You* 2022

Glenny, Misha, *The Fall of Yugoslavia* 1996

—, *The Balkans: Nationalism, War, and the Great Powers 1804–2012* 2017

Goodwin, Doris Kearns, *Lyndon Johnson & the American Dream* 1976

Gordon, Michael R., and Trainor, General Bernard E., *The Generals' War: The Inside Story of the Conflict in the Gulf* 1995

—, *COBRA II: The Inside Story of the Invasion and Occupation of Iraq* 2006

—, *The Endgame: The Inside Story of the Struggle for Iraq, from George W. Bush to Barack Obama* 2012

Grenier, Robert L., *88 Days to Kandahar: A CIA Diary* 2015

Halberstam, David, *The Best and the Brightest* 1968

Hanson, Victor Davis, *Carnage and Culture: Landmark Battles in the Rise of Western Power* 2001

—, *An Autumn of War: What America Learned from September 11 and the War on Terrorism* 2002

—, *The Savior Generals: How Five Great Commanders Saved Wars that Were Lost – from Ancient Greece to Iraq* 2013

Harding, Luke, *Invasion: Russia's Bloody War and Ukraine's Fight for Survival* 2022

Hastings, Max, *Warriors* 2005

—, *Vietnam: An Epic History of a Tragic War* 2018

—, *The Korean War* 2020

—, (ed.), *The Oxford Book of Military Anecdotes* 1985

—, and Jenkins, Simon, *The Battle for the Falklands* 1983

Helton, Arthur C., *The Price of Indifference: Refugees and Humanitarian Action in the New Century* 2002

Hennessy, Michael A., *Strategy in Vietnam: The Marines and Revolutionary Warfare in I Corps 1965–1972* 1997

Herr, Michael, *Dispatches* 1991

Herring, George C., *America's Longest War: The United States and Vietnam 1950–1975* 1986

Herzog, Chaim, *The Arab–Israeli Wars: War and Peace in the Middle East* 1982

—, *The War of Atonement* 1998

Hilsman, Roger, *To Move a Nation* 1967

Holbrooke, Richard, *To End a War* 1998

Horne, Alistair, *A Savage War of Peace: Algeria 1954–1962* 2006

—, *Kissinger's Year: 1973* 2009

—, *Hubris: The Tragedy of War in the Twentieth Century* 2015

Hosmer, Stephen T., et al., *Counterinsurgency: A Symposium, April 16–20, 1962* 2006

Howard, Michael, *War and the Nation State* 1978

—, *The Lessons of History* 1991

—, *The Invention of Peace* 2000

—, *War in European History* 2009

—, Andreopoulos, George, and Shulman, Mark R. (eds), *The Laws of War: Constraints on Warfare in the Western World* 1994

Ignatieff, Michael, *Virtual War: Kosovo and Beyond* 2000

Jackson, Julian, *A Certain Idea of France: The Life of Charles de Gaulle* 2018

Jackson, General Sir Mike, *Soldier: The Autobiography* 2007

Jankowicz, Nina, *How to Lose the Information War: Russia, Fake News, and the Future of Conflict* 2020

Jarrett, Philip (ed.), *Faster, Further, Higher: Leading-edge Aviation Technology since 1945* 2002

Johnson, Chalmers, *Autopsy on People's War* 1973

Johnson, David E., Markel, M. Wade, and Shannon, Brian, *The 2008 Battle of Sadr City: Reimagining Urban Combat* 2013

Johnson, Lyndon Baines, *The Vantage Point: Perspectives of the Presidency 1963–1969* 1971

Jones, Archer, *The Art of War in the Western World* 1987

Jones, Seth G., *Hunting in the Shadows: The Pursuit of al Qa'ida since 9/11* 2012

Judah, Tim, *Kosovo: War and Revenge* 2000

Kagan, Frederick W., *Choosing Victory: A Plan for Success in Iraq* 2006

—, *Finding the Target: Transformation of American Military Policy* 2007

Kagan, Kimberly, *The Surge: A Military History* 2009

Kakar, M. Hassan, *The Soviet Invasion and the Afghan Response 1979–1982* 1995

Kaplan, Fred, *The Insurgents: David Petraeus and the Plot to Change the American Way of War* 2013

Karnow, Stanley, *Vietnam: A History* 1997

Katz, Arthur M., *Life after Nuclear War: The Economic and Social Impacts of Nuclear Attacks on the United States* 1982

Keegan, John, *A History of Warfare* 2004

Kershaw, Ian, *To Hell and Back: Europe 1914–1949* 2015

Khalilzad, Zalmay, *The Envoy: From Kabul to the White House, My Journey through a Turbulent World* 2016

Kilcullen, David, The Accidental Guerrilla: Fighting Small Wars in the Midst of a Big One 2017

—, The Dragons and the Snakes: How the Rest Learned to Fight the West 2020

Kinnard, Douglas, The War Managers 1985

Kissinger, Henry, White House Years 1979

—, Years of Upheaval 1982

—, Diplomacy 1994

—, Leadership: Six Studies in World Strategy 2022

—, Schmidt, Eric, and Huttenlocher, Daniel, The Age of A.I. and Our Human Future 2021

Kitfield, James, Prodigal Soldiers: How the Generation of Officers Born of Vietnam Revolutionized the American Style of War 1995

Knorr, Klaus, and Morgan, Patrick (eds), Strategic Military Surprise 1983

Komer, Robert, Bureaucracy Does Its Thing 1972

Kostiner, Joseph, Conflict and Cooperation in the Gulf Region 2007

Krammer, Arnold, The Forgotten Friendship: Israel and the Soviet Bloc 1947–53 1974

Krepinevich Jr., Andrew F., The Army and Vietnam 1986

Kukielski, Philip, The U.S. Invasion of Grenada: Legacy of a Flawed Victory 2019

Kummer, David W., U.S. Marines in Afghanistan, 2001–2009: Anthology and Annotated Bibliography 2014

Lacouture, Jean, De Gaulle: The Ruler 1945–1970 1991

Lambeth, Benjamin S., NATO's Air War for Kosovo: A Strategic and Operational Assessment 2001

Lee, Wayne E., et al., The Other Face of Battle: America's Forgotten Wars and the Experience of Combat 2021

Lehman, John, On Seas of Glory: Heroic Men, Great Ships, and Epic Battles of the American Navy 2001

—, Oceans Ventured: Winning the Cold War at Sea 2018

LeoGrande, William M., Our Own Backyard: The United States in Central America 1977–1992 1998

Lewy, Guenter, America in Vietnam 1978

Logevall, Fredrik, Choosing War: The Lost Chance for Peace and the Escalation of War in Vietnam 1999

—, Embers of War: The Fall of an Empire and the Making of America's Vietnam 2012

Lowe, Keith, The Fear and the Freedom: How the Second World War Changed Us 2017

Lowrey, Nathan S., U.S. Marines in Afghanistan 2001–2002: From the Sea 2011

—, The Chairmanship of the Joint Chiefs of Staff 1949–2016 2016

Loyn, David, The Long War: The Inside Story of America and Afghanistan since 9/11 2021

Lynch, Michael, *The Chinese Civil War 1945–49* 2010

McChrystal, Stanley, *My Share of the Task: A Memoir* 2013

McClintock, Michael, *The American Connection: State Terror and Popular Resistance in El Salvador* 1985

—, *Instruments of Statecraft: U.S. Guerrilla Warfare, Counterinsurgency, and Counterterrorism 1940–1990* 1992

McConnell, Malcolm, *Just Cause: The Real Story of America's High-Tech Invasion of Panama* 1991

McCullough, David, *Truman* 1992

MacDonald, Michael, *Overreach: Delusions of Regime Change in Iraq* 2014

McMahan, Jeff, *Reagan and the World: Imperial Policy in the New Cold War* 1984

McMaster, H. R., *Dereliction of Duty: Lyndon Johnson, Robert McNamara, the Joint Chiefs of Staff, and the Lies that Led to Vietnam* 1997

—, *Battlegrounds: The Fight to Defend the Free World* 2020

MacMillan, Margaret, *War: How Conflict Shaped Us* 2020

McNamara, Robert, *In Retrospect: The Tragedy and Lessons of Vietnam* 1995

MacQueen, Norrie, *The United Nations since 1945* 1999

Malcolm, Noel, *Kosovo: A Short History* 1998

Malkasian, Carter, *War Comes to Garmser: Thirty Years of Conflict on the Afghan Frontier* 2013

—, *The American War in Afghanistan: A History* 2021

Mansoor, Peter R., *Baghdad at Sunrise: A Brigade Commander's War in Iraq* 2008

—, *Surge: My Journey with General David Petraeus and the Remaking of the Iraq War* 2013

Mansoor, Peter, and Murray, Williamson (eds), *The Culture of Military Organizations* 2019

Maraniss, David, *They Marched into Sunlight* 2004

Marshall, Tim, *Shadowplay: A Memoir from behind the Lines and under Fire* 2019

Matthews, Owen, *Overreach: The Inside Story of Putin's War against Ukraine* 2022

Mattis, Jim, and West, Bing, *Call Sign Chaos* 2019

Mauriac, François, *De Gaulle* 1965

Mayall, Simon, *Soldier in the Sand: A Personal History of the Modern Middle East* 2020

—, *Battlegrounds: The Fight to Defend the Free World* 2020

Michaels, Jim, *A Chance in Hell: The Men Who Triumphed over Iraq's Deadliest City and Turned the Tide of War* 2010

Mitter, Rana, *China's War with Japan 1937–1945* 2014

Moore, Harold G., and Galloway, Joseph L., *We Were Soldiers Once ... and Young: Ia Drang: The Battle that Changed the War in Vietnam* 1992

Morris, Benny, *Righteous Victims: A History of the Zionist–Arab Conflict, 1881–1998* 2002

Morris, Ian, *War: What Is It Good For?* 2014

Moyar, Mark, *Triumph Forsaken: The Vietnam War 1954–1965* 2006

—, *Phoenix and the Birds of Prey: Counterinsurgency and Counterterrorism in Vietnam* 2007

Mumford, Andrew, *Proxy Warfare* 2013

Mutawi, Samir A., *Jordan in the 1967 War* 1987

Nagl, John, *Learning to Eat Soup with a Knife: Counterinsurgency Lessons from Malaya and Vietnam* 2005

Naylor, Sean, *Not a Good Day to Die: The Untold Story of Operation Anaconda* 2005

Neumann, Brian, Mundey, Lisa, and Mikolashek, Jon, *The U.S. Army in Afghanistan: Operation Enduring Freedom, March 2002–April 2005* 2012

Nowrojee, B., *Shattered Lives: Sexual Violence during the Rwandan Genocide and Its Aftermath* 1996

Nutting, Anthony, *No End of a Lesson: The Story of Suez* 1967

Obama, Barack, *A Promised Land* 2020

O'Hanlon, Michael, *The Art of War in an Age of Peace: U.S. Strategy and Resolute Restraint* 2021

Oren, Michael, *Six Days of War: June 1967 and the Making of the Modern Middle East* 2003

Osborn, George K., et al., *Democracy, Strategy, and Vietnam: Implications for American Policymaking* 1987

O'Sullivan, John, *The President, the Pope, and the Prime Minister* 2006

Owen, Mark, with Kevin Maurer, *No Easy Day: The Firsthand Account of the Mission that Killed Osama Bin Laden* 2012

Pace, Lorin, *International Elements of Saddam Hussein's Conflicted Iraq: Shifting Allies, Resolute Foes* 2002

Packer, George, *The Assassins' Gate: America in Iraq* 2005

—, *Our Man: Richard Holbrooke and the End of the American Century* 2019

Palmer Jr., Bruce, *The 25-Year War: America's Military Role in Vietnam* 1984

Panetta, Leon, *Worthy Fights: A Memoir of Leadership in War and Peace* 2014

Paret, Peter (ed.), *Makers of Modern Strategy* 1986

Pavelec, Sterling Michael, *War and Warfare since 1945* 2017

Penfold, Peter, *Atrocities, Diamonds and Diplomacy: The Inside Story of the Conflict in Sierra Leone* 2012

Peres, Shimon, *No Room for Small Dreams* 2017

Petraeus, David, and Amos, James, *Counterinsurgency Field Manual FM 3-24* 2006

Pollack, Kenneth M., *Arabs at War: Military Effectiveness 1948–1991* 2002

Poole, Walter S., *The Effort to Save Somalia: August 1992–March 1994* 2005

Powell, Colin, *My American Journey* 1995

Pryce-Jones, David, *The War that Never Was: The Fall of the Soviet Empire 1985–1991* 1995

Puri, Samir, *Russia's Road to War with the West* 2022

Rabinovich, Abraham, *Yom Kippur War: The Epic Encounter that Transformed the Middle East* 2004

Race, Jeffrey, *War Comes to Long An: Revolutionary Conflict in a Vietnamese Province* 1972

Rayburn, Colonel Joel D., and Sobchak, Colonel Frank K. (eds), *The U.S. Army in the Iraq War*, vol. I: *Invasion–Insurgency–Civil War 2003–2006* 2019

—, *The U.S. Army in the Iraq War*, vol. 2: *Surge and Withdrawal 2007–2011* 2019

Report of the Independent Inquiry into the Actions of the United Nations during the 1994 Genocide in Rwanda, United Nations, 1999, S/1999/1257

Reynolds, David, *One World Divisible: A Global History since 1945* 2000

Rice, Susan, *Tough Love: My Story of the Things Worth Fighting For* 2019

Richardson, Louise, *What Terrorists Want: Understanding the Enemy, Containing the Threat* 2006

Ricks, Tom, *Fiasco: The American Military Adventure in Iraq* 2006

—, *The Gamble: General Petraeus and the American Military Adventure in Iraq* 2009

Rid, Thomas, *Active Measures: The Secret History of Disinformation and Political Warfare* 2021

Riedel, Bruce O., *Deadly Embrace: Pakistan, America, and the Future of the Global Jihad* 2012

Robbins, Keith, *The World since 1945* 1998

Roberts, Geoffrey, *Stalin's Wars: From World War to Cold War 1939–1953* 2006

Roberts, J. M., *The Penguin History of the Twentieth Century* 1999

Robinson, Linda, *Tell Me How This Ends: General David Petraeus and the Search for a Way Out of Iraq* 2008

—, *One Hundred Victories: Special Ops and the Future of American Warfare* 2013

Rose, Norman, *A Senseless, Squalid War: Voices from Palestine 1945–1948* 2009

Rosello, Victor M., *Lessons from El Salvador* 1993

Rudd, Gordon W., *Humanitarian Intervention: Assisting the Iraqi Kurds in Operation Provide Comfort 1991* 2004

—, *Reconstructing Iraq: Regime Change, Jay Garner, and the ORHA Story* 2011

Rumsfeld, Donald, *Known and Unknown: A Memoir* 2011

Rusk, Dean, *As I Saw It* 1990

Sanchez, Ricardo S., and Phillips, Donald T., *Wiser in Battle: A Soldier's Story* 2008

Sanger, David, *Confront and Conceal: Obama's Secret Wars and Surprising Use of American Power* 2012

—, *The Perfect Weapon: War, Sabotage, and Fear in the Cyber Age* 2018

Schadlow, Nadia, *War and the Art of Governance: Consolidating Combat Success into Political Victory* 2017

Schandler, Herbert Y., *Lyndon Johnson and Vietnam: The Unmaking of a President* 1977

Scharre, Paul, *Army of None: Autonomous Weapons and the Future of War* 2018

Schinella, Anthony M., *Bombs without Boots: The Limits of Airpower* 2019

Schofield, Victoria, *Kashmir in Conflict: India, Pakistan and the Unending War* 2010

Schroen, Gary C., *First In: An Insider's Account of How the CIA Spearheaded the War on Terror in Afghanistan* 2005

Schwarzkopf, H. Norman, with Peter Petre, *It Doesn't Take a Hero: The Autobiography of General H. Norman Schwarzkopf* 1993

Scott, James A., *Rampage: MacArthur, Yamashita, and the Battle of Manila* 2018

Scurr, John, *The Malayan Campaign 1948–60* 1982

Sebastyen, Victor, *Twelve Days: Revolution 1956* 2006

Sheehan, Neil, *A Bright Shining Lie: John Paul Vann and America in Vietnam* 1988

Shelton, General Hugh, *Without Hesitation: The Odyssey of an American Warrior* 2011

Shlaim, Avi, *Lion of Jordan: The Life of King Hussein in War and Peace* 2007

Silber, Laura, and Little, Allan, *Yugoslavia: Death of a Nation* 1997

Simpson, Emile, *War from the Ground Up: Twentieth-Century Combat as Politics* 2018

Singer, P. W., *Wired for War: The Robotics Revolution and Conflict in the 21st Century* 2009

Sky, Emma, *The Unravelling: High Hopes and Missed Opportunities in Iraq* 2015

Smith, Rupert, *The Utility of Force: The Art of War in the Modern World* 2005

Snepp, Frank, *Decent Interval: An Insider's Account of Saigon's Indecent End* 1977

Sorley, Lewis, *A Better War: The Unexamined Victories and Final Tragedy of America's Last Years in Vietnam* 1999

—, *Westmoreland: The General Who Lost Vietnam* 2011

—, (ed.), *The Vietnam War: An Assessment by South Vietnam's Generals* 2010

Stanick, Joseph T., *El Dorado Canyon: Reagan's Undeclared War with Qaddafi* 2003

Stanley, William, *The Protection Racket State: Elite Politics, Military Extortion, and Civil War in El Salvador* 1996

Stanton, Doug, *Horse Soldiers: The Extraordinary Story of a Band of U.S. Soldiers Who Rode to Victory in Afghanistan* 2009

Sterling, Brent L., *Other People's Wars: The U.S. Military and the Challenge of Learning from Foreign Conflicts* 2021

Stetler, Richard (ed.), *The Military Art of People's War: Selected Writings of General Vo Nguyen Giap* 1970

Stewart, Richard W., *The United States Army in Somalia 1992–1994* 2002

—, *The United States Army in Afghanistan: Operation Enduring Freedom, October 2001–March 2002* 2003

—, *United States Special Operations Command History 1987–2007* 2007

—, *Operation Urgent Fury: The Invasion of Grenada October 1983* 2013

Strachan, Hew, and Scheipers, Sibylle (eds), *The Changing Character of War* 2011

Strohn, Matthias (ed.), *The Long Shadow of World War Two: The Legacy of the War and its Impact on Political and Military Thinking since 1945* 2021

Subtelny, Orest, *Ukraine: A History* 1988

Symonds, Craig L., *Nimitz at War* 2022

Tapper, Jake, *The Outpost: An Untold Story of American Valor* 2012

Taylor, Maxwell D., *Swords and Plowshares* 1972

Tenet, George, with Bill Harlow, *At the Center of the Storm: My Years at the CIA* 2007

Thompson, Julian, *No Picnic: 3 Commando Brigade in the South Atlantic: 1982* 1992

Thompson, Robert, *Defeating Communist Insurgency: The Lessons of Malaya and Vietnam* 2005

Trauschweizer, Ingo, *The Cold War U.S. Army: Building Deterrence for Limited War* 2008

Tsai, C. C. (ed.), *Sunzi: The Art of War* 2018

Tucker, Spencer C., *The Encyclopedia of Middle East Wars: The United States in the Persian Gulf, Afghanistan, and Iraq Conflicts* 2010

Tyson, Ann Scott, *American Spartan: The Promise, the Mission, and the Betrayal of Special Forces Major Jim Gant* 2014

Vadney, T. E., *The World since 1945* 1998

Van Creveld, Martin, *Command in War* 1985

Van de Ven, Hans, *China at War: Triumph and Tragedy in the Emergence of the New China 1937–1952* 2017

Vulliamy, Ed, *Seasons in Hell: Understanding Bosnia's War* 1994

—, *The War Is Dead, Long Live the War: Bosnia, the Reckoning* 2012

Waldman, Thomas, *Vicarious Warfare: American Strategy and the Illusion of War on the Cheap* 2021

Walker, General Sir Walter, *Fighting On* 1997

Weigley, Russell F., *The Age of Battles* 1991

West, Bing, *No True Glory: A Frontline Account of the Battle for Fallujah* 2005

—, *The Strongest Tribe: War, Politics, and the Endgame in Iraq* 2008

Westad, Odd Arne, *Decisive Encounters: The Chinese Civil War 1946–1950* 2003

Westermeyer, Paul, with Christopher N. Blaker, *U.S. Marines in Afghanistan, 2010–2014: Anthology and Annotated Bibliography* 2017

Westmoreland, William, *A Soldier Reports* 1976

Whitlock, Craig, *The Afghanistan Papers: A Secret History of the War* 2021

Willbanks, James H., *Vietnam War Almanac* 2013

Wiest, Andrew, *Vietnam's Forgotten Army: Heroism and Betrayal in the ARVN* 2008

Woods, Kevin M., *Iraqi Perspectives Project: A View of Operation Iraqi Freedom from Saddam's Senior Leadership* 2006

Woodward, Bob, *Obama's Wars* 2010

Wright, Donald P., and Reese, Colonel Timothy R., *On Point II: Transition to the New Campaign: The United States Army in Operation Iraqi Freedom*, May 2003–January 2005 2008

—, et al., *A Different Kind of War: The U.S. Army in Operation Enduring Freedom, October 2001–September 2005* 2010

Zacke-Williams, Tunde, *When the State Fails: Studies on Intervention in the Sierra Leone Civil War* 2011

Zelensky, Volodymyr, *A Message from Ukraine: Speeches 2019–2022* 2022

Zetter, Kim, *Countdown to Zero Day: Stuxnet and the Launch of the World's First Digital Weapon* 2014

Articles

Adelman, Howard, "Chaos in the Camps," *Bulletin of the Atomic Scientists*, 58(6), 2016

Allen, William J., "Intervention in Panama: Operation Just Cause," in A. Timothy Warnock (ed.), *Short of War: Major USAF Contingency Operations 1947–1997*, Air Force History and Museums Program, 2000

Altman, Howard, "Why dissolving the Afghan Local Police program troubles its American architects," *Military Times*, 27 May 2020

Andrade, Dale, and Willbanks, James H., "CORDS/Phoenix: Counterinsurgency Lessons from Vietnam for the Future," *Military Review*, March–April 2006 pp. 9–23

Arkin, William M., "The Difference was in the Details," *Washington Post*, 17 January 1999

Ball, Captain Gregory, "1998 – Operation Desert Fox," Air Force Historical Support Division, https://www.afhistory.af.mil/FAQs/Fact-Sheets/Article/458976/1998-operation-desert-fox/

Balz, Dan, "President Warns Iraq of War Crimes Trials," *Washington Post*, 16 October 1990

Boone, Jon, "The financial scandal that broke Afghanistan's Kabul Bank," *Guardian*, 16 June 2011, https://www.theguardian.com/world/2011/jun/16/kabul-bank-afghanistan-financial-scandal

Bowden, Mark, "When Walter Cronkite Pronounced the War a 'Stalemate,'" *New York Times*, 26 February 2018

Carver, Michael, "Conventional Warfare in the Nuclear Age," in Peter Paret (ed.), *Makers of Modern Strategy*, 1986

Chotiner, Isaac, "Is the Russian Military a Paper Tiger?," *New Yorker*, 22 April 2022

Clines, Francis X., "Conflict in the Balkans: The Rescue," *New York Times*, 9 June 1995

Coffey, Ross, "Revisiting CORDS: The Need for Unity of Effort to Secure Victory in Iraq," *Military Review*, March–April 2006 pp. 24–34

Dawkins, Peter M., "The United States Army and the 'Other' War in Vietnam: A Study of the Complexity of Implementing Organizational Change," doctoral dissertation, Princeton University 1977

Drell, Sidney D., "The Shadow of the Bomb," *Policy Review*, 136, April–May 2006 p. 68

Drumbl, Mark A., "She Makes Me Ashamed to Be a Woman: The Genocide Conviction of Pauline Nyiramasuhuko, 2011," *Michigan Journal of International Law*, 2013

Dwyer, Jim, "A Nation at War: In the Field – V Corps Commander," *New York Times*, 28 March 2003

Endicott, Judy G., "Raid on Libya: Operation El Dorado Canyon," in A. Timothy Warnock (ed.), *Short of War: Major USAF Contingency Operations 1947–1997*, Air Force History and Museums Program, 2000

Feaver, Peter D., "The Right to Be Right: Civil–Military Relations and the Iraq Surge Decision," *International Security*, 35(4), Spring 2011

Fedarko, Kevin, and Thompson, Mark, "Rescuing Scott O'Grady: All for One," *Time*, 19 June 1995

Fein, Helen, "Discriminating Genocide from War Crimes: Vietnam and Afghanistan Reexamined," *Denver Journal of International Law & Policy*, 22(1), 1993

Filkins, Dexter, "Last Exit from Afghanistan," *New Yorker*, 8 March 2021

Freedman, Lawrence, "The First Two Generations of Nuclear Strategists," in Peter Paret (ed.), *Makers of Modern Strategy* 1986

Galvin, John R., "Uncomfortable Wars: Toward a New Paradigm," *Parameters*, 16(1), 1986 pp. 2–8

Gaouette, Nicole, et al., "The last US military planes have left Afghanistan, marking the end of the United States' longest war," CNN, 31 August 2021, https://www.cnn.com/2021/08/30/politics/us-military-withdraws-afghanistan/index.html

Handel, Michael, "Crisis and Surprise in Three Arab–Israeli Wars," in Klaus Knorr and Patrick Morgan (eds), *Strategic Military Surprise*, 1988

Hanson, Victor Davis, "Uses and Abuses of Military History," *New Criterion*, January 2023

Hastings, Michael, "The Runaway General," *Rolling Stone*, 22 June 2010

Haulman, Daniel L., "Crisis in Grenada: Operation Urgent Fury," in A. Timothy Warnock (ed.), *Short of War: Major USAF Contingency Operations 1947–1997*, Air Force History and Museums Program, 2000

—, "Crisis in Somalia: Operations Provide Relief and Restore Hope," in A. Timothy Warnock (ed.), *Short of War: Major USAF Contingency Operations 1947–1997*, Air Force History and Museums Program, 2000

—, "Resolution of Bosnia Crisis: Operation Deny Flight," in A. Timothy Warnock (ed.), *Short of War: Major USAF Contingency Operations 1947–1997*, Air Force History and Museums Program, 2000

Hess, Pamela, "Rumsfeld: Looting is transition to freedom," UPI, 11 April 2003

Hoffman, Frank G., and Mattis, James N., "Future Warfare: The Rise of Hybrid Wars," *Naval Institute Proceedings*, 132(11), November 2005

Holoboff, Elaine M., "Bad Boy or Good Business?: Russia's Use of Oil as a Mechanism of Coercive Diplomacy," in Lawrence Freedman (ed.), *Strategic Coercion*, 1998

Hopkins, Benjamin, "Afghanistan's Present Failure Lies in its Past Design," *Middle East Research and Information Project*, 18 January 2022, https://merip.org/2022/01/afghanistans-present-failure-lies-in-its-past-design/

—, "The Problem with 'Hearts and Minds' in Afghanistan," *Middle East Research and Information Project*, Summer 2010, https://merip.org/2010/05/the-problem-with-hearts-and-minds-in-afghanistan/

International Crisis Group, "Georgia: Avoiding War in South Ossetia," ICG *Europe Report*, No. 159, 26 November 2004

Kotkin, Stephen, "Freedom at Stake: How Did Russia and the West Fall Out?," *Times Literary Supplement*, 11 March 2022 p. 3

Lambeth, Benjamin S., "Operation Allied Force: Lessons for the Future," https://www.rand.org/pubs/research_briefs/RB75.html

Leinster, Colin, "The Two Wars of General Walt," *Life*, 26 May 1967

Matthews, Owen, "The red line: Biden and Xi's secret Ukraine talks revealed," *Spectator*, 26 November 2022 p. 13

Mearsheimer, John, "Back to the Future: Instability in Europe after the Cold War," *International Security*, 15(1), Summer 1990

Mills, Nicolaus, "Punished for telling truth about Iraq war," 20 March 2013, https://www.cnn.com/2013/03/20/opinion/mills-truth-teller-iraq/index.html

Neiberg, Michael S., "Total War, Total Victory: World War II in American Memory and Strategy," in Matthias Strohn (ed.), *The Long Shadow of World War Two: The Legacy of the War and its Impact on Political and Military Thinking since 1945*, 2021

Ong Keng Yong, "Lee Kuan Yew's Role in Singapore–U.S. Relations," *Asia Society*, 25 March 2015

Palermo, Joseph A., "40 Years Ago Today: Robert F. Kennedy's Most Important Speech on the Vietnam War," HuffPost.com, 8 February 2008

Pancevski, Bojan, "Europe Is Rushing Arms to Ukraine, But Running Out of Ammo," *Wall Street Journal*, 22 December 2022

Payne, Kenneth, "Artificial Intelligence as a revolution in military affairs," *Survival*, 60(5), October–November 2018

Peoples, Curtis, "The Use of the British Village Resettlement Model in Malaya and Vietnam," Texas Tech University, 4th Triennial Vietnam Symposium, 11–13 April 2002

Petraeus, David H., "The American Military and the Lessons of Vietnam," doctoral dissertation, Princeton University 1987

—, "Learning Counterinsurgency: Observations from Soldiering in Iraq," *Military Review Counterinsurgency Reader*, October 2006

—, and Serchuk, Vance, "Counter Russia's and China's Playbook," in "Lessons for the Next War," *Foreign Affairs*, 5 January 2023

Pipes, Richard, "Why the Soviet Union Thinks It Could Fight and Win a Nuclear War," *Commentary*, July 1977

Rempfer, Kyle, "Soldiers recall combat jumps into Panama on 30th anniversary," *Army Times*, 20 December 2019

Rezaei, Farhad, "The Islamic Revolution and the Bomb: Quran Meets Realism," in *Iran's Nuclear Program: A Study in Proliferation and Rollback*, 2017

Rice, Condoleezza, "The Making of Soviet Strategy," in Peter Paret (ed.), *Makers of Modern Strategy*, 1986

Sammut, Dennis, and Cvetkovski, Nikola, "The Georgia–South Ossetia Conflict," *Verification Technology Information Center*, 1996

Sennott, Charles M., "The First Battle of the 21st Century," *Atlantic*, 5 May 2015, https://www.theatlantic.com/international/archive/2015/05/war-afghanistan-spann-qala-i-jangi/392402/

Shaw Jr., Frederick J., "Crisis in Bosnia: Operation Provide Promise," in A. Timothy Warnock (ed.), *Short of War: Major USAF Contingency Operations 1947–1997*, Air Force History and Museums Program, 2000

Stewart, Richard W., "The United States Army in Somalia 1992–1994," US Army Center of Military History, Publication 70-81-1, 2003

Thompson, Mark, "The Fall of the Green Berets' *Lawrence of Afghanistan*," *Time*, 25 June 2014, https://time.com/2921469/the-fall-of-the-green-berets-lawrence-of-afghanistan/

Tirpak, John A., "Deliberate Force," *Air Force Magazine*, 1 October 1997

Ucko, David H., "Can Limited Intervention Work? Lessons from Britain's Success Story in Sierra Leone," *Journal of Strategic Studies*, 39(5 & 6), 2016

Van Vien, Gen. Cao, and Van Khuyen, Lt. Gen. Dong, "Reflections on the Vietnam War," in Sorley, Lewis (ed.), *The Vietnam War: An Assessment by South Vietnam's Generals*, 2010

Warner, Denis, "The Morning after the War Before," *Atlantic*, December 1972

Washington Post Editorial Board, "In Afghanistan, the Lights Go Out for Women," *Washington Post*, 26 December 2022

Watling, Jack, and Reynolds, Nick, "Operation Z: The Death Throes of an Imperial Delusion," Royal United Services Institute Special Report, 22 April 2022

White, Paul K., "Airpower and a Decade of Containment," *Joint Force Quarterly*, Winter 2000/1

Wilkins, Sam, "The Rise and Fall of Village Stability Operations in Afghanistan: Lessons for Future Irregular Warfare Campaigns," *Modern War Institute*, 9 August 2022

Wilkins, Warren, "When Strategy Isn't Enough: William Westmoreland and the War in Vietnam," *On Point*, December 2020

Government sources

Committee on Armed Services, US House of Representatives, "Security and Stability in Afghanistan: Status of U.S. Strategy and Operations and the Way Ahead," 110th Cong., 1st sess., 11 December 2007

Department of the Army, Field Manual 23-4, *Counterinsurgency* (December 2006)

Department of the Army, *Vietnam Studies: Airmobility 1961–1971* (Washington, DC, 1973)

Foreign Relations of the United States, 1961–1963, vol. III: *Vietnam, January–August 1963* (Washington, DC, 1991)

National Commission on Terrorist Attacks upon the United States, *The 9/11 Commission Report* (Washington, DC, 2004)

Special Inspector-General for Afghanistan Reconstruction, *What We Need to Learn: Lessons from Twenty Years of Afghanistan Reconstruction* (August 2021)

Special Inspector-General for Afghanistan Reconstruction, *Why the Afghan Government Collapsed* (November 2022)

Subcommittee on National Security and Foreign Affairs, Committee on Oversight and Government Reform, *Warlord, Inc.: Extortion and Corruption Along the U.S. Supply Chain in Afghanistan* (US House of Representatives, June 2010)

NOTES

Introduction

1. Howard, *War and the Nation State* p. 6
2. The intellectual construct for strategic leadership used throughout *Conflict* derives from that developed by General Petraeus and his team at the Belfer Center at the Harvard Kennedy School: https://www.belfercenter.org/publication/david-petraeus-strategic-leadership

One: The Death of the Dream of Peace

1. Lowe, *The Fear and the Freedom* p. 188
2. Howard, *The Invention of Peace* p. 1
3. Archer et al., *World History of Warfare* p. 550
4. Kissinger, *Leadership* p. 410
5. Kissinger et al., *The Age of A.I.* p. 143
6. Ibid.
7. Black, *A Short History of War* p. 203
8. Freedman and Michaels, *The Evolution of Nuclear Strategy* p. viii
9. Ibid., p. 36
10. Katz, *Life after Nuclear War* p. 3
11. Ibid., p. 4
12. Scharre, *Army of None* pp. 1–2
13. Freedman and Michaels, *The Evolution of Nuclear Strategy* p. 668
14. Ibid.
15. MacMillan, *War* p. 284
16. Keegan, *A History of Warfare* p. 56

17. Freedman and Michaels, *The Evolution of Nuclear Strategy* p. ix
18. Pavelec, *War and Warfare since 1945* p. 9
19. Dikötter, *The Tragedy of Liberation* p. 11
20. Ibid., p. 13
21. Lynch, *Chinese Civil War* p. 18
22. Westad, *Decisive Encounters* p. 7; Mitter, *China's War with Japan* p. 370
23. Mitter, *China's War with Japan* p. 370
24. Van de Ven, *China at War* p. 2
25. Westad, *Decisive Encounters* p. 7
26. Ibid., p. 8
27. Archer et al., *World History of Warfare* p. 558
28. Lynch, *Chinese Civil War* p. 91
29. Westad, *Decisive Encounters* p. 9
30. Mitter, *China's War with Japan* pp. 377–8
31. Lynch, *Chinese Civil War* p. 24
32. Archer et al., *World History of Warfare* p. 558
33. Tsai (ed.), *Sunzi: The Art of War* p. viii
34. Lynch, *Chinese Civil War* p. 26
35. Ibid., p. 28
36. Van de Ven, *China at War* p. 5
37. Mumford, *Proxy Warfare* p. 12
38. Archer et al., *World History of Warfare* p. 559
39. Ibid., p. 558
40. Lynch, *Chinese Civil War* p. 22
41. Ibid.
42. Ibid., p. 23
43. Dikötter, *The Tragedy of Liberation* p. 18
44. Lynch, *Chinese Civil War* p. 21
45. "China is Wordless on Traumas of Communists' Rise," *New York Times*, 1 October 2009
46. Ibid.
47. Dikötter, *The Tragedy of Liberation* pp. 17–18
48. Van de Ven, *China at War* p. 250
49. Ibid.
50. Ibid., p. 251
51. Lynch, *Chinese Civil War* p. 7
52. Gaddis, *We Now Know* p. 71
53. Horne, *Hubris* p. 282
54. Ibid.
55. Gaddis, *We Now Know* p. 75
56. Pavelec, *War and Warfare since 1945* p. 2
57. Neiberg, "Total War, Total Victory" p. 135

58. McCullough, *Truman* pp. 780–1
59. Horne, *Hubris* p. 291
60. Hastings, *Korean War* Appendix B
61. Ibid., p. 503
62. Gaddis, *We Now Know* p. 72
63. Roberts, *Stalin's Wars* p. 369
64. *Financial Times*, 16 September 2022, https://on.ft.com/3BReSKs
65. Neiberg, "Total War, Total Victory" p. 134
66. Symonds, *Nimitz at War passim*
67. Scott, *Rampage* p. 21
68. Neiberg, "Total War, Total Victory" p. 135
69. Ibid., p. 134
70. Horne, *Hubris* p. 287
71. Ibid., p. 288
72. Hastings, *The Korean War* p. 503
73. Pavelec, *War and Warfare since 1945* p. 20
74. Horne, *Hubris* p. 296
75. Freedman (ed.), *War* p. 43
76. McCullough, *Truman* p. 800
77. Ibid., p. 801
78. Ibid.
79. Horne, *Hubris* p. 294
80. McCullough, *Truman* p. 802
81. Ibid., p. 804
82. Horne, *Hubris* p. 296
83. McCullough, *Truman* p. 806
84. Ibid.
85. Rusk, *As I Saw It* p. 169
86. Archer et al., *World History of Warfare* p. 555
87. Neiberg, "Total War, Total Victory" p. 134
88. Hanson, *Savior Generals* p. 157
89. McCullough, *Truman* p. 815
90. Ibid.
91. Hanson, *Savior Generals* p. 141
92. Ibid., p. 148
93. Ibid., p. 166
94. Black, *War since 1945* p. 29
95. MacMillan, *War* p. 61
96. Archer et al., *World History of Warfare* p. 556
97. Hastings (ed.), *Oxford Book of Military Anecdotes* p. 474
98. Archer, *World History of Warfare* p. 556
99. Paret (ed.), *Makers of Modern Strategy* p. 780

100. Hastings, *The Korean War* p. xv
101. Ibid., p. xxi
102. Horne, *Hubris* p. 313

Two: Wars of Decolonization

1. Clausewitz, *On War*, 1984 edn p. 99
2. Black, *A Short History of War* p. 209
3. Boot, *Invisible Armies* p. 50
4. Richardson, *What Terrorists Want* passim
5. Schofield, *Kashmir in Conflict* p. 27
6. Ibid., p. 58
7. Black, *Air Power* p. 181
8. Krammer, *The Forgotten Friendship* p. 103
9. Schofield, *Kashmir in Conflict* p. 258
10. Bose, *Kashmir at the Crossroads* p. 291
11. Herzog, *The Arab–Israeli Wars* p. 338
12. Paret (ed.), *Makers of Modern Strategy* pp. 790–1
13. Gilbert, *Israel* p. 192
14. Peres, *No Room for Small Dreams* p. 49
15. Herzog, *The Arab–Israeli Wars* p. 339
16. Pavelec, *War and Warfare since 1945* p. 14
17. Peres, *No Room for Small Dreams* p. 48
18. Gilbert, *Israel* p. 201
19. Peres, *No Room for Small Dreams* p. 47
20. Gilbert, *Israel* p. 202
21. Ibid., p. 205
22. Ibid., p. 208
23. Ibid., p. 218
24. Ibid., p. 219
25. Ibid., p. 220
26. Ibid.
27. Oren, *Six Days of War* p. 5
28. Gilbert, *Israel* p. 249
29. Herzog, *The Arab–Israeli Wars* p. 338
30. Ibid., pp. 338–9
31. Scurr, *Malayan Campaign* p. 3
32. Boot, *Invisible Armies* p. 388
33. Scurr, *Malayan Campaign* p. 4
34. Boot, *Invisible Armies* p. 380
35. Ibid., p. 385
36. Scurr, *Malayan Campaign* p. 14

37. Boot, *Invisible Armies* p. 382
38. Ibid.
39. Churchill, *Malakand Field Force* p. 311
40. Boot, *Invisible Armies* p. 565
41. Scurr, *Malayan Campaign* p. 9
42. Ibid., pp. 15–17
43. Ibid., p. 18
44. Ibid.
45. Boot, *Invisible Armies* p. 384
46. Ibid.
47. Petraeus and Amos, *Counterinsurgency Field Manual* pp. 6–23
47. Ibid.
49. Horne, *A Savage War of Peace* p. 205
50. Boot, *Invisible Armies* p. 386
51. Ibid.
52. Ibid., p. 388
53. Ibid.; Scurr, *Malayan Campaign* p. 31
54. https://web.archive.org/web/20130922173616/; http://www.chinhphu.vn/portal/page/portal/English/TheSocialistRepublicOfVietnam/AboutVietnam/AboutVietnamDetail?categoryId=10000103&articleId=10002648
55. Barnet, *Intervention and Revolution* p. 185
56. Freedman (ed.), *War* p. 331
57. Ibid.
58. Ibid.
59. Clodfelter, *Warfare and Armed Conflicts* p. 657
60. Boot, *Invisible Armies* p. 392
61. Hastings (ed.), *The Oxford Book of Military Anecdotes* p. 475
62. Fall, *Street without Joy* p. 6
63. Ibid., p. viii
64. Howard et al. (eds), *The Laws of War* p. 205
65. Galula, *Counterinsurgency Warfare* p. 20
66. Ibid., p. 40
67. Horne, *A Savage War of Peace* p. 197
68. Howard et al. (eds), *The Laws of War* p. 206
69. Ibid.
70. Horne, *A Savage War of Peace* p. 199
71. Ibid., p. 204
72. Ibid., pp. 201–4
73. Ibid., p. 172
74. Cormac, *How to Stage a Coup* p. 18
75. Horne, *A Savage War of Peace* p. 195

76. Ibid., pp. 200–1
77. Ibid., p. 201
78. Ibid., p. 207
79. Howard et al. (eds), *The Laws of War* p. 207
80. Robbins, *The World since 1945* p. 87
81. Lacouture, *De Gaulle: The Ruler* p. 248
82. Vadney, *The World since 1945* p. 232
83. Boot, *Invisible Armies* p. 392
84. Simpson, *War from the Ground Up* p. 165
85. Walker, *Fighting On* p. 159
86. Ibid., p. 150
87. Simpson, *War from the Ground Up* p. 164
88. Ibid., p. 170
89. Ibid., p. 164
90. Ibid., p. 169
91. Ibid., p. 168
92. Walker, *Fighting On* p. 148
93. Akehurst, *We Won a War* p. 12
94. Ibid., p. 13
95. Ibid., p. 5
96. Ibid., p. 15
97. Ibid., p. 8
98. Ibid., pp. 25, 27
99. Ibid., p. 29
100. Ibid., p. 62
101. Ibid., p. 63
102. Ibid., p. 71
103. Ibid., p. 65
104. Ibid., p. 64
105. Ibid., p. 3
106. Ibid., p. 72
107. Ibid., p. 22
108. Ibid., p. 21
109. Ibid., p. 27
110. Ibid., p. 70
111. Jones, *The Art of War in the Western World* p. 600

Three: America's War in Vietnam

1. Clausewitz, *On War*, 1984 edn pp. 88–9
2. Ferguson, *Kissinger* p. 665
3. Krepinevich, *The Army and Vietnam* p. 4

4. Petraeus, "The American Military and the Lessons of Vietnam"
5. Ferguson, *Kissinger* p. 668
6. General Bruce Palmer's book on Vietnam was in fact titled *The 25-Year War: America's Military Role in Vietnam*
7. Department of the Army, Field Manual 23-4, *Counterinsurgency*, p. 11; Krepinevich, *The Army and Vietnam* p. 7; Johnson, *Autopsy on People's War*; and Thompson, *Defeating Communist Insurgency*
8. Department of the Army, Field Manual 23-4, *Counterinsurgency*, pp. 38–9
9. Nagl, *Learning to Eat Soup with a Knife* p. 115
10. Krepinevich, *The Army and Vietnam* p. 24
11. Ibid.
12. Ibid., p. 25
13. Ibid., p. 26
14. Ibid.
15. Petraeus, "The American Military and the Lessons of Vietnam" p. 92. The term "best and brightest" comes from Halberstam, *The Best and Brightest*
16. https://www.jfklibrary.org/asset-viewer/archives/USG/USG-17/USG-17
17. Ibid.
18. Logevall, *Embers of War* p. 714
19. Gelb and Betts, *The Irony of Vietnam* p. 69. Lansdale's views carried considerable weight as he had played a key role in helping two Filipino presidents eliminate the Huk insurgency in the early 1950s before shifting his focus to Vietnam
20. Ibid., pp. 70–1
21. Ibid., p. 72
22. Taylor, *Swords and Plowshares* pp. 225–6
23. Ibid., pp. 228–9
24. Ibid., pp. 232–3
25. Herring, *America's Longest War* pp. 81–5
26. Peoples, "The Use of the British Village Resettlement Model in Malaya and Vietnam," http://www.vietnam.ttu.edu/events/2002_Symposium/2002Papers_files/peoples.php
27. Karnow, *Vietnam: A History* p. 274
28. Ibid.
29. Herring, *America's Longest War* p. 88
30. Ibid., pp. 92–3
31. Ibid., p. 93; Hilsman, *To Move a Nation* p. 464
32. Herring, *America's Longest War* p. 93
33. Krepinevich, *The Army and Vietnam* p. 78
34. Ibid., p. 79; Neil Sheehan's description of this battle is particularly noteworthy. See Sheehan, *A Bright Shining Lie* pp. 203–65, with the casualty figures on p. 263

35. Krepinevich, *The Army and Vietnam* p. 79
36. *New York Times*, 15 January 1963
37. Cosmas, *MACV: The Joint Command in the Years of Escalation 1962–1967* p. 80
38. *Foreign Relations of the United States, 1961–1963*, vol. III: *Vietnam, January–August 1963*, 26
39. Herring, *America's Longest War* pp. 103–6; also see Hilsman, *To Move a Nation* pp. 483–94 and 514–23
40. Herring, *America's Longest War* p. 107; see, most prominent among Kennedy's defenders, Hilsman, *To Move a Nation* pp. 524–37
41. Ford, *CIA and the Vietnam Policymakers* pp. 21–2
42. Asselin, *Vietnam's American War* pp. 107–9
43. Pavelec, *War and Warfare since 1945* p. 48. For a superb assessment which concludes that the US could have chosen *not* to escalate and even to withdraw is presented by Logevall in *Choosing War*
44. Herring, *America's Longest War* p. 111
45. Ibid., p. 112
46. Ibid., p. 116; Krepinivech, *The Army and Vietnam* pp. 93–4
47. Krepinivech, *The Army and Vietnam* p. 95
48. Ibid., p. 97
49. Herring, *America's Longest War* pp. 125–6; Krepinevich, *The Army and Vietnam* pp. 98–9
50. Herring, *America's Longest War* pp. 128–9
51. Ibid., pp. 130–1
52. Ibid., pp. 134–5
53. Ibid., pp. 136–7
54. Ibid., pp. 138–9
55. Krepinevich, *The Army and Vietnam* pp. 156–9
56. Ibid., pp. 164–5
57. Ibid., p. 165
58. The battle is recounted in a classic work of the war by Lieutenant General (Ret.) Harold G. Moore, who commanded the main effort battalion and Joseph L. Galloway, the journalist who was with the unit. Moore and Galloway, *We Were Soldiers Once ... and Young*
59. Ibid., p. 345
60. Ibid., p. 343
61. Ibid., p. 342
62. Westmoreland, *A Soldier Reports* p. 191
63. Cited in Krepinevich, *The Army and Vietnam* p. 172; on Marine Corps operations in Vietnam, see also the superb study Hennessy, *Strategy in Vietnam*, and the excellent profile of then-Lt. Gen. Lew Walt, the Marine Commander in I Corps, Leinster, "The Two Wars of General Walt" pp. 83–4

64. Leinster, "The Two Wars of General Walt" pp. 83–4
65. Ibid.; Krepinevich, *The Army and Vietnam* pp. 173–4
66. Westmoreland, *A Soldier Reports* pp. 201–2
67. Krepinevich, *The Army and Vietnam* pp. 176–7
68. Ibid., p. 179
69. Ibid., pp. 179–80
70. Ibid., p. 180
71. Herring, *America's Longest War* p. 154
72. Sorley, *Westmoreland* pp. 104–5; the full text of the report can be found at https://apps.dtic.mil/sti/pdfs/AD0377743.pdf
73. Text of PROVN report, p. 9
74. Quoted in Sorley, *Westmoreland* p. 105. Sorley also describes a Marine document published in Vietnam several months later, in October 1966, that echoed the emphases in the PROVN study: "The very base of the war is the people of Vietnam. The revolutionary development program is the most critical of all the tasks, with the others being ancillary or contributing"
75. Ibid., p. 101
76. Quoted in Krepinevich, *The Army and Vietnam* pp. 183–4
77. Herring, *America's Longest War* pp. 155–6
78. Ibid., pp. 155–3
79. Quoted in Krepinevich, *The Army and Vietnam* p. 190. An account with particularly valuable insights into the conduct of search-and-destroy operations can be found in David Maraniss' exceptional work, *They Marched into Sunlight*, especially the chapters on the operations of the 2nd Battalion, 28th Infantry of the 1st Infantry Division
80. Krepinevich, *The Army and Vietnam* p. 191
81. Asselin, *Vietnam's American War* p. 157
82. See the text of the speech at https://timesmachine.nytimes.com/timesmachine/1967/04/29/83118429.pdf?pdf_redirect=true&ip=0
83. Quoted in Sorley, *Westmoreland* p. 167. See also Herring, *America's Longest War* p. 183
84. Krepinevich, *The Army and Vietnam* p. 193
85. Pavelec, *War and Warfare since 1945* pp. 56–7
86. Asselin, *Vietnam's American War* p. 158
87. Warner, "The Morning after the War Before" pp. 117–18
88. Asselin, *Vietnam's American War* pp. 157–9
89. Palermo, "40 Years Ago Today: Robert F. Kennedy's Most Important Speech on the Vietnam War"
90. Bowden, "When Walter Cronkite Pronounced the War a 'Stalemate'"
91. Schandler, *Lyndon Johnson and Vietnam* p. 290
92. Ibid., p. 318

93. Willbanks, *Vietnam War Almanac* p. 256

94. A superb discussion of the Clifford Task Forces is in Schandler, *Lyndon Johnson and Vietnam* pp. 121–76

95. Willbanks, *Vietnam War Almanac* p. 258

96. Sorley, *A Better War* p. 17

97. Ibid., pp. 17–20

98. Ibid., pp. 18–22

99. Ibid., p. 28

100. Krepinevich, *The Army and Vietnam* p. 254

101. Komer's first deputy was then-Brig. Gen. William A. Knowlton, later the father-in-law to General David Petraeus

102. An excellent overview of the CORDS program can be found in the article by Dale Andrade and James H. Willbanks, "CORDS/Phoenix." Also see Coffey, "Revisiting CORDS: The Need for Unity of Effort to Secure Victory in Iraq"

103. Andrade and Willbanks, "CORDS/Phoenix" pp. 14–17. Komer never gained General Abrams' confidence and was replaced in late 1968 by Ambassador William Colby, who had been the CIA Station Chief in Saigon and who established in November 1968 the Accelerated Pacification Program that energized CORDS and Phoenix and helped them be much more effective. See also Sorley, *A Better War* pp. 59–79

104. Andrade and Willbanks, "CORDS/Phoenix" pp. 18–19. Considerably more detail can be found in Mark Moyar's superb, very thorough *Phoenix and the Birds of Prey*

105. On this, see Lewy, *America in Vietnam* pp. 279–85

106. Andrade and Willbanks, "CORDS/Phoenix" p. 20; also see the detailed assessments of the program by Moyar in *Phoenix and the Birds of Prey*. CORDS and its various components, including the Phoenix Program, would be studied carefully as the US Army and US Marine Corps worked together in 2006 to publish the Counterinsurgency Field Manual that would be the intellectual foundation for the subsequent Surge in Iraq and operations in Afghanistan

107. Herring, *America's Longest War* pp. 222–4

108. Ibid., pp. 224–5

109. Sorley, *A Better War* p. 128

110. Ibid., pp. 129–30

111. Ibid., pp. 155–7

112. Herring, *America's Longest War* pp. 226–7

113. Ibid., pp. 228–9

114. Ibid., pp. 235–9

115. Ibid., pp. 240–1

116. Ibid., pp. 246–7

117. Ibid., pp. 250–5
118. Former CIA officer Frank Snepp provides a detailed examination of the withdrawal agreement and its execution in *Decent Interval*
119. The text of the agreement can be found at https://treaties.un.org/doc/ Publication/UNTS/Volume%20935/volume-935-I-13295-English.pdf; see also the assessment by Gen. Bruce Palmer, *The 25-Year War* p. 130. Henry Kissinger's detailed account of the process, the agreement and the aftermath, as well as his reflections, can be found in his *White House Years*, pp. 230–311, 436–521, 968–1048 and 1301–1476 and his *Years of Upheaval*, pp. 9–12, 301–73
120. Willbanks, *Vietnam War Almanac* p. 195
121. Ibid., p. 438; Herring, *America's Longest War* pp. 260–1
122. Herring, *America's Longest War* p. 262
123. Ibid., pp. 262–3
124. Ibid., pp. 263–4
125. Black, *War since 1945* p. 71
126. Ibid., p. 72
127. Gen. Cao Van Vien and Lt. Gen. Dong Van Khuyen, "Reflections on the Vietnam War," in Sorley (ed.), *The Vietnam War* p. 842. This volume is a source of exceptional essays and reflections by select former senior South Vietnamese generals who made their way to the US and were convened by the US Army's Center of Military History from 1976 to 1978. See, in particular, pp. 831–44
128. Andrade and Willbanks, "CORDS/Phoenix" p. 44. Also see Komer, *Bureaucracy Does Its Thing*
129. See, for example, the superb Princeton University doctoral dissertation by Peter M. Dawkins, "The United States Army and the 'Other' War in Vietnam: A Study of the Complexity of Implementing Organizational Change"
130. Krepinevich, *The Army and Vietnam* esp. pp. 27–55
131. Kitfield, *Prodigal Soldiers* pp. 104–6
132. Krepenevich, *The Army and Vietnam* p. 206
133. The text of the study can be found at https://apps.dtic.mil/sti/pdfs/ ADA063748.pdf
134. Kitfield, *Prodigal Soldiers* pp. 109–10
135. Palmer, *The 25-Year War* p. 179
136. Ibid., pp. 175–6
137. Krepinevich, *The Army and Vietnam* p. 271
138. Palmer, *The 25-Year War* p. 173
139. https://web.archive.org/web/20100413071553/http://www.ford.utexas. edu/library/exhibits/vietnam/750512f.htm; Singapore's highly respected president Lee Kuan Yew echoed Kissinger's and Palmer's assessments.

See Ong Keng Yong, "Lee Kuan Yew's Role in Singapore–U.S. Relations," *Asia Society*, 25 March 2015

140. Palmer, *The 25-Year War* p. 21
141. Kitfield, *Prodigal Soldiers*

Four: From the Sinai to Port Stanley

1. Colbeck, *With 3 Para to the Falklands* p. 185
2. Paret (ed.), *Makers of Modern Strategy* p. 791
3. Oren, *Six Days of War* pp. 164, 168, 171
4. Herzog, *Arab–Israeli Wars* p. 149
5. Oren, *Six Days of War* p. 164
6. Herzog, *Arab–Israeli Wars* p. 148
7. Oren, *Six Days of War* p. 170
8. Herzog, *Arab–Israeli Wars* p. 153
9. Horne, *Kissinger's Year* p. 238
10. Herzog, *Arab–Israeli Wars* p. 153
11. Dunstan, *The Six Day War 1967* p. 125
12. Herzog, *Arab–Israeli Wars* p. 165
13. Mutawi, *Jordan in the 1967 War* p. 139
14. Shlaim, *Lion of Jordan* p. 244
15. Herzog, *Arab–Israeli Wars* p. 183
16. Ibid., pp. 189–90
17. Ibid., p. 189
18. Paret (ed.), *Makers of Modern Strategy* p. 794
19. Herzog, *Arab–Israeli Wars* p. 189
20. Herzog, *The War of Atonement* p. xii
21. Horne, *Kissinger's Year* p. 237
22. Herzog, *The War of Atonement* p. xiii
23. Morris, *Righteous Victims* p. 390
24. Herzog, *The War of Atonement* p. xiv
25. Horne, *Kissinger's Year* pp. 238–9
26. Herzog, *The War of Atonement* p. xiii
27. Horne, *Kissinger's Year* p. 230
28. Handel, "Crisis and Surprise in Three Arab–Israeli Wars" pp. 136–40
29. Ibid., p. 113
30. Mark Helprin in *Claremont Review of Books*, Spring 2022 p. 12
31. Jones, *The Art of War in the Western World* p. 604
32. Smith, *The Utility of Force* p. 1
33. Jones, *The Art of War in the Western World* p. 605
34. Rabinovich, *The Yom Kippur War* p. 401
35. Pollack, *Arabs at War: Military Effectiveness* p. 120

36. Hastings (ed.), *Oxford Book of Military Anecdotes* p. 485
37. Ibid.
38. Herzog, *The War of Atonement* p. xix
39. Ibid., p. xi
40. Ibid., p. xii
41. Freedman, *Cold War* p. 166
42. Ibid., p. 160; Paret (ed.), *Makers of Modern Strategy* p. 797
43. Sterling, *Other People's Wars* p. 189
44. Kagan, *Finding the Target* p. 19
45. Sterling, *Other People's Wars* p. 246
46. Freedman, *Cold War* p. 166
47. Sterling, *Other People's Wars* p. 247
48. Ibid., p. 191
49. Ibid., p. 249
50. Herzog, *The War of Atonement* p. xvi
51. Freedman, *Cold War* p. 181; Reynolds, *One World Divisible* p. 365
52. Roberts, *Penguin History of the Twentieth Century* p. 746
53. Reynolds, *One World Divisible* p. 397
54. Fein, "Discriminating Genocide from War Crimes: Vietnam and Afghanistan Reexamined" pp. 29–62
55. Reynolds, *One World Divisible* p. 365
56. "4 Soviet deserters tell of cruel Afghanistan war," *New York Times*, 3 August 1984
57. Kakar, *The Soviet Invasion and the Afghan Response*, epilogue
58. David Lyon in *The Critic*, July 2022 p. 65
59. Private information
60. Kissinger, *Leadership* p. 343
61. Ibid., p. 345
62. Mark Almond in *The Critic*, April 2022 p. 31
63. John Beales in *History Today*, June 2022 p. 72
64. David Blagden in *The Critic*, April 2022 p. 30
65. Ibid., p. 28
66. O'Sullivan, *The President, the Pope and the Prime Minister* p. 157
67. Paret (ed.), *Makers of Modern Strategy* p. 807
68. Lehman, *On Seas of Glory* p. 352
69. Hastings and Jenkins, *The Battle for the Falklands* p. 91
70. Ibid., p. 92
71. Colbeck, *With 3 Para to the Falklands* p. 46
72. Freedman and Gamba-Stonehouse, *Signals of War* p. 190
73. Paret (ed.), *Makers of Modern Strategy* p. 808
74. Hastings and Jenkins, *The Battle for the Falklands* p. 90
75. Kissinger, *Leadership* p. 351

76. Hastings and Jenkins, *The Battle for the Falklands* p. 90
77. Ibid., p. 93
78. Colbeck, *With 3 Para to the Falklands* p. 43
79. Hastings and Jenkins, *The Battle for the Falklands* p. 93
80. Freedman and Gamba-Stonehouse, *Signals of War* p. 383
81. John Beales in *History Today*, June 2022 p. 72
82. Ibid., p. 76
83. Ibid., p. 82
84. Colbeck, *With 3 Para to the Falklands* p. 80
85. Ibid., p. 85
86. Hastings and Jenkins, *The Battle for the Falklands* p. 184
87. David Blagden in *The Critic*, April 2022 p. 29
88. Hastings and Jenkins, *The Battle for the Falklands* p. 286
89. Ibid., pp. 286–7
90. Ibid., p. 96
91. Ibid., p. 286
92. O'Sullivan, *The President, the Pope and the Prime Minister* p. 158
93. Kissinger, *Leadership* p. 351
94. O'Sullivan, *The President, the Pope and the Prime Minister* p. 159
95. Thompson, *No Picnic* p. 125
96. Ibid., p. 86
97. Colbeck, *With 3 Para to the Falklands* p. 92
98. Thompson, *No Picnic* p. 92
99. Freedman and Gamba-Stonehouse, *Signals of War* p. 396
100. Thompson, *No Picnic* p. 126
101. Colbeck, *With 3 Para to the Falklands* p. 172
102. Hastings and Jenkins, *The Battle for the Falklands* p. 298
103. Ibid., p. 299
104. Freedman and Gamba-Stonehouse, *Signals of War* p. 397
105. Hastings and Jenkins, *The Battle for the Falklands* p. 312
106. O'Sullivan, *The President, the Pope and the Prime Minister* p. 158
107. Lehman, *On Seas of Glory* p. 356
108. John Beales in *History Today*, June 2022 p. 80
109. Lehman, *On Seas of Glory* p. 353
110. Ibid., p. 357
111. Ibid., p. 358
112. Ibid.
113. Kissinger, *Leadership* p. 351

Five: Cold War Denouement

1. Archer et al., *World History of Warfare* p. 555
2. Van Creveld, *Command in War* p. 267
3. Ibid.
4. McNamara to President Johnson in 1967, https://www.latimes.com/archives/la-xpm-1995-04-11-me-53252-story.html
5. Stanley, *The Protection Racket State passim*
6. McClintock, *Instruments of Statecraft* ch. 10
7. McClintock, *The American Connection* pp. 213–14
8. LeoGrande, *Our Own Backyard* p. 44
9. Ibid., p. 94
10. McMahan, *Reagan and the World* p. 123
11. Rosello, *Lessons from El Salvador* p. 102
12. Galvin, *Fighting the Cold War* pp. 319–30; see also Galvin, "Uncomfortable Wars: Toward a New Paradigm" pp. 2–8, https://press.armywarcollege.edu/cgi/viewcontent.cgi?article=1415&context=parameters
13. Reynolds, *One World Divisible* p. 392
14. Pace, *International Elements of Saddam Hussein's Conflicted Iraq: Shifting Allies, Resolute Foes*
15. https://www.nytimes.com/1986/11/27/world/48-are-reported-killed-52-wounded-iranian-missile-hits-capital-iraq.html
16. Black, *A Short History of War* p. 213
17. McMaster, *Battlegrounds* p. 235
18. Freedman, *Cold War* p. 178
19. Archer et al., *World History of Warfare* p. 556
20. Richard Overy in *Literary Review*, September 2022 p. 34
21. Freedman, *Cold War* p. 178
22. Reynolds, *One World Divisible* p. 597
23. Kukielski, *The U.S. Invasion of Grenada* p. 22
24. Ibid.
25. Ibid., p. 85
26. Ibid., p. 95
27. Cole, *Operation Urgent Fury: Grenada* p. 47
28. Stewart, *Operation Urgent Fury* p. 18
29. Kukielski, *The U.S. Invasion of Grenada* p. 187
30. Reagan, "Remarks at the Annual Convention of the Congressional Medal of Honor Society in New York City, 1983," https://www.reaganlibrary.gov/archives/speech/remarks-annual-convention-congressional-medal-honor-society-new-york-city
31. Stewart, *Operation Urgent Fury* p. 32
32. Kukielski, *The U.S. Invasion of Grenada* p. 65

33. Ibid., p. 214
34. Cheney, *In My Time* p. 175
35. Powell, *My American Journey* p. 426
36. Allen, "Intervention in panama" p. 177
37. Ibid., p. 174
38. Rempfer, "Soldiers recall combat jumps into Panama"
39. "Legal fight looms over Noriega as dictator prepares to leave prison," *The Times*, 20 July 2007, https://web.archive.org/web/20110604160413/; http://www.timesonline.co.uk/tol/news/world/us_and_americas/ article2106381.ece?print=yes&randnum=1151003209000
40. Cheney, *In My Time* p. 184
41. Corrigan, *Desert Storm Air War* p. 12
42. https://foreignpolicy.com/2013/09/03/applying-the-8-questions-of-the- powell-doctrine-to-syria/
43. Ibid., p. 489
44. Gordon and Trainor, *The Generals' War* p. 57
45. Ibid.
46. Atkinson, *Crusade* p. 109
47. Gordon and Trainor, *The Generals' War* p. xv
48. "President Warns Iraq of War Crimes Trials," *Washington Post*, 16 October 1990
49. Gordon and Trainor, *The Generals' War* p. 74
50. Ibid., p. 84
51. Freedman and Karsh, *The Gulf Conflict* p. 312
52. Corrigan, *Desert Storm Air War* p. 45
53. Ibid., p. 44
54. Kitfield, *Prodigal Soldiers* passim
55. Corrigan, *Desert Storm Air War* p. xviii
56. Ignatieff, *Virtual War* p. 166
57. Gordon and Trainor, *The Generals' War* p. 71
58. Powell, *My American Journey* p. 512
59. "Iraqi missiles strike Israel," *Guardian*, 18 January 1991, https://www. theguardian.com/world/1991/jan/18/iraq.davidfairhall
60. Atkinson, *Crusade* p. 77
61. Freedman and Karsh, *The Gulf Conflict* p. 311
62. Ibid., p. 313
63. Powell, *My American Journey* p. 489
64. Freedman and Karsh, *The Gulf Conflict* p. 362
65. Schwarzkopf, *It Doesn't Take a Hero* p. 430
66. Ibid., p. 443
67. Freedman and Karsh, *The Gulf Conflict* p. 385
68. Ibid., p. 394

69. Schwarzkopf, *It Doesn't Take a Hero* p. xxx
70. Gordon and Trainor, *The Generals' War* p. 290
71. Freedman and Karsh, *The Gulf Conflict* p. 391
72. Ibid., p. 388
73. Schwarzkopf, *It Doesn't Take a Hero* p. 453
74. Ibid., p. 462
75. Gordon and Trainor, *The Generals' War* p. 386
76. Pavelec, *War and Warfare since 1945* p. 2
77. Mayall, *Soldier in the Sand* p. 144
78. Atkinson, *Crusade* p. 453
79. *Stars and Stripes*, 23 February 2003, https://www.stripes.com/living/u-s-troops-revisit-scene-of-deadly-gulf-war-barrage-1.5305
80. Waldman, *Vicarious Warfare* p. 44
81. Gordon and Trainor, *The Generals' War* p. 416
82. Atkinson, *Crusade* p. 493
83. Tucker, *The Encyclopedia of Middle East Wars* p. 470
84. Cheney, *In My Time* p. 224
85. Sterling, *Other People's Wars* p. 245
86. Ibid.
87. Ibid.
88. Stewart, *The United States Army in Somalia* p. 9
89. Ibid., p. 11
90. Ibid., p. 13
91. "Fallout from Somalia still haunts US policy 20 years later," *Stars and Stripes*, 3 October 2013, https://www.stripes.com/news/fallout-from-somalia-still-haunts-us-policy-20-years-later-1.244957
92. United States Special Operations Command History Offices, *Task Force Ranger: Operations in Somalia 3–4 October 1993*, https://web.archive.org/web/20210322004056/https://www.esd.whs.mil/Portals/54/Documents/FOID/Reading%20Room/International_Security_Affairs/07-A-2365_Task_Force_Ranger_Report_Operations_in_Somalia_1993.pdf
93. https://www.asomf.org/the-battle-of-mogadishu/
94. Poole, *The Effort to Save Somalia* p. 69
95. "Black Hawk Down's Long Shadow," *Daily Beast*, 9 October 2013, https://www.thedailybeast.com/black-hawk-downs-long-shadow
96. McChrystal, *My Share of the Task* p. 60

Six: The New World Disorder

1. Mearsheimer, "Back to the Future" p. 192
2. Forsyth, *The Caucasus* p. 684

3. Asmus, *A Little War that Shook the World* p. 60
4. Ibid., p. 61
5. De Waal, *The Caucasus* p. 144
6. Forsyth, *The Caucasus* p. 687
7. International Crisis Group, "Georgia: Avoiding War in South Ossetia" p. 4
8. Human Rights Watch, "Bloodshed in the Caucasus," https://www.hrw. org/reports/pdfs/g/georgia/georgia.923/georgia923full.pdf
9. Sammut and Cvetkovski, "The Georgia–South Ossetia Conflict" p. 14
10. Judah, *Kosovo* p. 42
11. Packer, *Our Man* p. 256
12. Ferguson, *The War of the World* p. 627
13. Ibid.
14. Vulliamy, *Seasons in Hell* p. 73
15. https://www.theguardian.com/world/from-the-archive-blog/2018/jul/13/ siege-of-sarajevo-ian-traynor-maggie-okane-1993
16. Ferguson, *The War of the World* p. 627
17. Ibid., p. 629
18. Ibid., p. 630
19. Black, *A Short History of War* p. 226
20. Shaw, "Crisis in Bosnia" p. 200
21. Ibid., p. 201
22. Ibid.
23. Ibid., p. 207
24. Haulman, "Resolution of Bosnian Crisis" p. 222
25. Powell, *My American Journey* p. 576
26. Packer, *Our Man* p. 263
27. Ibid., p. 266
28. Clark, *Waging Modern War* p. 7
29. Glenny, *The Balkans* p. 640
30. Clark, *Waging Modern War* p. 49
31. Silber and Little, *Yugoslavia* pp. 345, 350
32. Ibid., p. 365
33. Schinella, *Bombs without Boots* p. 23
34. Ibid., p. 26
35. Ibid., p. 17
36. Ibid., p. 29
37. Ibid., p. 366
38. Ibid., p. 30
39. Shaw, "Crisis in Bosnia" p. 226
40. Tirpak, "Deliberate Force"
41. Glenny, *The Balkans* p. 651
42. Albright, *Madame Secretary* p. 148

43. The National Security Archive, "The Rwanda 'Genocide Fax': What We Know Now," 9 January 2014, George Washington University, Washington, DC
44. Ibid.
45. Ferguson, *The War of the World* p. 631
46. Albright, *Madame Secretary* p. 149
47. Ibid., p. 150
48. *Report of the Independent Inquiry into the Actions of the United Nations during the 1994 Genocide in Rwanda* p. 1
49. Clinton, *My Life* p. 782
50. "France accused on Rwanda killings," BBC News, 24 October 2006
51. Nowrojee, *Shattered Lives passim*
52. Drumbl, "She Makes Me Ashamed to Be a Woman: The Genocide Conviction of Pauline Nyiramasuhuko, 2011"
53. Albright, *Madame Secretary* p. 154
54. "Bystanders to Genocide," *Atlantic Monthly*, September 2001
55. Lambeth, "Operation Allied Force: Lessons for the Future" *passim*
56. Ignatieff, *Virtual War* p. 13
57. Ibid., p. 21
58. Packer, *Our Man* p. 402
59. Judah, *Kosovo* p. 193
60. Ibid., p. 224
61. Ibid., p. 227
62. Lambeth, *NATO's Air War for Kosovo* pp. 11, 29
63. Ibid., p. 20
64. Ignatieff, *Virtual War* p. 170
65. Lambeth, *NATO's Air War for Kosovo* p. 18
66. Ignatieff, *Virtual War* p. 62
67. Ibid., p. 172
68. Ibid., p. 176
69. Felix, *Wesley K. Clark* p. 138
70. Ignatieff, *Virtual War* p. 100
71. Lambeth, *NATO's Air War for Kosovo* p. 27
72. Ignatieff, *Virtual War* p. 41
73. Ibid., p. 52
74. Lambeth, *NATO's Air War for Kosovo* p. 26
75. Ignatieff, *Virtual War* p. 105
76. Black, *A Short History of War* p. 226
77. Ignatieff, *Virtual War* p. 105
78. Lambeth, *NATO's Air War for Kosovo* p. 33
79. Ibid., p. 35
80. Judah, *Kosovo* p. 261

81. Shelton, *Without Hesitation* p. 370; and personal recollections of General David Petraeus, who was the Executive Officer to Gen. Shelton from 1997 to 1999

82. Clark, *Waging Modern War* p. 303

83. Shelton, *Without Hesitation* pp. 371–3

84. Ibid., pp. 384–5

85. Lambeth, *NATO's Air War for Kosovo* p. 43

86. Ignatieff, *Virtual War* p. 163

87. Shelton, *Without Hesitation* p. 384, and recollections of Gen. Petraeus

88. Clark, *Waging Modern War* p. 403

89. Borger, *The Butcher's Trail* p. 234

90. Archer et al., *World History of Warfare* p. 553

Seven: The War in Afghanistan

1. Churchill, *My Early Life* p. 246

2. The best examination of the terrorist attacks on 9/11 remains National Commission on Terrorist Attacks upon the United States, *The 9/11 Commission Report*. The determination of the "longest war in American history" depends on when one begins the counting on Vietnam, of course

3. Unless otherwise cited, the sources consulted for the planning and execution of the invasion of Afghanistan are Degen and Reardon, *Modern War in an Ancient Land*, vol. I; Lowrey, *U.S. Marines in Afghanistan*; Stewart, *The United States Army in Afghanistan*; and Wright et al., *A Different Kind of War*

4. Bin Laden issued a second *fatwa* on 23 February 2008, providing religious authorization for indiscriminate killing of Americans and Jews anywhere in the world. Six months later truck bombs detonated outside US embassies in Dar es Salaam, Tanzania and Nairobi, Kenya, killing more than 200 people, most of them Africans

5. Coll, *Ghost Wars*

6. Ibid., pp. 457–8. Unfortunately, Massoud would not live to see the triumph of the United Front. Two days before the terrorist attacks on the United States, two suicide bombers posing as television journalists blew themselves up while interviewing him, killing the most charismatic and effective leader of the Afghan resistance

7. Bush, *Decision Points* pp. 187–91; Tenet, *At the Center of the Storm* pp. 268–74

8. Degen and Reardon, *The United States Army in Afghanistan*, vol. I, p. 40

9. Bush, *Decision Points* pp. 187–8; Tenet, *At the Center of the Storm* pp. 275–8. For an in-depth examination of US–Pakistani relations, see Riedel, *Deadly Embrace*

10. Degen and Reardon, *The United States Army in Afghanistan*, vol. I, pp. 52–3; Franks and McConnell, *American Soldier* pp. 271–2; Rumsfeld, *Known and Unknown* pp. 377, 684

11. Schroen, *First In*

12. The story of ODA 595 is recounted in Stanton, *Horse Soldiers*, upon which the 2018 movie *12 Strong* is based. Over the next few weeks eight additional ODA teams would link up with various United Front leaders in Afghanistan

13. Degen and Reardon, *The United States Army in Afghanistan*, vol. I, p. 68

14. Ibid., pp. 81–6

15. Four hundred foreign fighters (among them two Americans) who surrendered were trucked to a prison fortress near Mazar-e Sharif, where they rose up against United Front elements holding them. US and British Special Operations Forces, CIA operatives and United Front personnel supported by airstrikes ended the uprising after six days of fighting, killing more than 300 of the detainees. Sennott, "The First Battle of the 21st Century"

16. The history of the CIA-sponsored campaign to take Kandahar is told by the former CIA station chief in Islamabad, Robert L. Grenier, *88 Days to Kandahar*

17. For the operations of the Marines in Afghanistan during this period, see Lowrey, *U.S. Marines in Afghanistan*

18. Whitlock, *The Afghanistan Papers* pp. 38–9

19. Historian Carter Malkasian writes that the unwillingness of the Bush administration to bring Taliban representatives into negotiations in 2001 "must be regarded as one of the greatest mistakes of the Afghan War." Malkasian, *The American War in Afghanistan* p. 76

20. Khalilzad, *The Envoy* pp. 119–27. A day prior to receiving word from Bonn of his new status, Karzai was nearly killed by an errant B-52 strike that killed twenty Afghans and wounded fifty others, along with two American dead and nineteen wounded, among them Captain Amerine

21. Special forces – the US Army's Green Berets – train indigenous forces to conduct unconventional and guerrilla warfare. Special operations forces – among them US Army Ranger Regiment and US Navy SEAL teams – conduct reconnaissance and direct-action missions requiring highly specialized and trained forces

22. For the details of the battle of Tora Bora, see Stewart, *United States Special Operations Command History*, https://archive.org/details/united-states-special-operations-command/mode/2up, pp. 93–7; and Degen and Reardon, *The United States Army in Afghanistan*, vol. I, pp. 119–26. For Brig. Gen. Mattis' offer to use Marines to seal the Pakistani border, see Lowrey, *U.S. Marines in Afghanistan* p. 209

23. Hagenbeck was given a force cap of 160 personnel for his headquarters when deploying to Karshi Khanabad, a penny-wise but pound-foolish approach to providing command and control for Operation Enduring Freedom. Furthermore, roughly a third of the staff were augmentees, replacing members of the division staff deployed to Kosovo on a peacekeeping mission

24. Naylor, *Not a Good Day to Die* p. 88

25. Ibid., p. 89

26. For an excellent history of Operation Anaconda, see Naylor, *Not a Good Day to Die*

27. Unless otherwise cited, the sources consulted for the seven years after the fall of the Taliban are Degen and Reardon, *The United States Army in Afghanistan*, vols I and II; Loyn, *The Long War*; Malkasian, *The American War in Afghanistan*; Neumann, Mundey and Mikolashek, *The U.S. Army in Afghanistan*; and Wright et al., *A Different Kind of War*

28. Bush, *Decision Points* p. 197

29. Degen and Reardon, *The United States Army in Afghanistan*, vol. I, pp. 177–8

30. Historian Carter Malkasian attributes the inattentiveness of the Bush administration to the opportunities available to them in Afghanistan after the fall of the Taliban to overconfidence. Malkasian, *The American War in Afghanistan* p. 80

31. Whitlock, *The Afghanistan Papers* p. 28

32. Bush, *Decision Points* p. 205. In this regard Bush and his Secretary of Defense were not on the same page. Rumsfeld writes in his memoir, "If some later contended that we never had a plan for full-fledged nation building or that we under-resourced such a plan, they were certainly correct. We did not go there to try to bring prosperity to every corner of Afghanistan. I believed – and continue to believe – that such a goal would have amounted to a fool's errand." Rumsfeld, *Known and Unknown* p. 683

33. Whitlock, *The Afghanistan Papers* p. 41. The United States dispensed $13.3 billion in Marshall Plan aid between 1948 and 1951, equivalent to $164 billion in 2022 dollars

34. This was in contrast to what we later did in Iraq, where we sought to populate the PRTs with as many civilians as the State Department and AID could provide, using military personnel only to fill shortfalls in civilian manning

35. Bush, *Decision Points* p. 193

36. Loyn, *The Long War* pp. 52–3

37. Rumsfeld, *Known and Unknown* p. 207

38. Hopkins, "Afghanistan's Present Failure Lies in its Past Design"

39. Some policymakers were more explicit in their criticism of Karzai. Susan Rice, who served as national security advisor in the second Obama administration, wrote, "Karzai plainly hated the U.S., and increasingly many of us (myself included) reviled his vitriolic nationalism, incorrigibly corrupt governance, and pompous leadership style." Rice, *Tough Love* p. 393

40. According to the Special Inspector-General for Afghanistan Reconstruction, "Absorptive capacity is the amount of international aid that a country can receive before it causes significant economic, social, and political disruptions and becomes counterproductive. Developing countries with fragile economies, such as those experiencing prolonged conflict, are believed to have a lower threshold for aid saturation. International aid above the absorptive capacity threshold can lead to waste, fraud, and increased corruption." Special Inspector-General for Afghanistan Reconstruction, *What We Need to Learn* p. 26

41. For an excellent examination of the impact of national caveats, see Auerswald and Saideman, *NATO in Afghanistan*

42. Whitlock, *The Afghanistan Papers* p. 113

43. Degen and Reardon, *The United States Army in Afghanistan*, vol. II, p. 144

44. The battle is detailed in Bowden, *The Three Battles of Wanat*

45. Committee on Armed Services, US House of Representatives, "Security and Stability in Afghanistan"

46. Degen and Reardon, *The United States Army in Afghanistan*, vol. II, pp. 223–4

47. Unless otherwise cited, the sources consulted for the Surge in Afghanistan are Collins and Hooker, *Lessons Encountered*; Degen and Reardon, *The United States Army in Afghanistan*, vol. II; Loyn, *The Long War*; Malkasian, *The American War in Afghanistan*; and Wright et al., *A Different Kind of War*

48. Woodward, *Obama's Wars* pp. 96–110

49. Obama, *A Promised Land* p. 321

50. Gates, *Duty* pp. 344–6. McKiernan was the first commanding general relieved in war since Douglas MacArthur in Korea in 1951

51. McChrystal, *My Share of the Task* chs 17 and 18

52. Gates, *Duty* pp. 358–9. The administration's stance became public in early 2010 with the release of classified documents to Wikileaks

53. The story of COP Keating is told superbly by Tapper in *The Outpost*. Two soldiers, Staff Sergeant Clinton L. Romesha and Specialist Ty Carter, were awarded the Medal of Honor for their heroism in defending the outpost

54. Gates, *Duty* p. 353

55. The cable was leaked to the public a year later by the website Wikileaks and so exacerbated the relationship between Ambassador Eikenberry and President Karzai that it became increasingly difficult for me to take the Ambassador to meetings with Karzai, as I had done with Ambassador Crocker in Iraq

56. Panetta, *Worthy Fights* pp. 253–4

57. Gates, *Duty* p. 384

58. Ibid., pp. 367–86. Secretary of State Hillary Clinton sided with Gates, Mullen and Petraeus to support the Surge. Clinton, *Hard Choices* pp. 146–9

59. Woodward, *Obama's Wars* pp. 315–16

60. Malkasian, *The American War in Afghanistan* p. 225

61. Hastings, "The Runaway General"

62. Woodward, *Obama's Wars* p. 373

63. Subcommittee on National Security and Foreign Affairs, Committee on Oversight and Government Reform, *Warlord, Inc*

64. The NDN eventually closed down in 2015 as Russia closed the corridors running through its territory and US and NATO forces in Afghanistan declined in number

65. Degen and Reardon, *The United States Army in Afghanistan*, vol. II, p. 314

66. The paper, "One Tribe at a Time," was published on author Steven Pressfield's blog and then linked on the popular counter-insurgency website *Small Wars Journal*, where it was read by the special forces community and came to the attention of General McChrystal and me. Gant subsequently published the paper in book form: Gant, *One Tribe at a Time*, copy located at https://www.globalsecurity.org/military/library/report/2009/2009_one_tribe_at_a_time.pdf

67. Tyson, *American Spartan* pp. 85–6; Robinson, *One Hundred Victories* pp. 24–5

68. Tyson, *American Spartan* p. 359. Tyson informally embedded with Gant's team for nine months in Konar Province, where she and Gant engaged in an intimate relationship. In 2012 Gant was relieved of command for various improprieties including his relationship in a combat zone with Tyson and abuse of prescription drugs and alcohol to treat post-traumatic stress disorder. Gant retired after his return to the United States and married Tyson in 2013. We had, perhaps, asked too much of him over the years. Thompson, "The Fall of the Green Berets' *Lawrence of Afghanistan*"

69. The most comprehensive account of the failure of aid programs in Afghanistan is the Special Inspector-General for Afghanistan Reconstruction, *What We Need to Learn*

70. Boone, "The financial scandal that broke Afghanistan's Kabul Bank"

71. Clinton, *Hard Choices* pp. 150–61
72. Ibid., pp. 164–9
73. Degen and Reardon, *The United States Army in Afghanistan*, vol. II, p. 312
74. Confirmation was complicated by significant damage to bin Laden's face because of the rounds that had struck him there, so one of the verifications was of bin Laden's height, confirmed when Vice Admiral Bill McRaven, the commander who had overseen the operation, ordered a SEAL who was the same height as bin Laden to lie down next to the body. The lack of a tape measure was the only detail overlooked in the meticulously planned operation. DNA analysis would definitively confirm the identity in a couple of days
75. For the account of the war against al-Qaeda and the manhunt for bin Laden see Jones, *Hunting in the Shadows*; Bergen, *Manhunt*; and Panetta, *Worthy Fights*, ch. 13. The best account of the raid on bin Laden's compound is Mark Owen (pseudonym for Matt Bissonnette), *No Easy Day*
76. For a discussion of what the Pakistani government knew about bin Laden's whereabouts, see Riedel, *Deadly Embrace*, preface: The Mystery of Abbottabad
77. Unless otherwise cited, the sources consulted for this phase in Afghanistan are Collins and Hooker, *Lessons Encountered*; Degen and Reardon, *The United States Army in Afghanistan*, vol. II; Loyn, *The Long War*; Malkasian, *The American War in Afghanistan*; and Wright et al., *A Different Kind of War*
78. Karzai ultimately refused to sign the document, which was concluded under his successor, "the far more sober, responsible, and gracious Ashraf Ghani." Rice, *Tough Love* p. 392
79. For a narrative of the battle, see Robinson, *One Hundred Victories* pp. 103–16
80. Degen and Reardon, *The United States Army in Afghanistan*, vol. II, pp. 395–7: "In most cases, there was still serviceable equipment left even after disposing of excess personal property and defense articles. On 1 March 2012, the DoD granted its Defense Logistics Agency authority to destroy serviceable equipment that U.S. Transportation Command representatives jointly determined too costly to ship. Serviceable vehicles that were not required elsewhere or whose values were less than the cost to ship them back to the United States – the majority of items being in the latter category – were destroyed in theater." The footnote to that account adds that, "Almost 4,000 vehicles were destroyed in theater, including more than 1,000 MRAPs. The U.S. military possessed a total of 22,500 MRAPs worldwide with 11,000 deployed to Afghanistan"
81. Degen and Reardon, *The United States Army in Afghanistan*, vol. II, p. 353

82. Wilkins, "The Rise and Fall of Village Stability Operations in Afghanistan: Lessons for Future Irregular Warfare Campaigns"; Altman, "Why dissolving the Afghan Local police program troubles its American architects"

83. Malkasian, *The American War in Afghanistan* p. 457

84. Filkins, "Last Exit from Afghanistan"

85. For an analysis of the swift collapse of the Afghan state, see the Special Inspector-General for Afghanistan Reconstruction, *Why the Afghan Government Collapsed*

86. Gaouette et al., "The last US military planes have left Afghanistan, marking the end of the United States' longest war"

87. Washington Post Editorial Board, "In Afghanistan, the Lights Go Out for Women"

88. Clausewitz, *On War*, 1984 edn p. 88

89. For an excellent analysis of how one district unravelled during this period, see Malkasian, *War Comes to Garmser* ch. 5

90. Sheehan, *A Bright Shining Lie* p. 67

91. I had opposed the shift from Soviet/Eastern Bloc helicopters that the Afghans could maintain to US-provided Black Hawks while I was the commander, but the decision was nonetheless made subsequently under Congressional pressure to "buy American"

92. Tyson, *American Spartan* pp. 210–11. At least one area historian agrees with Gant; see Hopkins, "The problem with 'Hearts and Minds' in Afghanistan"

93. The White House, "Continued U.S. Support for a peaceful, Stable Afghanistan," 25 June 2021, https://www.whitehouse.gov/briefing-room/statements-releases/2021/06/25/fact-sheet-continued-u-s-support-for-a-peaceful-stable-afghanistan/

Eight: The Iraq War

1. In fact, Iraq had no weapons of mass destruction in 2003, but Saddam wanted that reality to remain ambiguous to deter Iranian and Israeli aggression and to retain Iraq's status within the Arab world. For this reason, he ordered his subordinates to hinder inspections even though the regime had nothing to hide. This possibility had not occurred to US intelligence analysts – another display of their lack of imagination in the post-Cold War era. See Woods, *Iraqi Perspectives Project* pp. 91–3

2. US-led coalitions had, in fact, enforced no-fly zones over northern Iraq for five years to protect the Iraqi Kurds and over southern Iraq for ten years to protect the Shi'a Arabs; these were known as Operation Northern Watch and Operation Southern Watch

3. For a deeper examination of this fundamental assumption, see MacDonald, *Overreach*
4. This operation was named Operation Provide Comfort, undertaken to prevent a massacre of the Kurds and to provide humanitarian assistance in the wake of the Gulf War. See Rudd, *Humanitarian Intervention*
5. The sources consulted on the planning, build-up of forces and execution of the invasion of Iraq are Gordon and Trainor, *COBRA II*; Fontenot et al., *On Point*; and Rayburn and Sobchak (eds), *The U.S. Army in the Iraq War*, vol. I: *Invasion–Insurgency–Civil War*. Rumsfeld's views can be found in Rumsfeld, *Known and Unknown*
6. Rumsfeld did not initially approve the deployment of all elements of the 101st Corps Support Group, which was configured specifically to support the 101st Airborne Division (Air Assault) and its enormous helicopter fleet, causing substantial – and unnecessary – challenges at the outset of the operation. Among the assets delayed were the large tankers needed to provide water for flushing out helicopter engines in the dusty conditions and for other purposes. We sought to remedy the situation by renting small water tankers used by camel herders, which came equipped with disco lighting underneath the truck, which we could not figure out how to disconnect before the vehicles had to convoy north, interspersed incongruously with various armored vehicles
7. Mills, "Punished for telling truth about Iraq war"
8. An attempt was made to kill Saddam Hussein early in the morning of 19 March by using F-117 stealth fighters and Tomahawk cruise missiles to target his supposed hideout at Dora Farms, a compound outside Baghdad. The attack hit the target, but Saddam was not there
9. For the history of ORHA, see Rudd, *Reconstructing Iraq*
10. Woods, *Iraqi Perspectives Project* pp. 25–31
11. A convoy from the 507th Maintenance Company from the 3rd Infantry Division took a wrong turn and blundered into an ambush in Nasiriyah, leading to the deaths of eleven soldiers and the wounding of nine others, with seven of the wounded soldiers captured by Iraqi forces. Private First Class Jessica Lynch was subsequently recovered from an Iraqi hospital by special operations forces on 1 April
12. Fontenot et al., *On Point* pp. 179–92
13. Dwyer, "A Nation at War: In the Field – V Corps Commander"
14. The bridge at Objective Peach had a northern span and a southern span, either of which could be used to cross the Euphrates River. Iraqi demolitions partially destroyed the northern span, but US forces used the three-lane southern span to cross to the east side of the river

15. For the history of US Army involvement in governance, see Schadlow, *War and the Art of Governance*

16. Fontenot et al., *On Point* p. 301

17. Sergeant First Class Paul R. Smith would be awarded a posthumous Medal of Honor for saving the lives of three of his soldiers and using a .50-caliber machine gun to kill fifty Iraqi soldiers attacking his unit's position in the fighting for the airport

18. Rayburn and Sobchak (eds), *The U.S. Army in the Iraq War*, vol. I, p. 101

19. Atkinson, *In the Company of Soldiers* p. 6

20. Hess, "Rumsfeld: Looting is transition to freedom"

21. To be fair, Garner had never been provided with the resources and authority to accomplish the enormous tasks inherent in reconstructing Iraq and was replaced just as he was getting his feet on the ground

22. Ambassador Bremer did subsequently allow me to conduct a local reconciliation process in Nineveh Province that proved highly successful in getting critical government officials, civil servants and over 110 tenured professors back to work. Ultimately well over 10,000 former Ba'athists were reconciled in Nineveh in 2003, only to have Chalabi refuse to approve any of the actions after my departure in February 2004, fatally undermining the achievements in Mosul and Nineveh

23. Ricks, *Fiasco* p. 159

24. Bremer and McConnell, *My Year in Iraq* pp. 53–8

25. President Bush in retrospect believed he should have insisted on a thorough review of these two orders. He might have decided to implement them anyway, "but the discussion would have better prepared us for what followed." Bush, *Decision Points* p. 260. Below the NSC level, the draft orders, prepared by the office of the Undersecretary of Defense for Policy Douglas J. Feith, were subject to next to no inter-agency review

26. For Sanchez's view of his time in command of CJTF-7, see Sanchez and Phillips, *Wiser in Battle*

27. The sources consulted for the transition from offensive to counter-insurgency operations are Ricks, *Fiasco*; Wright and Reese, *On Point II*; and Rayburn and Sobchak (eds), *The U.S. Army in the Iraq War*, vol. I. For an Iraqi perspective, see Allawi, *The Occupation of Iraq*

28. Rayburn and Sobchak (eds), *The U.S. Army in the Iraq War*, vol. I, p. 261

29. In Baghdad, the four largest of eight bases (down from forty-six in May 2003) were Camp Victory/Liberty near Baghdad International Airport, FOB Falcon in southern Baghdad, FOB Rustamiyah in south-east Baghdad and Camp Taji north of the city

30. For an examination of the battles for Fallujah in 2004, see West, *No True Glory*

31. For an examination of one of these fights, in Karbala, see Mansoor, *Baghdad at Sunrise* ch. 10. Mansoor's 1st Brigade, as well as the other maneuver vre brigades of the 1st Armored Division, would be awarded a Presidential Unit Citation for its actions during this period

32. Some national contingents fought heroically. A battalion of El Salvadoran infantry held its positions in Najaf in hand-to-hand combat, and some British units upheld their proud military tradition by engaging in close combat with the JAM even as senior British military leaders, seeking a political solution to the crisis, refrained from ordering an attack on Sadr's militia

33. For the shortcomings of CPA, see Chandrasekaran, *Imperial Life in the Emerald City*; and Packer, *The Assassins' Gate*

34. Maj. Gen. Anthony M. Taguba, Article 15-6 Investigation of the 800th Military Police Brigade, 27 May 2004, https://www.aclu.org/sites/default/files/torturefoia/released/TR3.pdf

35. The UN Mission did return, however, and played a particularly important role during the Surge in Iraq under the exceedingly capable Staffan de Mistura, who was the Special Representative of the UN Secretary-General from 2005 to 2009 and was a tremendous partner for Ambassador Ryan Crocker and me – and who subsequently would head the UN Mission in Afghanistan, where it was a privilege to serve with him again

36. Despite the cordon around the city, the terrorist Zarqawi and several other insurgent leaders were able to flee before the battle, reportedly dressed as women

37. For an account of the battle, see West, *No True Glory*. For one soldier's recollection of the fighting, see David Bellavia, *House to House*. Bellavia was subsequently awarded the Medal of Honor in 2019 for his actions in Fallujah. Rayburn and Sobchak (eds), *The U.S. Army in the Iraq War*, vol. I, p. 354

38. For General Casey's view of his time in command of MNF-I, see Casey, *Strategic Reflections*

39. Kaplan, *The Insurgents* pp. 178–80

40. Rayburn and Sobchak (eds), *The U.S. Army in the Iraq War*, vol. I, p. 576. For a graphic representation of the violence in Iraq from 2004 to 2008, see Appendix A to this chapter

41. Casey, *Strategic Reflections* pp. 207–8

42. The brigade was further reinforced by a battalion from the 101st Airborne Division, a Marine battalion and a large contingent of Navy SEALs, giving it a substantial amount of combat power with which to contend for control of Ramadi

43. The best sources on the fight for Ramadi are Michaels, *A Chance in Hell*; and Doyle, *A Soldier's Dream*. Captain Patriquin was a crucial interlocutor between MacFarland and the local tribes until killed by a roadside bomb on 6 December 2006

44. Rayburn and Sobchak (eds), *The U.S. Army in the Iraq War*, vol. I, pp. 611–12

45. The sources consulted for the Surge are Mansoor, *Surge*; Ricks, *The Gamble*; Robinson, *Tell Me How This Ends*; West, *The Strongest Tribe*; Gordon and Trainor, *The Endgame*; Kagan, *The Surge*; and Rayburn and Sobchak (eds), *The U.S. Army in the Iraq War*, vol. II: *Surge and Withdrawal*

46. Feaver, "The Right to Be Right: Civil–Military Relations and the Iraq Surge"; Kagan, *Choosing Victory*. Another report, by the Congressionally mandated Iraq Study Group, called for the gradual withdrawal of US forces, an increased emphasis on training and advising Iraqi security forces, and a diplomatic solution to the conflict, which was unachievable given the disparate goals of neighboring states such as Iran and Syria and the internal causes of the Iraqi Civil War, which outside actors could only influence on the margins

47. Bush, *Decision Points* p. 376

48. Department of the Army, Field Manual 23-4, *Counterinsurgency*. I took command of MNF-I on 10 February 2007. For the development of the counter-insurgency doctrine, see Crane, *Cassandra in Oz*

49. The final version of the COIN Guidance in Appendix B to this chapter

50. Mansoor, *Surge* p. 78

51. For Admiral Fallon's reservations about the Surge, see Rayburn and Sobchak (eds), *The U.S. Army in the Iraq War*, vol. II, pp. 130–1; and Mansoor, *Surge* pp. 178–9

52. These were known as the Taji documents, after the location where they were discovered. Rayburn and Sobchak (eds), *The U.S. Army in the Iraq War*, vol. II, pp. 58–62

53. Not to be confused with the Anaconda Plan used by the Lincoln administration to guide Union forces during the Civil War. For a graphic representation of the Iraq War Anaconda Plan, see Appendix C to this chapter

54. Maliki would, regrettably, interfere with the professionalization of the Iraqi officer corps after the withdrawal of US forces in late 2011, leading to the collapse of some Iraqi Army formations when confronted by ISIS militants in 2014

55. Gates details his struggles with the Pentagon bureaucracy in Gates, *Duty* ch. 4

56. Among the snipers was Navy SEAL Chris Kyle, later made famous by the 2012 memoir and 2014 movie, *American Sniper*

57. For a graphic representation of the assets in the battle of Sadr City, see Appendix D to this chapter

58. A superb study of the battle of Sadr City is Johnson et al., *The 2008 Battle of Sadr City*

59. Astonishingly, US Ambassador to Iraq Christopher Hill convinced President Obama to back Maliki for another term in office over the objections of Gen. Odierno, who had returned to Iraq in September 2008 to be commander of MNF-I following Gen. Petraeus' move to US Central Command. Maliki, who had proven his sectarian credentials by then, was Iran's choice as well. The Iraqi Supreme Court, which Maliki had stacked with his sympathizers, ruled that the government would be formed not by the candidate whose party had gained the most seats in the election, but by the candidate who could first generate the majority of support in the Council of Representatives required to become prime minister. This enabled Maliki to twist enough arms to throw the election to his Dawa Party

60. For the best discussion of how Maliki turned victory into defeat in Iraq, see Sky, *The Unravelling*

61. The Obama administration withdrew US combat forces from Iraq because Prime Minister Maliki would not commit to gaining parliamentary approval of a SOFA. Ironically, when President Obama left office in 2017, there were more than 5,000 Americans on the ground in Iraq without a parliamentary-approved SOFA

Nine: Vladimir Putin's Existential War against Ukraine

1. Clausewitz, *On War*, 1873 edn p. 99

2. Chinkin and Kaldor, *New Wars* pp. 5–6

3. Freedman, *Future of War* p. 281

4. Chinkin and Kaldor, *New Wars* p. 16

5. "America's New Sanctions Strategy: How the U.S. Can Stop the Russian War Machine and Strengthen the International Economic Order" (foreignaffairs.com)

6. Allen et al., *Future War and the Defence of Europe* p. 132

7. Opinion | "Ukraine's counteroffensive is more than just bravado" – *The Washington Post*

8. Private information, and Galeotti, *The Weaponisation of Everything* p. 41

9. "Preliminary Lessons in Conventional Warfighting from Russia's Invasion of Ukraine: February–July 2022" | Royal United Services Institute (rusi.org)

10. *Times Literary Supplement*, 17 February 2023 p. 8

11. Hanson, "Uses and Abuses of Military History" p. 10

12. Howard, *War in European History* p. 136

13. Timothy Snyder: "Ukraine Holds the Future" (foreignaffairs.com)
14. Ukraine invasion: "West predicted Russia's advance every step of the way" | *The Times*
15. Derix and Shelkunova, *Zelensky* p. 178
16. Kotkin, "Freedom at Stake: How Did Russia and the West Fall Out?" p. 3
17. Derix and Shelkunova, *Zelensky* p. 180
18. Ibid.
19. https://www.wired.com/story/notpetya-cyberattack-ukraine-russia-code-crashed-the-world/; and Allen et al., *Future War and the Defence of Europe* p. 27
20. Watling and Reynolds, "Operation Z" p. 2
21. https://www.thetimes.co.uk/article/ukraine-calls-up-200-000-reserve-troops-for-war-6f8bdqnfr
22. "Deep underground, Ukrainian hackers take aim at the Kremlin" | News | *The Times*
23. https://www.thetimes.co.uk/article/ukraine-calls-up-200-000-reserve-troops-for-war-6f8bdqnfr
24. Watling and Reynolds, "Operation Z" p. 1
25. "Vladimir Putin's mad dash for Kyiv shown in leaked Ukraine battle plan" | World | *The Times*
26. Interview with Dr. Condoleeza Rice, 4 August 2022; and "Vladimir Putin's mad dash for Kyiv shown in leaked Ukraine battle plan" | World | *The Times*
27. https://www.latimes.com/world-nation/story/2022-04-10/battered-ukraine-air-field-was-key-to-russian-plan-to-take-the-capital-the-airport-fell-but-resistance-continued
28. Derix and Shelkunova, *Zelensky* p. 183
29. *The Times*, 25 February 2023
30. Derix and Shelkunova, *Zelensky* pp. 9–10
31. Zelensky, *A Message from Ukraine* pp. 51–2
32. "President Zelensky of Ukraine vows to fight Russia in every square" | *The Times*
33. Derix and Shelkunova, *Zelensky* p. 182
34. Adam Tooze in "John Mearsheimer and the dark origins of realism" – *New Statesman*
35. Gideon Rachman in "Russia learns a hard lesson about the folly of war" | *Financial Times* (ft.com)
36. Matthews, "The red line: Biden and Xi's secret Ukraine talks revealed" p. 13
37. https://www.foreignaffairs.com/articles/ukraine/2022-04-29/how-not-invade-nation?utm_campaign=wp_todays_worldview&utm_medium=email&utm_source=newsletter&wpisrc=nl_todayworld

38. "Over 1,000 Companies Have Curtailed Operations in Russia – But Some Remain" | Yale School of Management
39. "Russia's war has forced brands to pick a side" | Comment | *The Times*
40. https://berlinpolicyjournal.com/the-gerasimov-doctrine/
41. Watling and Reynolds, "Operation Z" p. 4
42. "Lessons for the Next War From Russia's Invasion of Ukraine: 12 Experts Weigh In" (foreignpolicy.com)
43. Ibid.
44. *Spectator*, 16 April 2022
45. https://www.thetimes.co.uk/article/ukrainian-resistance-armed-with-app-outwit-russian-invaders-ks9mkf98q?shareToken=68a68cfea4eccb e13b66fa3564fea06d
46. "Gen. David Petraeus: How the war in Ukraine will end" | How smart Technology changing lives (voonze.com)
47. Gen. Petraeus quoted in Jonathan Tepperman in *The Octavian Report*, 23 March 2022
48. Derix and Shelkunova, *Zelensky* p. 13
49. Ibid., p. 14
50. *Spectator*, 29 October 2022 p. 38
51. "Elon Musk uses Starlink satellites to provide internet coverage in Ukraine" | News | *The Times*
52. POLITICO Dispatch: "How Elon Musk's Starlink changed the war in Ukraine" on Apple podcasts
53. "Don't let tech companies call the shots in war" | Comment | *The Times*
54. "Snake Island commander reveals what really happened after Russian flagship was told 'go f--- yourself'" (telegraph.co.uk)
55. "Ukraine's fierce resistance slows Russia as missiles hit Kyiv" | News | *The Times*
56. https://www.thetimes.co.uk/article/ukrainian-forces-repel-russians-in-battle-for-kharkiv-twvgw637j
57. https://www.thetimes.co.uk/article/allies-expel-russia-from-swift-banking-system-jq6ldk2xq
58. Ibid.
59. https://www.thetimes.co.uk/article/invader-is-accused-of-targeting-civilians-in-kharkiv-onslaught-0chcrgfgw
60. "Boris Johnson condemns 'barbari' strikes on Ukraine" | News | *The Times*
61. https://www.thetimes.co.uk/article/what-can-the-head-of-the-international-criminal-court-achieve-over-ukraine-smn32tjkl
62. Derix and Shelkunova, *Zelensky* pp. 187–8
63. "Putin 'hid Ukraine invasion plan from cabinet, who are now too scared to resign'" | News | *The Times*

64. https://www.reuters.com/business/energy/russia-replaces-iraq-top-oil-supplier-india-nov-2022-12-14/#:~:text=India's%20oil%20imports%20from%20Russia,from%20October%2C%20the%20data%20showed

65. Mark Helprin in *Claremont Review of Books*, Spring 2022 p. 12

66. Private information

67. https://apple.news/AQ-yzMjQhRZ2Pf1VmNul-Ng

68. https://unherd.com/thepost/the-russian-armys-number-one-problem-hazing/

69. "Underdog Ukrainian pilots fight high-tech Russian adversaries with skill" | News | *The Times*

70. Helprin in *Claremont Review of Books*, Spring 2022 p. 12

71. "Lack of air superiority baffles experts" | News | *The Times*

72. "Underdog Ukrainian pilots fight high-tech Russian adversaries with skill" | News | *The Times*

73. https://www.thetimes.co.uk/article/fighting-for-the-skies-on-a-wing-and-a-prayer-5lb5hxg80?shareToken=7e081e184b6bd9586dbf4b4122e96ea1

74. https://www.army-technology.com/comment/air-superiority-in-ukraine/

75. "Fighting for the skies on a wing and a prayer" | News | *The Times*

76. https://www.thetimes.co.uk/article/no-fly-zone-in-ukraine-would-favor-the-russians-says-defence-secertary-ben-wallace-hxtzjcxl2 and https://www.thetimes.co.uk/article/what-is-a-no-fly-zone-and-what-could-it-mean-for-nato-srdsfl8md

77. https://www.thetimes.co.uk/article/game-changing-drones-helping-ukraine-in-battle-for-the-skies-lg68l8xjw

78. https://www.washingtonpost.com/world/2022/12/02/drones-russia-ukraine-air-war/

79. "How Ukraine is using drones and AI to fight the first digital war" | World | *The Times*

80. https://www.thetimes.co.uk/article/up-to-15-000-russian-troops-in-convoy-approaching-kyiv-tmp5cqc9v

81. https://www.thetimes.co.uk/article/ukraine-invasion-russia-war-latest-lnws2h2rf

82. "Russian military convoy 'has made little progress for three days'" | News | *The Times*

83. Helprin in *Claremont Review of Books*, Spring 2022 p. 12

84. "Ukraine's resistance has been emboldened by Russian failures from the start" | News | *The Times*

85. "Why has the Russian advance slowed?" | News | *The Times*

86. Clausewitz, *On War*, 1873 edn p. 91

87. "Why has the Russian advance slowed?" | News | *The Times*

88. Ibid.

89. "A day without dead Russians isn't a complete day" | World | *The Times*

90. "Vladimir putin vows total victory in Ukraine" | News | *The Times*
91. https://www.thetimes.co.uk/article/ukraine-scuttles-its-flagship-frigate-as-the-russians-close-in-wtd07bbqp
92. https://www.thetimes.co.uk/article/russian-commanders-die-as-army-falters-gjzq0f2jr
93. https://www.thetimes.co.uk/article/fourth-russian-general-killed-claims-ukraine-fkfs7pwnb, Why are so many Russian generals dying? – The post (unherd.com)
94. "US military learning from Russian troops using cell phones in Ukraine" (taskandpurpose.com)
95. https://unherd.com/thepost/why-are-so-many-russian-generals-dying/
96. "Putin's conscripts won't fix the Russian army's big flaw" | *The Spectator*
97. "Ukraine issues iodine pills as Russia shells Zaporizhzhya nuclear plant again" | World | *The Times*
98. https://www.thetimes.co.uk/article/putin-blackmailing-world-with-threat-of-a-nuclear-catastrophe-qn3fwftbh
99. https://www.thetimes.co.uk/article/ukraine-invasion-latest-krcwx07p5
100. "Russia 'using banned landmines' in northeastern Ukraine" | News | The Times
101. https://www.thetimes.co.uk/article/russia-admits-firing-thermobaric-vacuum-weapons-in-ukraine-g6sl78zcb
102. https://www.thetimes.co.uk/article/russia-is-accused-of-returning-to-medieval-warfare-with-siege-tactics-pmsctpw9h
103. https://www.thetimes.co.uk/article/russia-accused-bombing-refugee-route-mariupol-9n0xpcmjg
104. "Merciless Vladimir Putin shells Ukrainian evacuees" | News | The *Sunday Times* (thetimes.co.uk)
105. "Russia is accused of returning to medieval warfare with siege tactics" | News | *The Times*
106. Gen. Petraeus quoted in Jonathan Tepperman in *The Octavian Report*, 23 March 2022
107. https://www.thetimes.co.uk/article/putins-army-leaderless-say-us-defence-officials-pgrx5mntc
108. https://www.thetimes.co.uk/article/we-will-not-give-up-on-ukraine-Zelensky-says-in-address-to-parliament-lqwdnrkv8
109. https://www.thetimes.co.uk/article/ukraine-to-receive-nato-fighter-jets-8fj5fh999
110. Ibid.
111. Matthews, "The red line: Biden and Xi's secret Ukraine talks revealed" p. 13
112. https://www.thetimes.co.uk/article/britain-may-send-starstreak-anti-aircraft-weapons-to-ukraine-9kndz9pqf

113. https://www.thetimes.co.uk/article/meet-the-ghost-of-kyiv-coming-for-russias-soul-0s9gn9t89

114. https://www.thetimes.co.uk/article/ukraine-trench-warfare-russia-fcz95j7kl?shareToken=7e10e5227a98b468f4d0e01e0df5148f

115. Freedman, *Ukraine and the Art of Strategy* p. 164

116. https://www.thetimes.co.uk/article/call-center-welcomes-foreign-legion-d03lf0wwb

117. https://www.thetimes.co.uk/article/secret-war-being-waged-against-russia-by-eyes-in-the-sky-3q562d85q

118. https://www.amnesty.org/en/latest/news/2022/06/ukraine-deadly-mariupol-theatre-strike-a-clear-war-crime-by-russian-forces-new-investigation/

119. https://www.thetimes.co.uk/article/russian-atrocities-in-borodyanka-much-worse-than-bucha-warns-Zelensky-tfmljqp9d?shareToken=3005e0d924bd2e56d8067bb03a84ee09

120. "Civilians taken into Russia and stripped of ID, says Ukraine" | News | *The Times*

121. General Petraeus quoted in Jonathan Tepperman in *The Octavian Report*, 23 March 2022

122. Ibid.

123. https://unherd.com/thepost/the-russian-armys-number-one-problem-hazing/

124. https://www.thetimes.co.uk/article/ukraine-invasion-going-worse-than-chechnya-officer-tells-moscow-t8tzd2tw5

125. https://www.thetimes.co.uk/article/russian-commander-run-over-by-tank-deliberately-killed-by-own-troops-ncdw9fmr9

126. https://www.thetimes.co.uk/article/russian-soldier-gives-up-tank-for-citizenship-and-7-500-gqpbrhr2w

127. "Ukrainians watch astonished at Groundhog Day blunders" | News | *The Sunday Times* (thetimes.co.uk)

128. https://www.thetimes.co.uk/article/war-in-ukraine-town-of-bucha-shattered-by-russian-invaders-atrocities-bqgj58pdg and https://www.ohchr.org/en/statements/2022/09/update-chair-independent-international-commission-inquiry-ukraine-51st-session

129. Derix and Shelkunova, *Zelensky* p. 189

130. "Eyes in the sky search for mass graves in Ukraine" | News | *The Times*

131. https://www.independent.co.uk/news/world/europe/ukraine-war-crimes-borodyanka-bucha-hostomel-makariv-b2054530.html?amp

132. https://www.thetimes.co.uk/article/maxar-satellites-boss-we-could-see-invasion-of-ukraine-coming-cvw2x8gc5?shareToken=a7655295689bab1f26d855b07aa4e5b9

133. Watling and Reynolds, "Operation Z" p. 8

134. "Ukraine using Clearview AI facial recognition to identify Russian war dead" – *The Washington Post*

135. 400. Russia-Ukraine Conflict: Sign post to the Future (Part 1) | Mad Scientist Laboratory (army.mil)

136. *"The Times* view on the destruction of Russia's Black Sea flagship: Sink the Moskva" | Comment | *The Times*

137. https://www.thetimes.co.uk/article/even-a-relic-of-the-true-cross-couldnt-keep-the-12-500-ton-moskva-afloat-xr6v05rj6?shareToken=3e5 9d2450945f21ae7a7d08299defc4b

138. "What should the US Navy learn from Moskva's demise?" – Breaking Defense

139. *Spectator*, 16 April 2022

140. Watling and Reynolds, "Operation Z" p. 10

141. Ibid., p. 11

142. Ibid., pp. 13–14

143. https://www.thetimes.co.uk/article/russia-suspected-of-buying-eu-fridges-to-strip-for-weapon-parts-p73sdtngc?shareToken=1a4369c70d7 adda86abb3af63db9b988

144. "American Defense priorities After Ukraine" – War on the Rocks

145. General Sir Richard Barrons in the *Economist*, 9 February 2023

146. https://www.theguardian.com/commentisfree/2022/may/04/us-lend-lease-act-ukraine-1941-second-world-war & U.S. Sends Ukraine $400 Million in Military Equipment > U.S. Department of Defense > Defense Department News

147. "The West is struggling to forge a new arsenal of democracy" | *The Economist*

148. Pancevski, "Europe Is Rushing Arms to Ukraine, But Running Out of Ammo"

149. Letter to MPs and peers from James Cleverly, 17 February 2023, and "Tanks will help Kyiv break the deadlock. But its partners now face a fork in the road" | Ukraine | *The Guardian*

150. Hansard, House of Lords, 21 December 2022 col. 1194

151. "Britain to boost Ukraine's defences with laser-guided missile system" | News | *The Times*

152. "Uber-style technology helped Ukraine to destroy Russian battalion" | News | *The Times*

153. "Escalators and Quagmires" – by Lawrence Freedman (substack.com)

154. Matthews, "The red line: Biden and Xi's secret Ukraine talks revealed" p. 13

155. https://apple.news/A8k9bXo-yTEmdRIoV9c8xyQ and https://www.thetimes.co.uk/article/russia-will-formally-annex-four-occupied-regions-of-ukraine-nk7ftn8mx?shareToken=cbaf784a2b925ea178980dd03443 9db9

156. Lawrence Freedman "Gradually, then suddenly," *Comment is Freed*
157. Niall Ferguson, "Ukraine's path to Victory," Bloomberg, 13 September 2022
158. Putin speech: "Russia announces immediate 'partial mobilization' of citizens for its offensive in Ukraine" | CNN
159. "Ukraine accuses Russia of blackmail with plan to halt gas to Poland" | Reuters
160. Paret (ed.), *Makers of Modern Strategy* p. 811
161. Hansard, House of Lords, 21 December 2022 col. 1193
162. https://apple.news/ArY_77BRUROGXzMTn2DhGJA
163. "Ukraine's cultural gems looted by Russia" | World | *The Times*
164. *Times Literary Supplement*, 2 December 2022 p. 16
165. https://www.theguardian.com/world/live/2023/mar/11/russia-ukraine-war-live-bakhmut-fighting-escalates-as-moscows-forces-try-to-break-through-ukrainian-defences
166. "How many Russians have been killed in Ukraine?" | *The Economist*
167. "Ukraine holds on, praying that new western kit will break the deadlock" (thetimes.co.uk)
168. "Ukraine's war secret: how many of their troops have died?" | News | *The Sunday Times* (thetimes.co.uk)
169. "How many Russians have been killed in Ukraine?" | *The Economist*
170. https://www.theguardian.com/world/live/2023/mar/11/russia-ukraine-war-live-bakhmut-fighting-escalates-as-moscows-forces-try-to-break-through-ukrainian-defences
171. https://www.washingtonpost.com/national-security/2022/04/11/putin-misjudged-ukraine-hubris-isolation/
172. Gen. David Petraeus: "How the war in Ukraine will end" | How smart Technology changing lives (voonze.com)
173. Ibid.
174. Russian Offensive Campaign Assessment, January 15, 2023 | Institute for the Study of War (understandingwar.org)
175. Chotiner, "Is the Russian Military a Paper Tiger?"

Ten: The Wars of the Future

1. "One Year In: What Are the Lessons from Ukraine for the Future of War?" (newamerica.org)
2. MacMillan, *War* p. 284
3. Ibid., p. 285
4. Ibid., p. 287
5. Galeotti, *Weaponisation of Everything* p. 5
6. Ibid.

7. https://www.statista.com/statistics/272698/global-market-share-held-by-mobile-operating-systems-since-2009/
8. Allen et al., *Future War and the Defence of Europe* p. 27
9. Ibid., p. 23
10. Lee et al., *The Other Face of Battle* p. 205
11. Hoffman and Mattis, "Future Warfare: The Rise of Hybrid Wars"
12. Freedman, *The Future of War* p. 223
13. Ibid.
14. https://foreignpolicy.com/2018/03/05/im-sorry-for-creating-the-gerasimov-doctrine/
15. https://berlinpolicyjournal.com/the-gerasimov-doctrine/
16. https://thecritic.co.uk/issues/may-2022/the-moral-blindness-of-putins-generals/
17. Kostiner, *Conflict and Cooperation in the Gulf Region* p. 175
18. Ibid.
19. Kilcullen, *The Dragons and the Snakes* p. 29
20. Ibid., p. 208
21. https://www.nytimes.com/2021/09/18/world/middleeast/iran-nuclear-fakhrizadeh-assassination-israel.html
22. Kissinger et al., *The Age of A.I.* p. 159
23. "One Year In: What Are the Lessons from Ukraine for the Future of War?" (newamerica.org)
24. Singer, *Wired for War* p. 430
25. Black, *A Short History of War* p. 230
26. Braun et al. (eds), *Robotics and Military Operations* p. 11
27. Strachan and Scheipers (eds), *The Changing Character of War* p. 333
28. Singer, *Wired for War* p. 21
29. Strachan and Scheipers (eds), *The Changing Character of War* p. 336
30. https://www.sofx.com/green-berets-weaponized-robots-team-up-for-offensive-operations-breaking-defense/?utm_source=SOFX+Newsletter&utm_campaign=f5648af752-EMAIL_CAMPAIGN_2019_02_20_11_10_COPY_04&utm_medium=email&utm_term=0_974be175b9-f5648af752-201631409
31. Strachan and Scheipers (eds), *The Changing Character of War* p. 338
32. Ibid.
33. Ibid., p. 339
34. Ibid., p. 338
35. Singer, *Wired for War* p. 10
36. Braun et al. (eds), *Robotics and Military Operations* p. 9
37. Ibid., p. 20
38. Kissinger et al., *The Age of A.I.* p. 140
39. Ibid., p. 172

40. Braun et al. (eds), *Robotics and Military Operations* p. 4
41. Freedman, *Command* pp. 505–6
42. Kissinger et al., *The Age of A.I.* p. 157
43. Payne, "Artificial Intelligence as a revolution in military affairs"
44. Scharre, *Army of None* p. 4
45. Ibid., p. 5
46. Black Stilwell, https://www.wearethemighty.com/popular/robot-machine-guns-guard-dmz/
47. Scharre, *Army of None* p. 6
48. Ibid., p. 15
49. Ibid., p. 367
50. Kilcullen, *The Dragons and the Snakes* p. 30
51. Rezaei, "The Islamic Revolution and the Bomb: Quran Meets Realism" pp. 13–34
52. https://www.cnas.org/publications/reports/dangerous-straits-wargaming-a-future-conflict-over-taiwans?utm_medium=email&utm_campaign=Press%20Release%20-%20Dangerous%20Straits%20Wargame%20Report&utm_content=Press%20Release%20-%20Dangerous%20Straits%20Wargame%20Report+CID_d726ab5d26f018f5eb3dd615634a258f&utm_source=Campaign%20Monitor&utm_term=Dangerous%20Straits%20Wargaming%20a%20Future%20Conflict%20over%20Taiwan
53. Freedman, *Future of War* p. 281
54. Drell, "The Shadow of the Bomb" p. 68
55. https://edition.cnn.com/europe/live-news/ukraine-russia-putin-news-03-10-22/h_ffb15927857812d909a0d6f4c9644f6e
56. Rid, *Active Measures* p. 423
57. Ibid., p. 433
58. Ibid., pp. 433–4
59. Ibid., pp. 360–3
60. Ibid., p. 433
61. Galeotti, *The Weaponisation of Everything* p. 160
62. https://amp.theguardian.com/technology/2020/jan/13/what-are-deepfakes-and-how-can-you-spot-them
63. https://apnews.com/article/ap-top-news-artificial-intelligence-social-platforms-think-tanks-politics-bc2f19097a4c4fffaa00de6770b8a60d
64. https://apnews.com/article/ap-top-news-artificial-intelligence-social-platforms-think-tanks-politics-bc2f19097a4c4fffaa00de6770b8a60d
65. https://www.theguardian.com/us-news/2019/may/18/ex-cia-officer-kevin-mallory-sentenced-to-20-years-for-spying-for-china
66. https://www.wsj.com/articles/fraudsters-use-ai-to-mimic-ceos-voice-in-unusual-cybercrime-case-11567157402

67. https://www.theguardian.com/technology/ng-interactive/2019/jun/22/the-rise-of-the-deepfake-and-the-threat-to-democracy
68. Galeotti, *The Weaponisation of Everything* p. 161
69. Ibid.
70. https://www.wilsoncenter.org/blog-post/operation-denver-kgb-and-stasi-disinformation-regarding-aids
71. https://web.archive.org/web/20100324175917/https://www.cia.gov/library/center-for-the-study-of-intelligence/csi-publications/csi-studies/studies/vol53no4/pdf/U-%20Boghardt-AIDS-Made%20in%20the%20USA-17Dec.pdf
72. Galeotti, *The Weaponisation of Everything* p. 164
73. Ibid., pp. 168–9
74. Ibid., p. 166
75. Freedman, *The Future of War* p. 225
76. https://www.usnews.com/news/world-report/articles/2022-06-28/russia-defends-strike-on-shopping-mall-in-ukraine-that-killed-civilians
77. Dustin Volz, https://www.wsj.com, 27 July 2022
78. "400. Russia-Ukraine Conflict: Sign post to the Future (Part 1)" | Mad Scientist Laboratory (army.mil)
79. https://www.navytimes.com/news/your-navy/2023/01/03/new-in-2023-here-comes-the-first-ever-surface-drone-fleet/?utm_campaign=dfn-ebb&utm_medium=email&utm_source=sailthru&SToverlay=2002c2d9-c344-4bbb-8610-e5794efcfa7d
80. Strachan and Scheipers (eds), *The Changing Character of War* p. 337; and Bierbauer and Cooter, *Never Mind, We'll Do It Ourselves*
81. "Ukraine's Drone Spotters on Front Lines Wage New Kind of War" – WSJ
82. https://www.nytimes.com/2022/05/30/opinion/drones-ukraine-war.html?te=1&nl=opinion-today&emc=edit_ty_20220531
83. Singer, *Wired for War* p. 35
84. Strachan and Scheipers (eds), *The Changing Character of War* p. 336
85. https://www.understandingwar.org/backgrounder/russian-offensive-campaign-assessment-august-5
86. https://www.dailymail.co.uk/news/article-10964103/Motorbike-riding-terrorist-minced-spinning-blade-missile-precision-drone-strike-Syria.html
87. https://taskandpurpose.com/tech-tactics/army-xm1211-high-explosive-proximity-round-fielding/
88. Braun et al. (eds), *Robotics and Military Operations* p. 14
89. https://hackaday.com/2022/08/30/militaries-are-rushing-to-get-anti-drone-lasers-operational/
90. https://www.defenseone.com/technology/2022/02/drones-shooting-microwave-rays-could-be-drone-killers-tomorrow/361933/

91. Black, A *Short History of War* p. 230
92. https://www.wsj.com/articles/inside-a-u-s-navy-maritime-drone-operation-aimed-at-iran-11661954273?mod=itp_wsj&mod=djemITP_h
93. https://www.washingtonpost.com/opinions/2022/12/19/palantir-algorithm-data-ukraine-war/?utm_campaign=wp_follow_david_ignatius&utm_medium=email&utm_source=newsletter&wpisrc=nl-davidignatius&carta-url=https%3A%2F%2Fs2.washingtonpost.com%2Fcar-ln-tr%2F389c920%2F63a083f9ef9bf67b23287c39%2F59 6e8405ade4e25e028e8c91%2F5%2F19%2F63a083f9ef9bf67b23287 c39&wp_cu=5423d870f68ec1275991f7ae52461c78%7C3e20aea4-1174-11e0-a478-1231380f446b
94. https://www.armyupress.army.mil/Journals/Military-Review/English-Edition-Archives/November-December-2020/Allen-Data-Swept-2035/
95. https://madsciblog.tradoc.army.mil/46-integrated-sensors-the-critical-element-in-future-complex-environment-warfare/
96. https://taskandpurpose.com/news/russia-ukraine-electronic-warfare-us-troops/
97. Gen. David Petraeus: "How the war in Ukraine will end" | How smart Technology changing lives (voonze.com)
98. Opinion | "Russia learns the perils of aggression in an age of defensive dominance" – *The Washington Post*
99. Galeotti, *The Weaponisation of Everything* p. 113
100. https://www.wired.com/2014/11/countdown-to-zero-day-stuxnet/
101. Zetter, *Countdown to Zero Day* p. 12
102. https://www.haaretz.com/2010-09-28/ty-article/computer-virus-in-iran-actually-targeted-larger-nuclear-facility/0000017f-f5dd-d460-afff-ffffd3b30000
103. Sanger, *Confront and Conceal passim*; and Sanger, *The Perfect Weapon passim*
104. "Preliminary Lessons in Conventional Warfighting from Russia's Invasion of Ukraine: February–July 2022" | Royal United Services Institute (rusi.org)
105. Ibid.
106. https://www.spectator.co.uk/article/farewell-to-arms-britains-depleted-military/
107. "A New Theory of American Foreign Policy" – *The Atlantic*
109. General Sir Richard Barrons in the *Economist*, 9 February 2023
109. "American Defense Priorities After Ukraine" – War on the Rocks
110. Opinion | "Russia learns the perils of aggression in an age of defensive dominance" – *The Washington Post*
111. Gen. David Petraeus: "How the war in Ukraine will end" | How smart Technology changing lives (voonze.com)

112. "What Is China Learning From Russia's War in Ukraine?" | Foreign Affairs

113. "The Coming War Over Taiwan" – WSJ

114. https://www.foxnews.com/politics/bidens-fumbled-afghanistan-withdrawal-propaganda-gift-china-defense-department-finds?test=a7718066b09a50c919201c0561555900

115. "What China Has Learned From the Ukraine War" | Foreign Affairs

116. https://www.nytimes.com/2023/02/27/opinion/a-war-with-china-would-reach-deep-into-american-society.html?smid=nytcore-ios-share&referringSource=articleShare

117. "Eight New Points on the Porcupine: More Ukrainian Lessons for Taiwan" – War on the Rocks

118. Pancevski, "Europe Is Rushing Arms to Ukraine, But Running Out of Ammo"

119. https://www.foreignaffairs.com/russian-federation/americas-new-sanctions-strategy?utm_medium=newsletters&utm_source=twofa&utm_campaign=Putin%E2%80%99s%20Last%20Stand&utm_content=20221223&utm_term=FA%20This%20Week%20-%20112017; https://som.yale.edu/story/2022/over-1000-companies-have-curtailed-operations-russia-some-remain

INDEX